Mathematical Foundations of Information Flow

Proceedings of Symposia in APPLIED MATHEMATICS

Volume 71

Mathematical Foundations of Information Flow

Clifford Lectures
Information Flow in Physics,
Geometry, Logic and Computation
March 12–15, 2008
Tulane University
New Orleans, Louisiana

Samson Abramsky
Michael Mislove
Editors

American Mathematical Society
Providence, Rhode Island

EDITORIAL COMMITTEE

Mary Pugh   Lenya Ryzhik   Eitan Tadmor (Chair)

2010 *Mathematics Subject Classification.* Primary 18D35, 22A15, 78A15, 81P10, 81P45, 83C99, 91A10.

Library of Congress Cataloging-in-Publication Data

Mathematical foundations of information flow : Clifford lectures on information flow in physics, geometry, logic and computation, March 12–15, 2008, Tulane University, New Orleans, Louisiana / Samson Abramsky, Michael Mislove, editors.
    p. cm. — (Proceedings of symposia in applied mathematics ; v. 71)
  Papers from the 2008 Clifford lectures, an annual series sponsored by the Tulane University Mathematics Department in honor of A.H. Clifford.
  Includes bibliographical references.
  ISBN 978-0-8218-4923-1 (alk. paper)
  1. Categories (Mathematics)—Congresses. 2. Topological semigroups—Congresses. I. Clifford, A. H. (Alfred Hoblitzelle), 1908– II. Abramsky, Samson, 1953– III. Mislove, Michael W. IV. Tulane University. Dept. of Mathematics.

QA169.M388 2012
512′.62—dc23
                                                                                      2012009053

Copying and reprinting. Material in this book may be reproduced by any means for educational and scientific purposes without fee or permission with the exception of reproduction by services that collect fees for delivery of documents and provided that the customary acknowledgment of the source is given. This consent does not extend to other kinds of copying for general distribution, for advertising or promotional purposes, or for resale. Requests for permission for commercial use of material should be addressed to the Acquisitions Department, American Mathematical Society, 201 Charles Street, Providence, Rhode Island 02904-2294, USA. Requests can also be made by e-mail to reprint-permission@ams.org.

Excluded from these provisions is material in articles for which the author holds copyright. In such cases, requests for permission to use or reprint should be addressed directly to the author(s). (Copyright ownership is indicated in the notice in the lower right-hand corner of the first page of each article.)

© 2012 by the American Mathematical Society. All rights reserved.
The American Mathematical Society retains all rights
except those granted to the United States Government.
Copyright of individual articles may revert to the public domain 28 years
after publication. Contact the AMS for copyright status of individual articles.
Printed in the United States of America.

∞ The paper used in this book is acid-free and falls within the guidelines
established to ensure permanence and durability.
Visit the AMS home page at http://www.ams.org/

10 9 8 7 6 5 4 3 2 1   17 16 15 14 13 12

Contents

Preface	vii
H*-algebras and Nonunital Frobenius Algebras: First Steps in Infinite-dimensional Categorical Quantum Mechanics SAMSON ABRAMSKY and CHRIS HEUNEN	1
Teleportation in General Probabilistic Theories HOWARD BARNUM, JONATHAN BARRETT, MATTHEW LEIFER, and ALEXANDER WILCE	25
Fixed Points in Epistemic Game Theory ADAM BRANDENBURGER, AMANDA FRIEDENBERG, and H. JEROME KEISLER	49
Spekkenss Toy Theory as a Category of Processes BOB COECKE and BILL EDWARDS	61
Categorical Traces From Single-photon Linear Optics PETER HINES and PHILIP SCOTT	89
Compact Affine Monoids, Harmonic Analysis and Information Theory KARL H. HOFMANN and MICHAEL MISLOVE	125
The Scope of a Quantum Channel KEYE MARTIN	183
Spacetime Geometry From Causal Structure and a Measurement KEYE MARTIN and PRAKASH PANANGADEN	213
Geometry of Abstraction in Quantum Computation DUSKO PAVLOVIC	233

Preface

This volume contains papers from the 2008 Clifford Lectures. The Clifford Lectures is an annual series sponsored by the Tulane University Mathematics Department in honor of A. H. CLIFFORD, the father of algebraic semigroup theory and a longtime member of the Tulane mathematics department. The 2008 Clifford Lectures were delivered by Samson Abramsky, with the theme of *Information Flow in Physics, Geometry, Logic and Computation* [1]. The Lectures included five talks by Professor Abramsky, as well as invited talks by twelve colleagues on topics ranging from mathematics, and in particular topology, to computer science, physics, classical and quantum information, systems biology and finite model theory. This broad range of topics was deliberate in the design of the lectures, the aim of which was to encourage collaboration among a group of researchers, all of whom were working on some aspect of information flow. Rather than comprising a proceedings of that meeting alone, this volume represents the culmination of a series of meetings on the same theme. Indeed, the 2008 Clifford Lectures provided the impetus for a series of meetings focused on information flow, and the authors of the papers in this volume have been participants in most of these meetings. The meetings included two Workshops on Informatic Phenomena held at Tulane in the fall of 2008 and 2009 [5, 6], a Seminar on the *Semantics of Information* held at Schloß Dagstuhl, the German International Meeting Center for Computer Science, in June 2010 [3], the 2011 Clifford Lectures [2] in March 2011 which featured talks by Dr. Christopher Fuchs (Perimeter Institute) on quantum information, and finally, the forthcoming *Seminar on Information Flow* [4], to be held again at Schloß Dagstuhl in August 2012. These meetings maintain the broad representation of topics of the initial 2008 Clifford Lectures, and several fruitful collaborations have sprung up among the researchers who participated in the meetings, some of which are represented in this volume. This volume represents a significant component of the research presented at the series of meetings just described.

While the theme of the 2008 Clifford Lectures was deliberately broad, the focus of the research presented in this volume is narrower. The principal theme represented in this volume is information flow in classical and quantum physics and its mathematical underpinnings. This is quite appropriate, since Professor Abramsky's research in this area was the impetus for the 2008 Clifford Lectures and the basis for casting the wide net of research interests featured at that and the subsequent meetings. More precisely, the focus of much of Professor Abramsky's recent research has been the application of ideas from theoretical computer science to develop a novel categorical formulation of quantum mechanics, as the basis of a new approach to quantum physics and quantum information. In keeping with this theme, all the papers in this volume focus on information flow in quantum and classical physics and its mathematical underpinnings.

The focus of the AMS *Proceedings* of Symposia in Applied Mathematics series is on the application of mathematics to other disciplines. In that spirit, the papers in this volume comprise a broad representation of applications of mathematics to quantum physics and to classical and quantum information: The research reported here utilizes category theory, domain theory, harmonic analysis, probability theory, Shannon information theory, as well as topology, as tools for modeling quantum physics and classical and quantum information.

The papers in this volume

We shall give a brief indication of the contents of the papers which appear in this volume.

(1) The paper by Abramsky and Heunen addresses the issue of extending the categorical quantum mechanics paradigm to the infinite-dimensional case. At the same time, it makes connections with some classic topics in operator algebras, notably the work by Ambrose on H^*-algebras and an infinite-dimensional extension of the Wedderburn structure theorem. It relates these to Frobenius algebras, which have been studied in categorical quantum mechanics as an algebraic way of capturing orthonormal bases and measurements. It also characterizes Frobenius algebras in various categories of relations.

(2) The paper by Barnum, Barrett, Leifer and Wilce considers which probability theories support teleportation. Previous work of the authors showed that phenomena associated to quantum mechanics such as no-cloning and no-broadcasting are generic in all non-classical probabilistic theories. On the other hand, teleportation is not supported in most such theories, leading the authors in the present paper to explore which probabilistic theories support this protocol. They isolate a natural class of composite systems which they term *regular* and establish necessary and sufficient conditions for a regular tripartite system to support a conclusive teleportation protocol. They also give sufficient conditions for deterministic teleportation, yielding a large supply of composite state spaces that are neither classical nor quantum, but that do support such a protocol

(3) The paper by Brandenburger, Friedenberg and Keisler looks at another fruitful source of ideas about modelling information flow, coming from game theory, and the interaction of rational agents. In particular, it looks at epistemic game theory, where there is an explicit formal representation of the belief states of the agents, in terms of type spaces. In this context, order-theoretic fixpoints play a prominent rôle, in contrast to the topological theorems of Brouwer and Kakutani, which are widely used in the study of Nash equilibria. Order-theoretic fixpoints are also widely used in theoretical computer science. An interesting point of difference is that the fixpoints used in epistemic game theory may come from non-monotonic functions; this leads to a number of interesting mathematical questions and results.

(4) The paper by Coecke and Edwards looks at the well-known 'toy model' of quantum mechanics developed by Rob Spekkens from the perspective

of categorical quantum mechanics. This model shows that many features held to be characteristic of quantum mechanics can be realized in an intriguingly simple model based on finite sets and relations. Previous work had shown that this model could be captured in an elegant fashion in the setting of categorical quantum mechanics. However, the constraints placed on the model by the 'knowledge balance principle' means that it is surprisingly difficult to give an explicit description of the full model, as opposed to an inductive construction. The present paper gives such a description, which can serve as a basis for further investigations.

(5) The paper by Hines and Scott uses the classical Sagnac interferometer as a thought experiment in single-photon linear optics, which leads to a general construction on Hilbert spaces. This construct has a close connection to constructions from algebraic and categorical program semantics, the so-called trace. The authors analyze their general construction in terms of a categorical trace which generalizes a 'particle-style' trace on Hilbert space they studied in an earlier paper. They show this general construction has a physical realization based on the thought experiment that motivated the work.

(6) The paper by Hofmann and Mislove has two aims. The first is to provide a self-contained, accessible account of some basic results in the theory of compact monoids and harmonic analysis, and to demonstrate how these results, when applied to the compact affine monoid of probability measures on a compact group, lead to Wendel's proof that such a group has a unique Haar measure. The second goal is to apply some of the same theory to analyze Shannon's classical information of discrete lossless noisy channels with finite inputs and outputs. Using domain theory as an additional tool, the main result generalizes work of Martin, Allwein and Moskowitz about the nature of Shannon capacity as a function on the family of binary channels.

(7) The paper by Martin introduces the notion of the *scope* of a unital quantum channel. Such a channel has a range of possible Shannon capacities for sending classical information, depending on the basis used to encode the information. The author calls this range the *scope* of the channel. He shows that, in the case of qubit channels, the scope is a compact interval, and he uses the algebraic structure of the family of channels as a monoid to characterize the scope of the channel, and how to calculate it. The author also presents an adaptive scheme for communication in which the participants can maximize the information transmitted after they first determine the state of the environment, for which a method is also presented. The author shows how this work can be applied in quantum cryptography to minimize the error rate over any time interval during which the environment remains stable.

(8) The paper by Martin and Panangaden is the second in which the authors have explored a partial order on the events in spacetime that is defined by the causal structure. In an earlier paper the authors used techniques from domain theory to show that the topology of globally hyperbolic space-times could be reconstructed from the causal structure. But the causal

structure determines the metric only up to a local rescaling (a conformal transformation); in a four-dimensional spacetime, the metric tensor has ten components, and thus effectively only nine are determined by the causal structure. To remedy this deficiency, the authors again apply domain theory. They first establish the relationship between measurement in domain theory, the concept of a global time function and the Lorentz distance. Then they are able to domain-theoretically recover the final tenth component of the metric tensor, thereby obtaining causal reconstruction of not only the topology of spacetime, but also its geometry.

(9) Finally, the paper by Pavlovic continues the theme of categorical models of quantum computation by considering how to identify classical data in a quantum computing setting. It is shown that polynomial extensions of (dagger-)monoidal closed categories capture exactly the classical data and admissible operations thereon, namely, copying, deleting and abstraction. A running example of Simon's algorithm is used to illustrate the results.

Acknowledgments

There are several organizations and people who deserve thanks for their support in helping with the organization and financial support for the 2008 Clifford Lectures, and the subsequent meetings described above, all of which contributed to the research reported in this volume. First and foremost, the Tulane Mathematics Department deserves thanks for inviting Professor Abramsky to be the 2008 Clifford Lecturer, as well as for its subsequent invitation to Dr. Christopher Fuchs to be the 2011 Clifford Lecturer. In addition, the department deserves thanks for hosting the two Workshops in Informatic Phenomena in 2008 and 2009. Dr. Keye Martin (NRL) also deserves thanks for having the vision to propose the 2008 and 2009 Workshops on Informatic Phenomena, both of which he helped to organize. Dr. Martin also helped to organize the 2010 Schloß Dagstuhl meeting. Finally, Schloß Dagstuhl, and in particular Professor Dr. Reinhard Wilhelm, its scientific director, deserve thanks for hosting two seminars on this topic, in 2010 and 2012.

Finally, thanks are due for the financial support that agencies provided in support of the meetings mentioned above. First and foremost, we thank the US Office of Naval Research, and especially Dr. Ralph Wachter, whose research program provided generous support to help underwrite the 2008 Clifford Lectures, as well as the other meetings listed above. Dr. Keye Martin (NRL) also is owed a debt of thanks for providing the funds to support the Workshops on Informatics Phenomena held in 2008 and 2009, and for providing support for the participants to attend the Seminar on the Semantics of Information at Schloß Dagstuhl in June 2010. The U.K. Engineering and Physical Sciences Research Council have supported Professor Abramsky's research through a Senior Fellowship.

Samson Abramsky
Michael Mislove
December 2011

Bibliography

[1] 2008 Clifford Lectures, http://www.math.tulane.edu/~mwm/clifford
[2] 2011 Clifford Lectures, http://tulane.edu/sse/math/news/clifford-lectures-2011.cfm
[3] Seminar on the Semantics of Information,
    http://www.dagstuhl.de/en/program/calendar/semhp/?semnr=10232
[4] Seminar on the Information Flow and Its Applications,
    http://www.dagstuhl.de/en/program/calendar/semhp/?semnr=12352
[5] 2008 Workshop on Informatic Phenomena http://www.math.tulane.edu/~mwm/WIP2008
[6] 2009 Workshop on Informatic Phenomena http://www.math.tulane.edu/~mwm/WIP2009

H*-algebras and nonunital Frobenius algebras: first steps in infinite-dimensional categorical quantum mechanics

Samson Abramsky and Chris Heunen

ABSTRACT. A certain class of Frobenius algebras has been used to characterize orthonormal bases and observables on finite-dimensional Hilbert spaces. The presence of units in these algebras means that they can only be realized finite-dimensionally. We seek a suitable generalization, which will allow arbitrary bases, and therefore observables with discrete spectra, to be described within categorical axiomatizations of quantum mechanics. We develop a definition of H*-algebra that can be interpreted in any symmetric monoidal dagger category, reduces to the classical notion from functional analysis in the category of (possibly infinite-dimensional) Hilbert spaces, and hence provides a categorical way to speak about orthonormal bases and quantum observables in arbitrary dimension. Moreover, these algebras reduce to the usual notion of Frobenius algebra in compact categories. We then investigate the relations between nonunital Frobenius algebras and H*-algebras. We give a number of equivalent conditions to characterize when they coincide in the category of Hilbert spaces. We also show that they always coincide in categories of generalized relations and positive matrices.

1. Introduction

The context for this paper comes from the ongoing work on *categorical quantum mechanics* [**AC04, AC09**]. This work has shown how large parts of quantum mechanics can be axiomatized in terms of monoidal dagger categories and structures definable within them. This axiomatization can be used to perform high-level reasoning and calculations relating to quantum information, using diagrammatic methods [**Sel10**]; and also as a basis for exploring foundational issues in quantum mechanics and quantum computation. In particular, a form of Frobenius algebras has been used to give an algebraic axiomatization of *orthonormal bases* and *observables* [**CP07, CPV09**].

1991 *Mathematics Subject Classification*. Primary 18D35, 81P15; Secondary 46K15, 46B28.

Key words and phrases. Orthonormal bases, Frobenius algebras, H*-algebras, monoidal categories, categorical quantum mechanics.

Supported by an EPSRC Senior Fellowship and by ONR.

Supported by the Netherlands Organisation for Scientific Research (NWO).

The structures used so far (*e.g.* compact closure, Frobenius algebras) have only finite-dimensional realizations in Hilbert spaces. This raises some interesting questions and challenges:

- Find a good general notion of Frobenius structure which works in the infinite-dimensional case in **Hilb**.
- Use this to characterize general bases and therefore general observables with discrete spectra.
- Similarly extend the analysis for other categories.
- Clarify the mathematics, and relate it to the wider literature.

As we shall see, an intriguing problem remains open, but much of this program of work has been accomplished.

The further contents of the paper are as follows. Section 2 recalls some background on monoidal dagger categories and Frobenius algebras, and poses the problem. Section 3 introduces the key notion of H*-algebra, in the general setting of symmetric monoidal dagger categories. In Section 4, we prove our results relating to **Hilb**, the category of Hilbert spaces (of unrestricted dimension). We show how H*-algebras provide exactly the right algebraic notion to characterize orthonormal bases in arbitrary dimension. We give several equivalent characterizations of when H*-algebras and nonunital Frobenius algebras coincide in the category of Hilbert spaces. Section 5 studies H*-algebras in categories of generalized relations and positive matrices. We show that in these settings, where no phenomena of 'destructive interference' arise, H*-algebras and nonunital Frobenius algebras always coincide. Finally, Section 6 provides an outlook for future work.

2. Background

The basic setting is that of *dagger symmetric monoidal categories*. We briefly recall the definitions, referring to [**AC09**] for further details and motivation.

A *dagger category* is a category **D** equipped with an identity-on-objects, contravariant, strictly involutive functor. Concretely, for each arrow $f \colon A \to B$, there is an arrow $f^\dagger \colon B \to A$, and this assignment satisfies
$$\mathrm{id}^\dagger = \mathrm{id}, \qquad (g \circ f)^\dagger = f^\dagger \circ g^\dagger, \qquad f^{\dagger\dagger} = f.$$
An arrow $f \colon A \to B$ is *dagger monic* when $f^\dagger \circ f = \mathrm{id}_A$, and a *dagger iso(morphism)* if both f and $f^\dagger$ are dagger monics.

A *symmetric monoidal dagger category* is a dagger category with a symmetric monoidal structure $(\mathbf{D}, \otimes, I, \lambda, \rho, \alpha, \sigma)$ such that
$$(f \otimes g)^\dagger = f^\dagger \otimes g^\dagger$$
and moreover the natural isomorphisms λ, ρ, α, σ are componentwise dagger isomorphisms.

Examples.

- The category **Hilb** of Hilbert spaces and continuous linear maps, and its (full) subcategory **fHilb** of finite-dimensional Hilbert spaces. Here the dagger is the adjoint, and the tensor product has its standard interpretation for Hilbert spaces. Dagger isomorphisms are *unitaries*, and dagger monics are *isometries*.
- The category **Rel** of sets and relations. Here the dagger is relational converse, while the monoidal structure is given by the cartesian product.

This generalizes to relations valued in a commutative quantale [**Ros90**], and to the category of relations of any regular category [**CKS84**]. This has a full sub-category **fRel**, of finite sets and relations.

- The category **lbfRel**, of *locally bifinite relations*. This is the subcategory of **Rel** comprising those relations which are image-finite, meaning that each element in the domain is related to only finitely many elements in the codomain, and whose converses are also image-finite. This forms a monoidal dagger subcategory of **Rel**. It serves as a kind of qualitative approximation of the passage from finite- to infinite-dimensional Hilbert spaces. For example, a set carries a compact structure in **lbfRel** if and only if it is finite.

- A common generalization of **fHilb** and **fRel** is obtained by forming the category $\mathbf{Mat}(S)$, where S is a commutative semiring with a specified involution [**Heu09b**]. Objects of $\mathbf{Mat}(S)$ are finite sets, and morphisms are maps $X \times Y \to S$, which we think of as 'X times Y matrices'. Composition is by matrix multiplication, while the dagger is conjugate transpose, where conjugation of a matrix means elementwise application of the involution on S. The tensor product of X and Y is given by $X \times Y$, with the action on matrices given by componentwise multiplication, corresponding to the 'Kronecker product' of matrices. If we take $S = \mathbb{C}$, this yields a category equivalent to **fHilb**, while taking S to be the Boolean semiring $\{0,1\}$, with trivial involution, gives **fRel**.

- An infinitary generalization of $\mathbf{Mat}(\mathbb{C})$ is given by $\mathbf{Mat}_{\ell^2}(\mathbb{C})$. This category has arbitrary sets as objects, and its morphisms $X \to Y$ are matrices $M \colon X \times Y \to \mathbb{C}$ such that for each $x \in X$, the family $\{M(x,y)\}_{y \in Y}$ is ℓ^2-summable; and for each $y \in Y$, the family $\{M(x,y)\}_{x \in X}$ is ℓ^2-summable. The category **Hilb** is equivalent to a (nonfull) subcategory of $\mathbf{Mat}_{\ell^2}(\mathbb{C})$ [**BEH08**, Theorem 3.1.7].

Graphical Calculus. We briefly recall the graphical calculus for symmetric monoidal dagger categories [**Sel10**]. This can be seen as a two-dimensional version of *Dirac notation*, which allows equational reasoning to be performed graphically in a sound and complete fashion. A morphism $f \colon X \to Y$ is represented pictorially as $\begin{smallmatrix}Y\\ \boxed{f}\\ X\end{smallmatrix}$, the identity on X simply becomes $\begin{smallmatrix}X\\ |\\ X\end{smallmatrix}$, and composition and tensor products appear as follows.

$$\begin{array}{c}Z\\ \boxed{g \circ f}\\ X\end{array} = \begin{array}{c}Z\\ \boxed{g}\\ \boxed{f}\\ X\end{array} \qquad \begin{array}{c}Y \otimes Z\\ \boxed{f \otimes g}\\ W \otimes X\end{array} = \begin{array}{cc}Y & Z\\ \boxed{f} & \boxed{g}\\ W & X\end{array}$$

The symmetry isomorphism σ is drawn as $\times$. The dagger is represented graphically by a horizontal reflection.

2.1. Dagger Frobenius algebras.

Frobenius algebras are a classic notion in mathematics [**Nak39**]. A particular form of such algebras was introduced in the general setting of monoidal dagger categories by Coecke and Pavlović in [**CP07**]. In their version, a *dagger Frobenius structure* on an object A in a dagger monoidal category is a commutative comonoid $(I \xleftarrow{\varepsilon} A \xrightarrow{\delta} A \otimes A)$ satisfying certain additional equations:

(A) $\qquad (\mathrm{id}_A \otimes \delta) \circ \delta = (\delta \otimes \mathrm{id}_A) \circ \delta,$

(U) $\qquad (\mathrm{id}_A \otimes \varepsilon) \circ \delta = \mathrm{id}_A,$

(C) $\qquad \sigma \circ \delta = \delta,$

(M) $\qquad \delta^\dagger \circ \delta = \mathrm{id}_A,$

(F) $\qquad \delta \circ \delta^\dagger = (\delta^\dagger \otimes \mathrm{id}_A) \circ (\mathrm{id}_A \otimes \delta).$

These equations become more perspicuous when represented diagrammatically, as below. Here, we draw the comultiplication δ as ⋎, and the counit ε as ⸰.

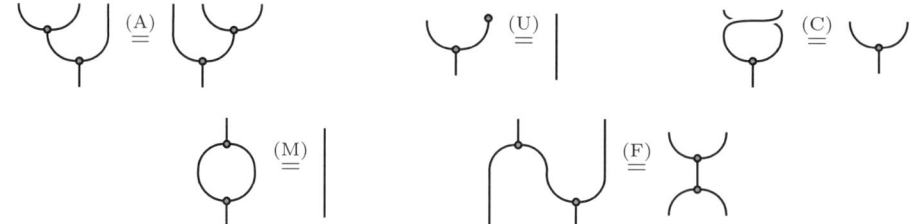

A 'right-handed version' of the Frobenius law (F) follows from (C); in the noncommutative case we should add this symmetric version (F') to axiom (F).

2.2. Dagger Frobenius algebras in quantum mechanics.

Frobenius algebras provide a high-level algebraic way of talking about *orthonormal bases*, and hence can be seen as modeling quantum mechanical *observables*.

To put this in context, we recall the *no-cloning theorem* [**WZ82**], which says that there is no quantum evolution (*i.e.* unitary operator) $f\colon H \to H \otimes H$ such that, for any $|\phi\rangle \in H$,

$$f|\phi\rangle = |\phi\rangle \otimes |\phi\rangle.$$

A general form of no-cloning holds for structural reasons in categorical quantum mechanics [**Abr10**]. In particular, there is no *natural*, *i.e.* uniform or basis-independent, family of diagonal morphisms in a compact closed category, unless the category collapses, so that endomorphisms are scalar multiples of the identity.

However, if we drop naturality, we *can* define such maps in **Hilb** in a basis-dependent fashion. Moreover, it turns out that such maps can be used to *uniquely determine* bases. Firstly, consider *copying maps*, which can be defined in arbitrary dimension: for a given basis $\{|i\rangle\}_{i \in I}$ of H, define $\delta\colon H \to H \otimes H$ by (continuous linear extension of) $|i\rangle \mapsto |ii\rangle$.

For example, consider the map $\delta_{\mathrm{std}}\colon \mathbb{C}^2 \to \mathbb{C}^2 \otimes \mathbb{C}^2$ defined by

$$|0\rangle \mapsto |00\rangle, \qquad |1\rangle \mapsto |11\rangle.$$

By construction, this copies the elements of the computational basis — and *only* these, as in general

$$\delta_{\mathrm{std}}(\alpha|0\rangle + \beta|1\rangle) = \alpha|00\rangle + \beta|11\rangle \neq (\alpha|0\rangle + \beta|1\rangle) \otimes (\alpha|0\rangle + \beta|1\rangle).$$

Next, consider *deleting maps* $\varepsilon\colon H \to \mathbb{C}$ by linearly extending $|e_i\rangle \mapsto 1$. In contrast to copying, these can be defined in *finite dimension only*. It is straightforward to verify that these maps define a dagger Frobenius structure on H. Moreover, the following result provides a striking converse.

THEOREM 1. [**CPV09**] *Orthonormal bases of a finite-dimensional Hilbert space H are in one-to-one correspondence with dagger Frobenius structures on H.* □

This result in fact follows easily from previous results in the literature on Frobenius algebras [**Abr97**]; we will give a short proof from the established literature in Section 4.4.

Another result provides a counterpart—at first sight displaying very different looking behaviour—in the category **Rel**.

THEOREM 2. [**Pav09**] *Dagger Frobenius structures in the category **Rel** correspond to disjoint unions of abelian groups.* □

We shall provide a different proof of this result in Section 5.1, which makes no use of units, and hence generalizes to a wide range of other situations, such as locally bifinite and quantale-valued relations, and positive ℓ_2-matrices.

2.3. The problem. The notion of Frobenius structure as defined above, which requires a unit, limits us to the *finite-dimensional case* in **Hilb**, as the following lemma shows.

LEMMA 3. *A Frobenius algebra in **Hilb** is unital if and only if it is finite-dimensional.*

PROOF. Sufficiency is shown in [**Koc03**, 3.6.9]. Necessity follows from [**Kap48**, Corollary to Theorem 4]. □

In fact, a Frobenius structure on an object A induces a *compact* (or *rigid*) structure on A, with A as its own dual (see [**AC09**]). Indeed, put $\eta = \delta \circ \varepsilon^\dagger \colon I \to A \otimes A$. In the category **fHilb**, for example, $\eta \colon \mathbb{C} \to \mathbb{C}^2 \otimes \mathbb{C}^2$ is an *entangled state preparation*:

$$\eta_{\text{std}} = \delta_{\text{std}} \circ \varepsilon^\dagger_{\text{std}} = (1 \mapsto \delta_{\text{std}}(|0\rangle + |1\rangle)) = (1 \mapsto |00\rangle + |11\rangle).$$

In general it is easy to see that η indeed provides a dagger compact structure on A, with $A^* = A$:

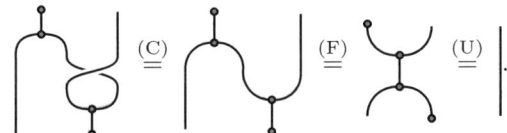

As is well-known, a compact structure exists only for finite-dimensional spaces in **Hilb**. Thus to obtain a notion capable of being extended beyond the finite-dimensional case, we need to drop the assumption of a unit.

3. H*-algebras

We begin our investigation of suitable axioms for a notion of algebra which can characterize orthonormal bases in arbitrary dimension by recalling the axioms for Frobenius structures.

(A) $\qquad (\text{id}_A \otimes \delta) \circ \delta = (\delta \otimes \text{id}_A) \circ \delta$

(U) $\qquad (\text{id}_A \otimes \varepsilon) \circ \delta = \text{id}_A$

(C) $\qquad \sigma \circ \delta = \delta$

(M) $\qquad \delta^\dagger \circ \delta = \text{id}_A$

(F) $\qquad \delta \circ \delta^\dagger = (\delta^\dagger \otimes \text{id}_A) \circ (\text{id}_A \otimes \delta)$

We note in passing that there is some redundancy in the definition of Frobenius structure.

LEMMA 4. *In any dagger monoidal category, (M), (F) and (F') imply (A).*

PROOF.

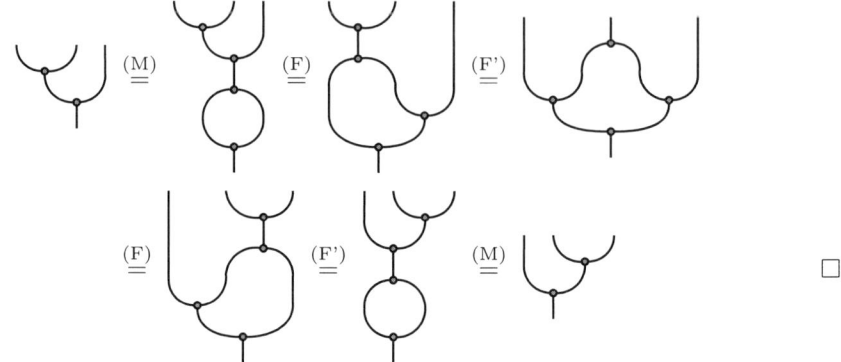

The axioms (U), (C), and (F) are independent:
- As we have seen, for an orthonormal basis $\{|n\rangle \mid n \in \mathbb{N}\}$ of a separable (infinite-dimensional) Hilbert space, the map $\delta(|n\rangle) = |nn\rangle$ satisfies everything except for (U).
- Group algebras of finite noncommutative groups [**Amb45**, Example 4] satisfy everything except for (C).
- Any nontrivial commutative (unital) Hopf algebra satisfies everything except for (F) by [**Koc03**, Proposition 2.4.10].

It is worth noting that under additional assumptions, such as unitality and enrichment in abelian groups, (A) and (M) are known to imply (F) [**LR97**, Section 6].

We shall now *redefine* a Frobenius algebra[1] in a dagger monoidal category to be an object A equipped with a comultiplication $\delta : A \to A \otimes A$ satisfying (A), (C), (M) and (F). A Frobenius algebra which additionally has an arrow $\varepsilon \colon A \to I$ satisfying (U) will explicitly be called *unital*.

3.1. Regular representation as pointwise abstraction. As we have seen, unital Frobenius algebras allow us to define compact, and hence closed, structure. How much of this can we keep in key examples such as **Hilb**?

The category **Hilb** has well-behaved duals, since $H \cong H^{**}$, and indeed there is a conjugate-linear isomorphism $H \cong H^*$. However, it is *not* the case that the tensor unit $\mathbb{C}$ is exponentiable in **Hilb**, since if it was, we would have a bounded linear evaluation map
$$H \otimes H^* \to \mathbb{C},$$
and hence its adjoint $\mathbb{C} \to H \otimes H^*$, and a compact structure.

We shall now present an axiom which captures what seems to be the best we can do in general in the way of a 'transfer of variables'. It is, indeed, a general form,

---

[1]In the literature the unital version is more specifically termed a special commutative dagger Frobenius algebra (sometimes also called a separable algebra, or a Q-system). As we will only be concerned with these kinds of Frobenius algebras, we prefer to keep terminology simple and dispense with the adjectives.

meaningful in any monoidal dagger category, of a salient structure in functional analysis.

Suppose we have a comultiplication $\delta \colon A \to A \otimes A$, and hence a multiplication $\mu = \delta^\dagger \colon A \otimes A \to A$. We can *curry* the multiplication (this process is also called λ-abstraction [**Bar01**]) for *points*—this is just the regular representation![2] Thus we have a function $R \colon \mathbf{D}(I, A) \to \mathbf{D}(A, A)$ defined by

$$R(a) = \mu \circ (\mathrm{id} \otimes a) = \quad .$$

If μ is associative, this is a semigroup homomorphism.

3.2. Axiom (H). An endomorphism homset $\mathbf{D}(A, A)$ in a dagger category $\mathbf{D}$ is not just a monoid, but a *monoid with involution*, because of the dagger. We say that (A, μ) *satisfies axiom (H)* if there is an operation $a \mapsto a^*$ on $\mathbf{D}(I, A)$ such that R becomes a homomorphism of involutive semigroups, *i.e.*

$$R(a^*) = R(a)^\dagger$$

for every $a \colon I \to A$. This unfolds to

(H) $$\mu \circ (a^* \otimes \mathrm{id}) = (a^\dagger \otimes \mathrm{id}) \circ \mu^\dagger;$$

or diagrammatically:

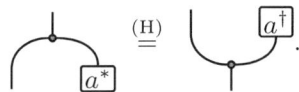

Thus $a \mapsto a^*$ is indeed a 'transfer of variables'.

3.3. Relationships between axioms (F) and (H). The rest of this section compares axioms (F) and (H) at the abstract level of monoidal dagger categories.

The following observation by Coecke, Pavlović and Vicary is the central idea in their proof of Theorem 1.

LEMMA 5. *In any dagger monoidal category, (F) and (U) imply (H).*

PROOF. Define $a^* = (a^\dagger \otimes \mathrm{id}) \circ \delta \circ \varepsilon^\dagger$.

This indeed satisfies (H).

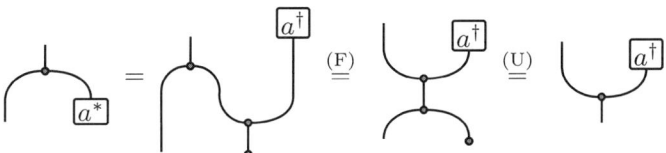

□

Recall that a category is *monoidally well-pointed* if the following holds:

$$f = g \colon A \otimes A' \to B \otimes B' \iff \forall x \colon I \to A, y \colon I \to A'.\ f \circ (x \otimes y) = g \circ (x \otimes y).$$

All the categories listed in our Examples are monoidally well-pointed in this sense.

---

[2]As we are in a commutative context, there is no need to distinguish between left and right regular representations.

LEMMA 6. *In a monoidally well-pointed dagger monoidal category, (H) and (A) imply (F).*

PROOF. For any $a\colon I \to A$ we have the following.

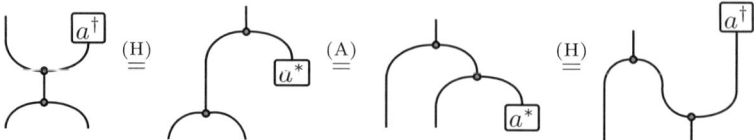

Then (F) follows from monoidal well-pointedness. □

Lemma 5 is strengthened by the following proposition, which proves that compactness implies unitality.

PROPOSITION 7. *Any Frobenius algebra in a dagger compact category is unital.*

PROOF. [**Car91**, Remark (1) on page 503] Suppose that $\delta\colon A \to A \otimes A$ is a nonunital Frobenius algebra in a compact category. Define $\varepsilon\colon A \to I$ as follows.

$$\varepsilon = \;\big|\; = \;\;\text{(diagram)}$$

Then the following holds, where we draw the unit and counit of compactness by caps and cups (without dots).

That is, (U) holds. □

Thus in the unital, monoidally well-pointed case, (F) and (H) are essentially equivalent. Our interest is, of course, in the nonunital case. To explain the provenance of the (H) axiom, and its implications for obtaining a correspondence with orthonormal bases in **Hilb** in arbitrary dimension, we shall now study the situation in the concrete setting of Hilbert spaces.

4. H*-algebras in Hilb

We begin by revisiting Theorem 1. How should the correspondence between Frobenius algebras and orthonormal bases be expressed mathematically? In fact, the content of this result is really a *structure theorem* of a classic genre in algebra [**Alb39**]. The following theorem, the *Wedderburn structure theorem*, is the prime example; it was subsequently generalized by Artin, and there have been many subsequent developments.

THEOREM 8 (Wedderburn, 1908). *Every finite-dimensional semisimple algebra is isomorphic to a product of full matrix algebras. In the commutative case over the complex numbers, this has the form: the algebra is isomorphic to a product of one-dimensional complex algebras.* □

To see the connection between the Wedderburn structure theorem and Theorem 1, consider the coalgebra A determined by an orthonormal basis $\{|i\rangle\}$ on a Hilbert space:
$$\delta \colon |i\rangle \mapsto |ii\rangle.$$
This is isomorphic as a coalgebra to a direct sum of one-dimensional coalgebras
$$\delta_{\mathbb{C}} \colon \mathbb{C} \to \mathbb{C} \otimes \mathbb{C}, \qquad 1 \mapsto 1 \otimes 1.$$
To say that a Frobenius algebra corresponds to an orthonormal basis is exactly to say that it is isomorphic as a coalgebra to a Hilbert space direct sum of one-dimensional coalgebras:
$$A \cong \bigoplus_I (\mathbb{C}, \delta_{\mathbb{C}}),$$
where the cardinality of I is the dimension of H. Applying dagger, this is equivalent to A being isomorphic as an *algebra* to the direct sum of one-dimensional *algebras*
$$A \cong \bigoplus_I (\mathbb{C}, \mu_{\mathbb{C}}), \qquad \mu_{\mathbb{C}} \colon 1 \otimes 1 \mapsto 1.$$

In this case, we say that the Frobenius algebra *admits the structure theorem*, making the view of bases as (co)algebras precise.

4.1. H*-algebras. There is a remarkable generalization of the Wedderburn structure theorem to an infinite-dimensional setting, in a classic paper from 1945 by Warren Ambrose on 'H*-algebras' [**Amb45**]. He defines an H*-algebra[3] as a (not necessarily unital) Banach algebra based on a Hilbert space H, such that for each $x \in H$ there is an $x^* \in H$ with
$$\langle xy \mid z \rangle = \langle y \mid x^* z \rangle$$
for all $y, z \in H$, and similarly for right multiplication. Note that
$$\langle xy \mid z \rangle = (\mu \circ (x \otimes y))^\dagger \circ z = (x^\dagger \otimes y^\dagger) \circ \mu^\dagger \circ z,$$
where we identify points $x \in H$ with morphisms $x \colon \mathbb{C} \to H$, and similarly
$$\langle y \mid x^* z \rangle = y^\dagger \circ \mu \circ (x^* \otimes z).$$
Using the monoidal well-pointedness of **Hilb**, it is easy to see that this is equivalent to the (H) condition![4] The following two lemmas show that the assumptions (A), (C), (M) and (H) indeed result in an H*-algebra.

LEMMA 9. *A monoid in* **Hilb** *satisfying (M) is a Banach algebra.*

---

[3]The notion termed 2-H*-algebra in [**Bae97**] was inspired by Ambrose's notion of H*-algebra. The former could be seen as a categorification of the latter; the two notions should not be confused.

[4]Notice that neither axiom (H) nor Ambrose's definition of H*-algebra requires the operation $a \mapsto a^*$ to be continuous. However, in the setting of **Hilb**, continuity follows automatically from axiom (H) [**Amb45**, Theorem 2.3].

PROOF. The condition (M) implies that $P = \mu^\dagger \circ \mu$ is a projector:
$$P^2 = \mu^\dagger \circ \mu \circ \mu^\dagger \circ \mu = \mu^\dagger \circ \mu = P$$
and clearly $P = P^\dagger$. Hence a monoid in **Hilb** satisfying (M) is a Banach algebra:
$$\begin{aligned}
\|xy\|^2 &= \langle xy \mid xy \rangle \\
&= (x^\dagger \otimes y^\dagger) \circ \mu^\dagger \circ \mu \circ (x \otimes y) \\
&= \langle x \otimes y \mid P(x \otimes y) \rangle \\
&\leq \langle x \otimes y \mid x \otimes y \rangle \\
&= \langle x \mid x \rangle \langle y \mid y \rangle \\
&= \|x\|^2 \|y\|^2.
\end{aligned}$$
□

Remark. In fact, it can be shown that the multiplication of a semigroup in **Hilb** satisfying (H) is automatically continuous, so that after adjusting by a constant, the semigroup is a Banach algebra [**Ing65**, Corollary 2.2].[5]

The following lemma establishes *properness*, which corresponds to x^* being the unique vector with the property defining H*-algebras. It follows that $R(x^*)$ is the adjoint of $R(x)$.

LEMMA 10. *Suppose $\delta \colon A \to A \otimes A$ in **Hilb** satisfies (A) and (H). Then (M) implies properness, i.e. $aA = 0 \Rightarrow a = 0$. Hence (M) holds if and only if the regular representation is monic.*

PROOF. By [**Amb45**, Theorem 2.2], A is the direct sum of its trivial ideal A' and a proper H*-algebra A''. Here, the trivial ideal is $A' = \{a \in A \mid aA = 0\}$. Since the direct sum of Hilbert spaces is a dagger biproduct, we can write δ as $\delta' \oplus \delta'' \colon A \to A \otimes A$, where $\delta' \colon A' \to A' \otimes A'$ and $\delta'' \colon A'' \to A'' \otimes A''$. The latter two morphisms are again dagger monic as a consequence of (M). So the multiplication $\delta'^\dagger$ of A' is epic, which forces $A' = 0$. □

The following proposition summarizes the preceding discussion.

PROPOSITION 11. *Any structure (A, μ) in **Hilb** satisfying (A), (H) and (M) is an H*-algebra (and also satisfies (F)); and conversely, an H*-algebra satisfies (A), (H) and (M), and hence also (F).* □

Ambrose proved a complete structure theorem for H*-algebras, of which we now state the commutative case.

THEOREM 12 (Ambrose, 1945). *Any proper commutative H*-algebra (of arbitrary dimension) is isomorphic to a Hilbert space direct sum of one-dimensional algebras.* □

---

[5]Hence Proposition 11 and Theorem 12 can be altered to show that a monoid in **Hilb** satisfying properness, (A), (C), and (H) (but not necessarily (M)!), corresponds to an *orthogonal basis*. This may have consequences for attempts to classify multipartite entanglement according to various Frobenius structures [**CK10**]. Compare also the second entry in the table on page 11 of [**CPV09**]: in finite dimension, δ is monic by (U), but in infinite dimension, one has to explicitly postulate δ to be monic to prevent *e.g.* the trivial algebra $\delta(a) = 0$ and obtain a correspondence with orthogonal bases.

This is equivalent to asserting isomorphism qua coalgebras. So it is exactly the result we are after! Rather than relying on Ambrose's results, we now give a direct, conceptual proof, using a few notions from Gelfand duality for commutative Banach algebras.

4.2. Copyables and semisimplicity. A *copyable element* of a semigroup $\delta \colon A \to A \otimes A$ in a monoidal category is a semigroup homomorphism to it from the canonical semigroup on the monoidal unit. More precisely, a copyable element is a morphism $a \colon I \to A$ such that $(a \otimes a) \circ \delta = \delta \circ a$. In a monoidally well-pointed category such as **Hilb**, we can speak of a copyable element of δ as a point $a \in A$ with $\delta(a) = a \otimes a$.[6]

PROPOSITION 13. *Assuming only (A), nonzero copyable elements are linearly independent.*

PROOF. [**Hof70**, Theorem 10.18(ii)] Suppose that $\{a_0, \ldots, a_n\}$ is a minimal nonempty linearly dependent set of nonzero copyables. Then we can write a_0 as $\sum_{i=1}^{n} \alpha_i a_i$ for a suitable choice of coefficients $\alpha_i \in \mathbb{C}$. So

$$\sum_{i=1}^{n} \alpha_i (a_i \otimes a_i) = \sum_{i=1}^{n} \alpha_i \delta(a_i)$$
$$= \delta(a_0)$$
$$= \Big(\sum_{i=1}^{n} \alpha_i a_i\Big) \otimes \Big(\sum_{j=1}^{n} \alpha_j a_j\Big)$$
$$= \sum_{i,j=1}^{n} \alpha_i \alpha_j (a_i \otimes a_j).$$

By minimality, $\{a_1, \ldots, a_n\}$ is linearly independent. Hence $\alpha_i^2 = \alpha_i$ for all i, and $\alpha_i \alpha_j = 0$ for $i \neq j$. So $\alpha_i = 0$ or $\alpha_i = 1$ for all i. If $\alpha_j = 1$, then $\alpha_i = 0$ for all $i \neq j$, so $a_0 = a_j$. By minimality, then $j = 1$ and $\{a_0, a_j\} = \{a_0\}$, which is impossible. So we must have $\alpha_i = 0$ for all i. But then $a_0 = 0$, which is likewise a contradiction. □

PROPOSITION 14. *Assuming only (M), nonzero copyable elements have unit norm.*

PROOF. Let a be a copyable element. Then $\delta(a) = a \otimes a$. Hence

$$\|a\| = \|\delta(a)\| = \|a \otimes a\| = \|a\|^2.$$

It follows that $\|a\|$ is either 0 or 1. Therefore, if a is a nonzero, then $\|a\| = 1$. □

PROPOSITION 15. *Assuming only (F), copyable elements are pairwise orthogonal.*

---

[6] Copyable elements are also called *primitive* in the context of C*-bigebras [**Hof70**], and *grouplike* in the study of Hopf algebras [**Swe69, Kas95**].

PROOF. [**CPV09**, Corollary 4.7] Let a, b be copyables. Then:

$$\begin{aligned}
\langle a \,|\, a \rangle \cdot \langle a \,|\, a \rangle \cdot \langle b \,|\, a \rangle &= \langle a \otimes a \otimes b \,|\, a \otimes a \otimes a \rangle \\
&= \langle (\delta \otimes \mathrm{id})(a \otimes b) \,|\, (\mathrm{id} \otimes \delta)(a \otimes a) \rangle \\
&= \langle a \otimes b \,|\, (\delta^\dagger \otimes \mathrm{id}) \circ (\mathrm{id} \otimes \delta)(a \otimes a) \rangle \\
&= \langle a \otimes b \,|\, (\mathrm{id} \otimes \delta^\dagger) \circ (\delta \otimes \mathrm{id})(a \otimes a) \rangle \\
&= \langle (\mathrm{id} \otimes \delta)(a \otimes b) \,|\, (\delta \otimes \mathrm{id})(a \otimes a) \rangle \\
&= \langle a \otimes b \otimes b \,|\, a \otimes a \otimes a \rangle \\
&= \langle a \,|\, a \rangle \cdot \langle b \,|\, a \rangle \cdot \langle b \,|\, a \rangle.
\end{aligned}$$

Analogously $\langle b \,|\, b \rangle \langle b \,|\, b \rangle \langle a \,|\, b \rangle = \langle b \,|\, b \rangle \langle a \,|\, b \rangle \langle a \,|\, b \rangle$. Hence, if $\langle a \,|\, a \rangle$ and $\langle b \,|\, a \rangle$ are both nonzero, then $\langle a \,|\, a \rangle = \langle b \,|\, a \rangle$ and $\langle b \,|\, b \rangle = \langle a \,|\, b \rangle$. So $\langle a \,|\, a \rangle, \langle b \,|\, b \rangle \in \mathbb{R}$ and $\langle a \,|\, a \rangle = \langle a \,|\, b \rangle = \langle b \,|\, a \rangle = \langle b \,|\, b \rangle$. Now suppose $\langle a \,|\, b \rangle \neq 0$. Then we can conclude $\langle a - b \,|\, a - b \rangle = \langle a \,|\, a \rangle - \langle a \,|\, b \rangle - \langle b \,|\, a \rangle + \langle b \,|\, b \rangle = 0$. So $a - b = 0$, *i.e.* $a = b$. Hence the copyables are pairwise orthogonal. □

Applying dagger, a copyable element of A corresponds exactly to a comonoid homomorphism $(\mathbb{C}, \delta_\mathbb{C}) \to (A, \delta)$:

$$\begin{array}{ccc}
1 & \longmapsto & 1 \otimes 1 \\
\downarrow & & \downarrow \\
a & \longmapsto & a \otimes a.
\end{array}$$

We have already seen that copyable elements correspond exactly to algebra homomorphisms

$$(A, \mu) \to (\mathbb{C}, \mu_\mathbb{C}),$$

i.e. to *characters* of the algebra—the elements of the Gelfand spectrum of A [**Ped89**]. This leads to our first characterization of when a (nonunital) Frobenius algebra in **Hilb** corresponds to an orthonormal basis.

THEOREM 16. *A Frobenius algebra in* **Hilb** *admits the structure theorem and hence corresponds to an orthonormal basis if and only if it is semisimple.*

PROOF. We first consider sufficiency. Form a direct sum of one-dimensional coalgebras indexed by the copyables of (A, δ). This will have an isometric embedding as a coalgebra into (A, δ):

$$e \colon \bigoplus_{\{a \mid \delta(a) = a \otimes a\}} (\mathbb{C}, \delta_\mathbb{C}) \to (A, \delta).$$

The image S of e is a closed subspace of A, and has an orthonormal basis given by the images of the characters of A qua copyables. The structure theorem holds if the image of e spans A.

Given $a \in A$ and a character c, the evaluation $c(a)$ gives the Fourier coefficient of a at the basis element of S corresponding to c. Now S will be the whole of A if and only if distinct vectors have distinct projections on S, *i.e.* if and only if distinct vectors have distinct Gelfand transforms $\hat{a} \colon c \mapsto c(a)$. Hence the Ambrose structure theorem holds when the Gelfand representation is injective, which holds if and only if the algebra is semisimple.

Necessity is easy to see from the form of a direct sum of one-dimensional algebras, as the lattice of ideals is a complete atomic boolean algebra, where the atoms are the generators of the algebras. □

We shall restate the previous theorem in terms of axiom (H), so that we have a characterization that lends itself to categories other than **Hilb**.

PROPOSITION 17. *A Frobenius algebra in* **Hilb** *satisfies (H) if and only if it is semisimple, and hence admits the structure theorem.*

PROOF. Semisimplicity of proper H*-algebras follows from results in [**Amb45**]. Conversely, $\bigoplus_I(\mathbb{C}, \mu_\mathbb{C})$ is easily seen to satisfy (H); we can define x^* by taking conjugate coefficients in the given basis. □

4.3. Categorical formulation. We can recast these results into a categorical form. Recall that there is a functor $\ell^2 \colon \mathbf{PInj} \to \mathbf{Hilb}$ on the category of sets and partial injections [**Bar92**, **Heu09a**]. It sends a set X to the Hilbert space $\ell^2(X) = \{\varphi \colon X \to \mathbb{C} \mid \sum_{x \in X} |\varphi(x)|^2 < \infty\}$, which is the free Hilbert space on X that is equipped with an orthonormal basis, *i.e.* an H*-algebra, in a sense we will now make precise. First, we make Frobenius algebras and H*-algebras into categories. While other choices of morphisms can fruitfully be made [**Heu10**], the following one suits our current purposes.

DEFINITION 18. Let **D** be a symmetric monoidal dagger category. We denote by **HStar(D)** the category whose objects are H*-algebras in **D**, and by **Frob(D)** the category whose objects are Frobenius algebras in **D**. A morphism $(A, \delta) \to (A', \delta')$ in both categories is a morphism $f \colon A \to A'$ in **D** satisfying $(f \otimes f) \circ \delta = \delta' \circ f$ and $f^\dagger \circ f = \mathrm{id}$.

PROPOSITION 19. *Every object in* **PInj** *carries a unique H*-algebra structure, namely $\delta(a) = (a, a)$.*

PROOF. Let $\delta = (A \twoheadleftarrow_{\delta_1}\!\!\rightarrowtail D \leftarrowtail\!\!_{\delta_2}\twoheadrightarrow A \times A)$ be an object of **HStar(PInj)**. Because of (M), we may assume that $\delta_1 = \mathrm{id}$. By (C), we find that δ_2 is a tuple of some $d \colon A \to A$ with itself. It follows from (A) that $d = d \circ d$. Finally, since **PInj** is monoidally well-pointed, δ satisfies (F) by Lemma 6. Writing out what (F) means gives

$$\{((d(b), b), (b, d(b))) \mid b \in A\} = \{((c, d(c)), (d(c), c)) \mid c \in A\}.$$

Hence for all $b \in A$, there is $c \in A$ with $b = d(c)$ and $d(b) = c$. Taking $b = d(a)$ we find that $c = a$, so that for all $a \in A$ we have $d \circ d(a) = a$. Therefore $d = d \circ d = \mathrm{id}$. We conclude that δ is the diagonal function $a \mapsto (a, a)$. □

As a corollary one finds that an object in **PInj** with its unique H*-algebra structure is unital if and only if it is a singleton set, which is another good argument against demanding (U).

If we drop the condition $f^\dagger \circ f = \mathrm{id}$ on morphisms in Definition 18, the previous proposition can also be read as saying that the categories **HStar(PInj)**, **Frob(PInj)**, and **PInj** are isomorphic.

Since the Hilbert space $\ell^2(X)$ comes with a chosen basis induced by X, the ℓ^2 construction is in fact a functor $\ell^2 \colon \mathbf{HStar(PInj)} \to \mathbf{HStar(Hilb)}$. Conversely, there is a functor U in the other direction taking an H*-algebra to the set of its copyables; this is functorial by [**Amb45**, Example 3]. These two functors are adjoints:

$$\mathbf{HStar}(\mathbf{PInj}) \underset{U}{\overset{\ell^2}{\rightleftarrows}} \mathbf{HStar}(\mathbf{Hilb}).$$

The Ambrose structure theorem, Theorem 12, can now be restated as saying that this adjunction is in fact an equivalence.

Similarly, there is an adjunction between **Frob**(**PInj**) and **Frob**(**Hilb**), but it is not yet clear if this is an equivalence, too, *i.e.* if **Frob**(**Hilb**) and **HStar**(**Hilb**) are equivalent categories. In fact, this question is the central issue of the rest of this paper, and will lead to the main open question in Section 4.5 to follow. In the meantime, we shall use the categorical formulation to give different characterizations of when Frobenius algebras in **Hilb** admit the structure theorem.

4.4. Further conditions. There are in fact a number of conditions on Frobenius algebras in **Hilb** which are equivalent to admitting the structure theorem. This section gives two more.

THEOREM 20. *A Frobenius algebra in* **Hilb** *is an H*-algebra, and hence corresponds to an orthonormal basis, if and only if it is a directed colimit (in* **Frob**(**Hilb**)*) of unital Frobenius algebras.*

PROOF. Given an orthonormal basis $\{|i\rangle\}_{i \in I}$ for A, define $\delta \colon A \to A \otimes A$ by (continuous linear extension of) $\delta|i\rangle = |ii\rangle$. For finite subsets F of I, define $\delta_F \colon \ell^2(F) \to \ell^2(F) \otimes \ell^2(F)$ by $\delta_F|i\rangle = |ii\rangle$. These are well-defined objects of **Frob**(**Hilb**) by Theorem 1. Since F is finite, every δ_F is a unital Frobenius algebra in **Hilb**. Together they form a (directed) diagram in **Frob**(**Hilb**) by inclusions $i_{F \subseteq F'} \colon \ell^2(F) \hookrightarrow \ell^2(F')$ if $F \subseteq F'$; the latter are well-defined morphisms since $\delta_{F'} \circ i_{F \subseteq F'}|i\rangle = |ii\rangle = (i_{F \subseteq F'} \otimes i_{F \subseteq F'}) \circ \delta_F|i\rangle$. Finally, we verify that δ is the colimit of this diagram. The colimiting cocone is given by the inclusions $i_F \colon \ell^2(F) \hookrightarrow A$; these are morphisms $i_F \colon \delta_F \to \delta$ in **Frob**(**Hilb**) since $\delta \circ i_F = (i_F \otimes i_F) \circ \delta_F$, that are easily seen to form a cocone. Now, if $f_F \colon \delta_F \to (A', \delta')$ form another cocone, define $m \colon X \to X'$ by (continuous linear extension of) $m|i\rangle = f_{\ell^2(\{|i\rangle\})}|i\rangle$. Then $m \circ i_F|i\rangle = f_{\ell^2(\{|i\rangle\})}|i\rangle = f_F|i\rangle$ for $i \in F$, so that indeed $m \circ i_F = f_F$. Moreover, m is the unique such morphism. Thus δ is indeed a colimit of the δ_F.

Conversely, suppose (A, δ) is a colimit of some diagram $d \colon \mathbf{I} \to \mathbf{Frob}(\mathbf{Hilb})$. We will show that the nonzero copyables form an orthonormal basis for A. By Lemma 13 and Proposition 15, it suffices to prove that they span a dense subspace of A. Let $a \in A$ be given. Since the colimiting cocone morphisms $c_i \colon A_i \to A$ are jointly epic, the union of their images is dense in A, and therefore a can be written as a limit of $c_i(a_i)$ with $a_i \in A_i$ for some of the $i \in \mathbf{I}$. These a_i, in turn, can be written as linear combinations of elements of copyables of A_i by Theorem 1. Now, c_i maps copyables into copyables, and so we have written a as a limit of linear combinations of copyables of A. Hence the copyables of A spans a dense subspace of A, and therefore form an orthonormal basis.

Finally, we verify that these two constructions are mutually inverse. Starting with a δ, one obtains $E = \{e \mid \delta(e) = e \otimes e\}$, and then $\delta' \colon A \to A \otimes A$ by (continuous linear extension of) $\delta'(e) = e \otimes e$ for $e \in E$. The definition of E then gives $\delta' = \delta$.

Conversely, starting with an orthonormal basis $\{|i\rangle\}_{i \in I}$, one obtains a map $\delta \colon A \to A \otimes A$ by (continuous linear extension of) $\delta|i\rangle = |ii\rangle$, and then it follows that $E = \{a \in A \mid \delta(a) = a \otimes a\}$. It is trivial that $\{|i\rangle \mid i \in I\} \subseteq E$. Moreover, we know that E is linearly independent by 13. Since it contains a basis, it must therefore be a basis itself. Hence indeed $E = \{|i\rangle \mid i \in I\}$. □

For *separable* Hilbert spaces, there is also a characterization in terms of approximate units as follows.

THEOREM 21. *A Frobenius algebra on a separable Hilbert space in* **Hilb** *is an H*-algebra, and hence corresponds to an orthonormal basis, if and only if there is a sequence* e_n *such that* $e_n a$ *converges to* a *for all* a, *and* $(\mathrm{id} \otimes a^\dagger) \circ \delta(e_n)$ *converges.*

PROOF. Writing $a_n^* = (\mathrm{id} \otimes a^\dagger) \circ \delta(e_n)$, by assumption $a^* = \lim_{n \to \infty} a_n^*$ is well-defined. Since morphisms in **Hilb** are continuous functions and composition preserves continuity, (H) holds by the following argument.

Hence approximate units imply (H). Conversely, using the Ambrose structure theorem, Theorem 12, we can always define e_n to be the sum of the first n copyables. □

Summarizing, we have the following result.

THEOREM 22. *For a Frobenius algebra in* **Hilb**, *the following are equivalent:*
  (a) *it is induced by an orthonormal basis;*
  (b) *it admits the structure theorem;*
  (c) *it is semisimple;*
  (d) *it satisfies axiom (H);*
  (e) *it is a directed colimit (with respect to isometric homomorphisms) of finite-dimensional unital Frobenius algebras.*

Moreover, if the Hilbert space is separable, these are equivalent to:
  (f) *it has a suitable form of approximate identity.* □

We see that the finite-dimensional result follows immediately from our general result and Lemma 5, which shows that the algebra is C* and hence semisimple. In fact, the influential thesis [**Abr97**] (see also [**Abr00**, **Koc03**]) already observes explicitly (and in much wider generality) that:
- If (M) holds, a unital Frobenius algebra is semisimple [**Abr97**, Theorem 2.3.3].
- A commutative semisimple unital Frobenius algebra is a direct sum of fields [**Abr97**, Theorem 2.2.5].

Thus the only additional ingredient required to obtain Theorem 1 is the elementary Proposition 15.

4.5. The main question. The main remaining question in our quest for a suitable notion of algebra to characterize orthonormal bases in arbitrary dimension is the following.

> In the presence of (A), (C), and (M), does (F) imply (H)?

We can ask this question for the central case of **Hilb**, and for monoidal dagger categories in general.

If the answer is positive, then nonunital Frobenius algebras give us the right notion of observable to use in categorical quantum mechanics. If it is negative, we may consider adopting (H) as the right axiom instead of (F).

At present, these questions remain open, both for **Hilb** and for the general case. However, we have been able to achieve positive results for a large family of categories; these will be described in the following section. We shall conclude this section by further narrowing down the question in the category **Hilb**.

Recall that the *Jacobson radical* of a commutative ring is the intersection of all its maximal regular ideals, and that a ring is called *radical* when it equals its Jacobson radical.

PROPOSITION 23. *Frobenius algebras A in* **Hilb** *decompose as a direct sum*

$$A \cong S \oplus R$$

of (co)algebras, where S is an H*-algebra and R is a radical algebra.

PROOF. Let a be a copyable element of a Frobenius algebra A in **Hilb**. Consider the embedding into A of a as a one-dimensional algebra. This embedding is a kernel, since it is isometric and its domain is finite-dimensional. Observe that this embedding is an algebra homomorphism as well as a coalgebra homomorphism, because copyables are idempotents by (M). Now it follows from [**Heu10**, Lemma 19] that also the orthogonal complement of the embedding is both an algebra homomorphism and a coalgebra homomorphism. Finally, Frobenius algebra structure restricts along such embeddings by [**Heu10**, Proposition 9].

We can apply this to the embedding of the closed span of all copyables of A, and conclude that A decomposes (as a (co)algebra) into a direct sum of its copyables and the orthogonal subspace. By definition, the former summand is semisimple, and is hence a H*-algebra by Proposition 17. The latter summand by construction has no copyables and hence no characters, and is therefore radical. □

This shows how the Jacobson radical of a Frobenius algebra sits inside it in a very simple way. Indeed, we are left with not just a nonsemisimple algebra, but a radical one, which is the opposite of a semisimple algebra—an algebra is semisimple precisely when its Jacobson radical is zero. Therefore, in the category **Hilb**, our main remaining question above reduces to finding out whether R must be zero, as follows.

> Does there exist a nontrivial radical Frobenius algebra?

Although there is an extensive literature about commutative radical Banach algebras, including a complete classification that in fact ties in with approximate units [**Est83**], this question seems to be rather difficult.

5. H*-algebras in categories of relations and positive matrices

We have been able to give a complete analysis of nonunital Frobenius algebras in several (related) cases, including:

- categories of relations, and locally bifinite relations, valued in cancellative quantales;
- nonnegative matrices with ℓ^2-summable rows and columns.

The common feature of these cases can be characterized as *the absence of destructive interference*.

The main result we obtain is as follows.

THEOREM 24. *Nonunital Frobenius algebras in all these categories decompose as direct sums of abelian groups, and satisfy (H).*

The remainder of this section is devoted to the proof of this theorem. Our plan is as follows. First, we shall prove the result for **Rel**, the category of sets and relations. In this case, our main question is already answered directly by Proposition 7 and Theorem 22. Moreover, the result in this case has appeared in [**Pav09**]. However, our proof is quite different, and in particular makes no use of units. This means that it can be carried over to the other situations mentioned above.

5.1. Frobenius algebras in Rel and lbfRel. We assume given a set A, and a Frobenius algebra structure on it given by a relation $\Delta \subseteq A \times (A \times A)$. We shall write ∇ for $\Delta^\dagger$.

DEFINITION 25. Define $x \sim y$ if and only if $(x, y)\nabla z$ for some z. By (M), the relation ∇ is single-valued and surjective. Therefore, we may also use multiplicative notation xy (suppressing the ∇), and write $x \sim y$ to mean that xy is defined.

LEMMA 26. *The relation $\sim$ is reflexive.*

PROOF. Let $a \in A$. By (M), we have $a = a_1 a_2$ for some $a_1, a_2 \in A$. Then $(a_2, a)(\text{id} \otimes \Delta)(a_2, a_1, a_2)$ and $(a_2, a_1, a_2)(\nabla \otimes \text{id})(a, a_2)$ by (C), so by (F) we have $(a_2, a)\Delta \circ \nabla(a, a_2)$, so that aa_2 is defined. Diagrammatically, we annotate the lines with elements to show they are related by that morphism.

Also $(a, a)(\Delta \otimes \text{id})(a_1, a_2, a)$ and $(a_1, a_2, a)(\text{id} \otimes (\Delta \circ \nabla))(a_1, a, a_2)$, so by (F) we have $(a, a)(\text{id} \otimes \Delta) \circ \Delta \circ \nabla(a_1, a, a_2)$, so that a^2 is defined.

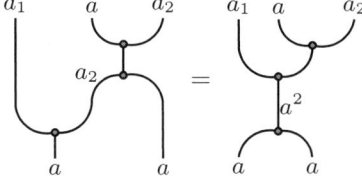

That is, $a \sim a$. □

LEMMA 27. *The relation $\sim$ is transitive.*

PROOF. Suppose that $a \sim b$ and $b \sim c$. Then $d = ab$ is defined. By Lemma 26, then $(ab)d = d^2$ is defined. Hence by (A), also $a(bd)$ is defined. Applying (F) now yields $\bar{b}$ such that $a = d\bar{b}$.

$$\begin{array}{c} \text{[diagram]} \end{array}$$

It now follows from (C) that $a = d\bar{b} = a\bar{b}b$; in particular $a\bar{b}$ is defined. But then also $ac = (a\bar{b}b)c = (a\bar{b})(bc)$ is seen to be defined by the assumption $b \sim c$ and another application of (F).

$$\begin{array}{c} \text{[diagram]} \end{array}$$

Hence $a \sim c$. □

PROPOSITION 28. *The Frobenius algebra A is a disjoint union of totally defined commutative semigroups, each satisfying (F).*

PROOF. By the previous two lemmas and (C), the relation $\sim$ is a equivalence relation. Hence A is a disjoint union of the equivalence classes under $\sim$. By definition of $\sim$, the multiplication ∇ is totally defined on these equivalence classes. Moreover, they inherit the properties (M), (C), (A) and (F) from A. □

LEMMA 29. [**Hun02**] *A semigroup S is a group if and only if $aS = S = Sa$ for all $a \in S$.*

PROOF. The condition $aS = S$ means $\forall b \exists c [b = ac]$. If S is a group, this is obviously fulfilled by $c = a^{-1}b$. For the converse, fix $a \in S$. Applying the condition with $b = a$ yields c such that $a = ac$. Define $e = c$, and let $x \in S$. Then applying the condition with $b = x$ gives c with $x = ac$. Hence $ex = eac = ac = x$. Thus S is a monoid with (global) unit e. Applying the condition once more, with $a = x$ and $b = e$ yields x^{-1} with $xx^{-1} = e$. □

THEOREM 30. *A is a disjoint union of commutative groups.*

PROOF. Let A' be one of the equivalence classes of A, and consider $a, b \in A'$. This means that $a \sim b$. As in the proof of Lemma 27, there is a $\bar{b}$ such that $a = \bar{b}ba$. Putting $c = \bar{b}a$ thus gives $\forall a, b \in A' \exists c \in A'[a = cb]$. In other words, $aA' = A'$ and similarly $A' = A'a$ for all $a \in A$. Hence A' is a (commutative) group by Lemma 29. □

The following theorem already follows from Proposition 7 and Theorem 22, but now we have a direct proof that also carries over to the theorem after it, which does not follow from the earlier results.

THEOREM 31. *In **Rel**, Frobenius algebras satisfy (H), and the conditions (F) and (H) are equivalent in the presence of the other axioms for Frobenius algebras.*

PROOF. This follows directly from Lemma 30, since we can define $a^* = a^{-1}$, where a^{-1} is the inverse in the disjoint summand containing a. More precisely, a point of A in **Rel** will be a subset of A, and we apply the definition $a^* = a^{-1}$ pointwise to this subset. This assignment is easily seen to satisfy (H). □

THEOREM 32. *In* **lbfRel**, *Frobenius algebras are disjoint unions of abelian groups and hence satisfy (H), and the conditions (F) and (H) are equivalent in the presence of the other axioms for Frobenius algebras.*

PROOF. The proof above made no use of units, and is equally valid in **lbfRel**. □

5.2. Quantale-valued relations. We shall now consider categories of the form $\mathbf{Rel}(Q)$, where Q is a commutative, cancellative quantale. Recall that a commutative quantale [**Ros90**] is a structure $(Q, \cdot, 1, \leq)$, where $(Q, \cdot, 1)$ is a commutative monoid, and $(Q, \leq)$ is a partial order which is a complete lattice, *i.e.* it has suprema of arbitrary subsets. In particular, the supremum of the empty set is the least element of the poset, written 0. The multiplication is required to distribute over arbitrary joins, *i.e.*

$$x \cdot (\bigvee_{i \in I} y_i) = \bigvee_{i \in I} x \cdot y_i, \qquad (\bigvee_{i \in I} x_i) \cdot y = \bigvee_{i \in I} x_i \cdot y.$$

The quantale is called *cancellative* if

$$x \cdot y = x \cdot z \implies x = 0 \vee y = z.$$

An example is given by the extended nonnegative reals $[0, \infty]$ with the usual ordering, and multiplication as the monoid operation. Note that the only nontrivial example when the monoid operation is idempotent, *i.e.* when the quantale is a locale, is the two-element boolean algebra $\mathbf{2} = \{0, 1\}$, since in the idempotent case $x \cdot 1 = x \cdot x$ for all x. We write **canQuant** for the category of cancellative quantales which are nontrivial, *i.e.* in which $0 \neq 1$.

PROPOSITION 33. *The two element boolean algebra is terminal in* **canQuant**.

PROOF. The unique homomorphism $h \colon Q \to \mathbf{2}$ sends 0 to itself, and everything else to 1. Preservation of sups holds trivially, and cancellativity implies that multiplication is preserved. □

The category $\mathbf{Rel}(Q)$ has sets as objects; morphisms $R \colon X \nrightarrow Y$ are Q-valued matrices, *i.e.* functions $X \times Y \to Q$. Composition is relational composition evaluated in Q, *i.e.* if $R \colon X \nrightarrow Y$ and $S \colon Y \nrightarrow Z$, then

$$S \circ R(x, z) = \bigvee_{y \in Y} R(x, y) \cdot S(y, z).$$

It is easily verified that this yields a category, with identities given by diagonal matrices; that it has a monoidal structure induced by cartesian product; and that is has a dagger given by matrix transpose, *i.e.* relational converse. Thus $\mathbf{Rel}(Q)$ is a symmetric monoidal dagger category, and the notion of Frobenius algebra makes sense in it. Note that $\mathbf{Rel}(\mathbf{2})$ is just **Rel**.

A homomorphism of quantales $h \colon Q \to R$ induces a (strong) monoidal dagger functor $h^* \colon \mathbf{Rel}(Q) \to \mathbf{Rel}(R)$, which transports Frobenius algebras in $\mathbf{Rel}(Q)$ to Frobenius algebras in $\mathbf{Rel}(R)$. In particular, by Proposition 33, a Frobenius algebra $\Delta \colon A \nrightarrow A \times A$ in $\mathbf{Rel}(Q)$ has a reduct $h^* \Delta \colon A \nrightarrow A \times A$ in **Rel**. Hence Theorems 30 and 31 apply to this reduct. The remaining degree of freedom in the Frobenius algebra in $\mathbf{Rel}(Q)$ is which elements of Q can be assigned to the elements of the matrix.

Suppose that we have a Frobenius algebra $\Delta\colon A \rightarrowtail A \times A$ in $\mathbf{Rel}(Q)$. We write $M\colon (A \times A) \times A \to Q$ for the matrix function corresponding to the converse of Δ, and we write $M(a,b,c)$ rather than $M((a,b),c)$.

Because the unique homomorphism $Q \to \mathbf{2}$ reflects 0, an entry $M(a,b,c)$ is nonzero if and only if the corresponding relation $(a,b)(h^*\nabla)c$ holds. Applying Theorem 30, this immediately implies that for each $a,b \in A$, there is exactly one $c \in A$ such that $M(a,b,c) \neq 0$.

We can use this observation to apply similar diagrams to those used in our proofs for $\mathbf{Rel}$ to obtain constraints on the values taken by the matrix in Q.

PROPOSITION 34. *With notation as above:*

(a) *If e is an identity element in one of the disjoint summands, then for all a,b in that disjoint summand we have $M(a,e,a) = M(b,e,b)$. We write q_e for this common value.*

(b) *For all $a,b \in A$, we have $M(a,b,ab)^2 = 1$.*

PROOF. For (a), consider the diagram

This implies the equation
$$M(a,b,ab) \cdot M(ab,e,ab) = M(a,b,ab) \cdot M(b,e,b)$$
and hence, by cancellativity, $M(ab,e,ab) = M(b,e,b)$. Hence for any c, taking $b = a^{-1}c$, $M(c,e,c) = M(b,e,b)$.

For (b), consider the diagram

This implies the equation $M(a,b,ab)^2 = q_e^2$. Now applying (M), for each c we obtain that
$$\bigvee \{M(a,b,ab)^2 \mid ab = c\} = 1.$$
As all the terms in this supremum are the same, $M(a,b,ab)^2 = 1$. □

Thus if $q^2 = 1$ implies $q = 1$ in Q, the matrix M is in fact valued in $\mathbf{2}$. Otherwise, we can choose square roots of unity for the entries.

THEOREM 35. *Let Q be a cancellative quantale. Suppose that $q^2 = 1$ implies $q = 1$ in Q. Then every Frobenius algebra in $\mathbf{Rel}(Q)$ satisfies (H).* □

5.3. Positive ℓ^2 matrices. We now consider the case of matrices in $\mathbf{Mat}_{\ell^2}(\mathbb{C})$ valued in the non-negative reals. These form a monoidal dagger subcategory of $\mathbf{Mat}_{\ell^2}(\mathbb{C})$, which we denote by $\mathbf{Mat}_{\ell^2}(\mathbb{R}^+)$. Note that the semiring $(\mathbb{R}^+, +, 0, \times, 1)$ has a unique 0-reflecting semiring homomorphism to $\mathbf{2}$. Hence a Frobenius algebra in $\mathbf{Mat}_{\ell^2}(\mathbb{R}^+)$ has a reduct to one in $\mathbf{Rel}$ via this homomorphism. Just as before, we can apply Theorem 30 to this reduct.

We have the following analogue to Proposition 34, where $M\colon (A\times A)\times A \to \mathbb{R}^+$ is the matrix realizing the Frobenius algebra structure.

PROPOSITION 36. *The function M is constant on each disjoint summand of A.*

PROOF. We can use the same reasoning as in Proposition 34(a) to show that, if e is an identity element in one of the disjoint summands, then for all a, b in that disjoint summand, $M(a,e,a) = M(b,e,b)$. We write r_e for this common value.

Using the same reasoning as in Proposition 34(b) one finds $M(a,b,ab)^2 = r_e^2$. Since we are in $\mathbb{R}^+$, this implies $M(a,b,ab) = r_e$, so that M is constant on each disjoint summand. $\square$

PROPOSITION 37. *Each disjoint summand is finite, and the common value of M on that summand is $1/\sqrt{d}$, where d is the cardinality of the summand.*

PROOF. Applying (M), for each c in the summand we obtain that

$$\sum_{ab=c} M(a,b,c)^2 = 1.$$

Since the summand is a group, for each c and a there is a unique b such that $ab = c$. Moreover, by Proposition 36, all the terms in this sum are equal. Thus the sum must be finite, with the number of terms d the cardinality of the summand. We can therefore rewrite the equation as $dr_e^2 = 1$, and hence $r_e = 1/\sqrt{d}$. $\square$

THEOREM 38. *Every Frobenius algebra in $\mathbf{Mat}_{\ell^2}(\mathbb{R}^+)$ satisfies (H).*

PROOF. We make the same pointwise assignment $x^* = x^{-1}$ on the elements of A as in the proof of Theorem 31, with the weight $1/\sqrt{d}$ determined by the summand. $\square$

In the case when the matrix represents a bounded linear map in **Hilb**, we can apply Theorem 22, and obtain the following.

PROPOSITION 39. *If a Frobenius algebra in **Hilb** can be represented by a nonnegative real matrix, then it corresponds to a direct sum of one-dimensional algebras, and hence to an orthonormal basis.* $\square$

Conversely, if a Frobenius algebra in **Hilb** satisfies (H), it is induced by an orthonormal basis, and hence has a matrix representation with nonnegative entries. Therefore we have found another equivalent characterization of when (F) implies (H) in **Hilb** to add to our list in Theorem 22.

PROPOSITION 40. *A Frobenius algebra A in **Hilb** satisfies (H) if and only if there is a basis of A such that the matrix of the comultiplication has nonnegative entries when represented on that basis.*

5.4. Discussion. How different is the situation with Frobenius algebras in these matrix categories from **Hilb**? In fact, it is not as different as it might at first appear.

- The category **Hilb** is equivalent to a full subcategory of the dagger monoidal category of complex matrices with ℓ^2-summable rows and columns. The 'only' assumption needed but not satisfied is positivity.

- The result 'looks' different, but beware. Consider *group rings* (or algebras) over the complex numbers, for finite abelian groups. They can easily be set up to fulfil all our axioms, including (U), so that Theorem 1 applies, and they decompose as direct sums of one-dimensional algebras. But the isomorphism which gives this decomposition may be quite non-obvious.[7] Note that copyable elements are idempotents, so the only group element which is copyable is the identity.
- This decomposition result indeed shows that group rings over the complex numbers are very weak invariants of the groups. The group rings of two finite abelian groups will be isomorphic if the groups have the same order [**Str66**]!
- However, this is highly sensitive to which field we are over. Group algebras over the *rationals* are isomorphism invariants of groups [**Ing65**].

6. Outlook

We are still investigating our main question, of whether (F) implies (H), in **Hilb** and elsewhere.

Beyond this, we see the following main lines for continuing a development of categorical quantum mechanics applicable to infinite-dimensional situations.

- We are now able to consider observables with infinite discrete spectra. Beyond this lie continuous observables and projection-valued measures; it remains to be seen how these can be analyzed in the setting of categorical quantum mechanics.
- Complementary observables should be studied in this setting. The bialgebra approach studied in [**CD08**] is based on axiomatizing *mutually unbiased bases*, and does not extend directly to the infinite-dimensional case. However, complementary observables are studied from a much more general perspective in works such as [**BGL95**], and this should provide a good basis for suitable categorical axiomatizations.
- This leads on to another point. There may be other means, within the setting of categorical quantum mechanics, of representing observables, measurements and complementarity, which may be more flexible than the Frobenius algebra approach, and in a sense more natural, since tensor product structure is not inherent in the basic notion of measurement. Methodologically, one should beware of concluding over-hastily that a particular approach is canonical, simply on the grounds that it captures the standard notion in finite-dimensional Hilbert spaces. There may be several ways of doing this, and some more definitive characterization would be desirable.
- A related investigation to the present one is the work on nuclear and traced ideals in [**ABP99**]. It seems likely that some combination of the ideas developed there, and those we have studied in this paper, will prove fruitful.

Acknowledgement. We thank Rick Blute for stimulating the early stages of the research that led to this article.

---

[7] It would be interesting, for example, to know the computational complexity of determining this isomorphism, given a presentation of the group. As far as we know, this question has not been studied.

References

[ABP99] Samson Abramsky, Richard Blute, and Prakash Panangaden, *Nuclear and trace ideals in tensored *-categories*, Journal of Pure and Applied Algebra **143** (1999), 3–47.

[Abr97] Lowell Abrams, *Frobenius algebra structures in topological quantum field theory and quantum cohomology*, Ph.D. thesis, John Hopkins University, 1997.

[Abr00] _____, *The quantum Euler class and the quantum cohomology of the Grassmannians*, Israel Journal of Mathematics **117** (2000), no. 1, 335–352.

[Abr10] Samson Abramsky, *No-cloning in categorical quantum mechanics*, Semantic Techniques in Quantum Computation (Simon Gay and Raja Nagarajan, eds.), Cambridge University Press, 2010, pp. 1–28.

[AC04] Samson Abramsky and Bob Coecke, *A categorical semantics of quantum protocols*, Logic in Computer Science 19, IEEE Computer Society, 2004, pp. 415–425.

[AC09] _____, *Categorical quantum mechanics*, Handbook of Quantum Logic and Quantum Structures (K. Engesser, D. Gabbay, and D. Lehmann, eds.), vol. 2, Elsevier, 2009, pp. 261–325.

[Alb39] A. Adrian Albert, *Structure of algebras*, Colloquium Publications, vol. 26, American Mathematical Society, 1939.

[Amb45] Warren Ambrose, *Structure theorems for a special class of Banach algebras*, Transactions of the American Mathematical Society **57** (1945), no. 3, 364–386.

[Bae97] John C. Baez, *Higher-dimensional algebra II: 2-Hilbert spaces*, Advances in Mathematics **127** (1997), 125–189.

[Bar92] Michael Barr, *Algebraically compact functors*, Journal of Pure and Applied Algebra **82** (1992), 211–231.

[Bar01] Henk P. Barendregt, *The lambda calculus: its syntax and semantics*, North-Holland, 2001.

[BEH08] Jiří Blank, Pavel Exner, and Miloslav Havlíček, *Hilbert space operators in quantum physics*, Springer, 2008.

[BGL95] Paul Busch, Marian Grabowski, and Pekka J. Lahti, *Operational Quantum Physics*, Springer, 1995.

[Car91] Aurelio Carboni, *Matrices, relations and group representations*, Journal of algebras **136** (1991), 497–529.

[CD08] Bob Coecke and Ross Duncan, *Interacting quantum observables: Categorical algebra and diagrammatics*, Automata, Languages and Programming, ICALP 2008, Lecture Notes in Computer Science, no. 5126, Springer, 2008, pp. 298–310.

[CK10] Bob Coecke and Aleks Kissinger, *The compositional structure of multipartite quantum entanglement*, International Colloquium on Automata, Languages and Programming, Volume II, Lecture Notes in Computer Science, no. 6199, Springer, 2010, pp. 297–308.

[CKS84] Aurelio Carboni, Stefano Kasangian, and Ross Street, *Bicategories of spans and relations*, Journal of Pure and Applied Algebra **33** (1984), 259–267.

[CP07] Bob Coecke and Duško Pavlović, *Quantum measurements without sums*, Mathematics of Quantum Computing and Technology, Taylor and Francis, 2007.

[CPV09] Bob Coecke, Duško Pavlović, and Jamie Vicary, *A new description of orthogonal bases*, Mathematical Structures in Computer Science (2009).

[Est83] Jean Esterle, *Radical Banach algebras and automatic continuity*, Lectures Notes in Mathematics, no. 975, ch. Elements for a classification of commutative radical Banach algebras, pp. 4–65, Springer, 1983.

[Heu09a] Chris Heunen, *Categorical quantum models and logics*, Ph.D. thesis, Radboud University Nijmegen, 2009.

[Heu09b] _____, *An embedding theorem for Hilbert categories*, Theory and Applications of Categories **22** (2009), no. 13, 321–344.

[Heu10] _____, *Complementarity in categorical quantum mechanics*, To appear in Foundations of Physics (2010).

[Hof70] Karl H. Hofmann, *The duality of compact semigroups and C*-bigebras*, Lecture Notes in Mathematics, vol. 129, Springer, 1970.

[Hun02] Edward V. Huntington, *Simplified definition of a group*, Bulletin of the American Mathematical Society **8** (1902), no. 7, 296–300.

[Ing65] Lars Ingelstam, *Real algebras with a Hilbert space structure*, Arkiv för Matematik **6** (1965), no. 24, 459–465.

[Kap48] Irving Kaplansky, *Dual rings*, Annals of mathematics **49** (1948), no. 3, 689–701.

[Kas95] Christian Kassel, *Quantum groups*, Springer, 1995.

[Koc03] Joachim Kock, *Frobenius algebras and 2-D topological quantum field theories*, London Mathematical Society Student Texts, no. 59, Cambridge University Press, 2003.

[LR97] Roberto Longo and John E. Roberts, *A theory of dimension*, K-Theory **11** (1997), 103.

[Nak39] Tadasi Nakayama, *On Frobeniusean algebras. I*, Annals of Mathematics (second series) **40** (1939), no. 3, 611–633.

[Pav09] Duško Pavlović, *Quantum and classical structures in nondeterministic computation*, Third International symposium on Quantum Interaction (P. Bruza et al., ed.), Lecture Notes in Artificial Intelligence, vol. 5494, Springer, 2009, pp. 143–157.

[Ped89] Gert K. Pedersen, *Analysis Now*, Springer, 1989.

[Ros90] Kimmo I. Rosenthal, *Quantales and their applicatoins*, Pitman Research Notes in Mathematics, Longman Scientific & Technical, 1990.

[Sel10] Peter Selinger, *A survey of graphical languages for monoidal categories*, New Structures for Physics, Lecture Notes in Physics, Springer, 2010.

[Str66] Robert S. Strichartz, *Isomorphisms of group algebras*, Proceedings of the American Mathematical Society **17** (1966), 858–862.

[Swe69] Moss E. Sweedler, *Hopf algebras*, W.A. Benjamin, 1969.

[WZ82] William K. Wootters and Wojciech H. Zurek, *A single quantum cannot be cloned*, Nature **299** (1982), 802–803.

Oxford University Computing Laboratory, Wolfson Building, Parks Road, Oxford OX1 3QD, United Kingdom
E-mail address: samson.abramsky@comlab.ox.ac.uk

Oxford University Computing Laboratory, Wolfson Building, Parks Road, Oxford OX1 3QD, United Kingdom
E-mail address: chris.heunen@comlab.ox.ac.uk

Teleportation in General Probabilistic Theories

Howard Barnum, Jonathan Barrett, Matthew Leifer, and Alexander Wilce

ABSTRACT. In a previous paper, we showed that many important quantum information-theoretic phenomena, including the no-cloning and no-broadcasting theorems, are in fact generic in all non-classical probabilistic theories. An exception is teleportation, which most such theories do not support. In this paper, we investigate which probabilistic theories, and more particularly, which composite systems, *do* support a teleportation protocol. We isolate a natural class of composite systems that we term *regular*, and establish necessary and sufficient conditions for a regular tripartite system to support a conclusive, or post-selected, teleportation protocol. We also establish a sufficient condition for deterministic teleportation that yields a large supply of composite state spaces, neither classical nor quantum, that support such a protocol.

1. Introduction

The standard quantum teleportation protocol [8] allows two parties, Alice and Bob, to transmit an unknown quantum state from Alice's site to Bob's; in compliance with the no-cloning theorem, Alice's copy is destroyed in the process. The protocol assumes that Alice and Bob have access to the two wings, A and B, of a bipartite system $A \otimes B$ in a maximally entangled state, which serves as a kind of quantum channel. The state to be teleported belongs to an auxiliary system A' at Alice's site, which is coupled to her half of the shared system. Alice measures an observable corresponding to the Bell basis on the combined system $A' \otimes A$. Depending upon the result, she instructs Bob (via purely classical signaling) to perform a particular unitary correction on his wing, B, of the shared $A \otimes B$ system. With certainty, Alice now knows that the state of Bob's system is identical to the state (whatever it was) of her ancillary system A'.

The possibility of teleportation is surprising, in view of the no-cloning and no-broadcasting theorems, which prohibit the copying of quantum information. In a

---

2010 *Mathematics Subject Classification*. Primary 81P16; Secondary 51A20.

At the Institute for Quantum Computing, ML was supported in part by MITACS and OR-DCF. ML was supported in part by grant RFP1-06-006 from FQXi. Research at Perimeter Institute for Theoretical Physics is supported in part by the Government of Canada through NSERC and by the Province of Ontario through MRI. This work was also carried out partially under the auspices of the US Department of Energy through the LDRD program at Los Alamos National Laboratory under Contract No. DE-AC52-06NA25396.

previous paper [**6**], we have shown that both no-cloning and no-broadcasting theorems are in fact quite generic features of essentially any *non-classical* probabilistic theory, and not specifically quantum at all. However, as pointed out in [**4, 5**], most such theories *do not* allow for teleportation. Since classical theories do allow for it, the possibility of teleportation can be regarded, in some very rough qualitative sense, as a measure of the relative *classicality* (or at any rate, *tameness*) of quantum theory.

In this note, we make some precise statements about *which* probabilistic theories – and more particularly, which tripartite systems – admit teleportation. For simplicity, consider the case in which the three component systems, A', A and B, in the protocol described above are identical. Then an obvious necessary condition for the protocol to succeed is that the cone of unnormalized states in A be isomorphic to the *dual* cone of unnormalized *effects* in A^* – a strong condition that is nevertheless satisfied by both quantum and classical systems. As we shall see, this is sufficient to ground conclusive (or one-outcome post-selected) teleportation. To obtain deterministic teleportation appears to be more difficult; however, where the state space has sufficient symmetry, a sort of deterministic teleportation can always be achieved with respect to a possibly continuously-indexed observable. Specializing to the case in which the state space is symmetric under the action of a finite group, we obtain a wealth of examples of state spaces that are neither classical nor quantum-mechanical, but nevertheless support a genuine deterministic teleportation protocol.

2. Probabilistic Models

This section assembles the necessary machinery of generalized probability theory, using a standard mathematical framework deriving from the work of Mackey [**18**] and subsequently refined by many authors, notably Davies and Lewis [**9**], C. M. Edwards [**10**] and Ludwig [**17**]. We use more or less the same notation as in [**5, 6**]; as in the latter, we consider only probabilistic models having finite-dimensional state spaces.

Abstract State Spaces We recall that an ordered vector equipped with a distinguished *positive cone*, A_+ – a convex set closed under multiplication by positive scalars, and also satisfying $A_+ \cap -A_+ = \{0\}$. Elements of A_+ are termed *positive* vectors. The cone A_+ induces a partial ordering on A, given by setting $a \leq b$ iff $b - a \in A_+$. We shall also assume here that A_+ is closed (recall here the standing assumption that all spaces are finite dimensional, so there is no ambiguity about the topology), (spans) A, so that every $a \in A$ can be written — in general, not uniquely – as a difference of positive vectors.

We model a physical system by an ordered vector space A with a (closed, pointed, generating) positive cone A_+, which we regard as consisting of un-normalized "states". We also posit a distinguished order unit, that is, a linear functional u_A that is *strictly* positive on non-zero positive elements of A. This defines a compact convex set $\Omega_A = u_A^{-1}(1)$ of *normalized* states. We refer to the extreme points of Ω_A as *pure states*; non-pure (normalized) states are *mixed*.

We shall call an ordered linear space, equipped with such a functional – more formally: a pair (A, u_A) – an *abstract state space*. If (A, u_A) and (B, u_B) are abstract state spaces, we write $A \leq B$ to indicate that (i) A is a subspace of B; (ii) $A_+ \subseteq B_+$; and (iii) u_A is the restriction of u_B to A_+. Similarly, $A \simeq B$, read "A is isomorphic to B", means that there exists an invertible, positive linear mapping $A \to B$, with a positive inverse, and taking the order unit of A to that of B. Equivalently, such a mapping takes A's normalized state space Ω_A bijectively (and affinely) onto B's normalized state space Ω_B. We refer to an isomorphism $A \to A$ as a *symmetry* of A. A positive linear mapping with positive inverse, but not necessarily preserving the order unit, we refer to as an *order isomorphism* between A and B, and we say that A and B are order-isomorphic if such a map exists.

By way of illustration, discrete classical probability theory concerns the case in which A is the space $\mathbb{R}^E$ of all real-valued functions α on a finite set E of measurement outcomes, in the natural point-wise ordering. The order unit is the functional $u_A(\alpha) := \sum_{x \in E} \alpha(x)$, hence the normalized state space Ω_A consists of all probability weights on E. In elementary quantum probability theory, A is the space of Hermitian operators on a complex Hilbert space $\mathbf{H}$, ordered in the usual way; the order unit is the trace, so that Ω_A is the set of density operators.

EXAMPLE 2.1. For a simple example of a probabilistic system that is neither classical nor quantum consider a pair of measurements with outcome-sets $\{x, x'\}$ and $\{y, y'\}$, respectively. A simultaneous assignment of probabilities to the outcomes of these two measurements amounts to a pair of real numbers $0 \leq p(x), p(y) \leq 1$; these can be chosen freely, so the set Ω of all such probability assignments can be identified with the unit square $[0,1]^2$. The set Ω spans a three-dimensional subspace, A, of $\mathbb{R}^X$. Ordered pointwise, and with $u(\alpha) = 1$ for every $\alpha \in \Omega$, A is an abstract state space. In some sense, this is the simplest possible non-classical probabilistic system – more complicated than a classical bit, but much simpler than the simplest quantum system, the qubit. Accordingly, we'll refer to this example (of which we'll make further use below) as a *square bit* or *squit*, for short.

Effects Physical events (e.g., measurement outcomes) associated with an abstract state space A are represented by *effects*, that is, positive linear functionals $f \in A^*$ with $f(\alpha) \leq 1$ for all $\alpha \in \Omega_A$, or, equivalently, $f \leq u_A$. The understanding is that $f(\alpha)$ represents the *probability* that the event in question will occur when the system's state is α. As indicated above, we wish to restrict our attention here to cases in which the space A is finite-dimensional. Thus we may identify A with A^{**}, so that, for $\alpha \in A$ and $f \in A^*$, we may write $f(\alpha)$ as $\alpha(f)$ whenever it suits us. In the sequel, we shall continue always to denote states by lower case Greek letters, and effects, by lower case Roman letters.

It is helpful to note that the set Ω_A of normalized states actually determines both the ordered space A and the order-unit u_A: one can take A to be the dual of the space of affine real-valued functionals on Ω_A, ordered by the cone of non-negative affine functionals; in this representation, u_A is simply the constant affine functional on Ω_A with value 1. When describing a particular abstract state space, it is often easiest simply to specify the convex set Ω_A.

As an example, consider the region $\Omega = \{(x, y, 1) | |x|, |y| \leq 1\}$ in $\mathbb{R}^3$ – a square, parallel to the x-y plane and displaced from the origin by one unit in the z direction. This is the base of a "pyramidal" cone, $\{(x, y, z) | |x|, |y| \leq z\}$. Ordered by this cone, and with u the unique functional taking the value 1 on Ω, we have an abstract state space equivalent to that of the "square bit" discussed in Example

2.1. In fact, the two basic measurements $\{x, x'\}$ and $\{y, y'\}$ in that example can be recovered from the geometry of the state space: simply identify x with the unique functional taking the value 1 on some face of Ω, and 0 on the opposite face, and let $x' = u - x$; similarly for y and y', using the other two faces.

Observables Let $(X, \mathcal{B})$ be a measurable space: an X-*valued observable* on a state space A is a weakly countably additive vector measure $F : \mathcal{B} \to A^*$ with $F(X) = u$. This guarantees that if $\alpha \in \Omega$, $B \mapsto F(B)(\alpha)$ is a (finitely additive) probability measure on $\mathcal{B}$. If μ is a given measure on $(X, \mathcal{B})$, we shall call $f : X \to A^*$ a *density* for F with respect to μ iff, for every $\alpha \in \Omega$ and every set $B \in \mathcal{B}$,

$$\int_B f(x)(\alpha) d\mu(x) = F(B)(\alpha).$$

In the simplest case, where X is a finite set and μ is the counting measure, an X-valued observable amounts to a list $(f_1, ..., f_n)$ of effects with $\sum_i f_i = u$. In the sequel, when we speak of *an observable*, without specifying the value space, this is what we have in mind.

The Base Norm Note that the point-wise ordering of functionals in A^* on Ω_A is exactly the usual dual ordering. There is a natural norm on A^*, namely the supremum norm $\|f\| = \sup_{\alpha \in \Omega_A} |f(\alpha)|$; this gives rise in turn to a norm on A, called the *base norm*, with respect to which $\|\alpha\| = u_A(\alpha)$ for $\alpha \in A_+$. In particular, every normalized state has norm 1, and conversely, a positive element of A having norm 1 is a normalized state (so that the two meanings of "normalized" coincide). In the sequel, we shall write $\widetilde{\alpha}$ for the normalized version of a positive weight $\alpha \in A_+$, i.e.,

$$\widetilde{\alpha} := \frac{\alpha}{\|\alpha\|} = \frac{\alpha}{u_A(\alpha)}$$

It will be convenient to stipulate that $\widetilde{0} = 0$.

Processes and Dynamics We represent physical *processes* involving an initial system with state space A and a final system with state space B by positive linear mappings $\phi : A \to B$ having the property that $\|\phi(\alpha)\| \leq 1$ for all $\alpha \in \Omega_A$, which is just to say that ϕ is norm contractive, or, equivalently, that $\|\phi\| \leq 1$. In this case, we understand that $\|\phi(\alpha)\| = u(\phi(\alpha))$ represents the probability that the process occurs when the input is α; indeed, we can regard the effect $u \circ \phi \in A^*$ as recording precisely this occurrence. Thus, a family $\{\phi_i | i \in I\}$ of positive linear mappings with $\|\phi_i\| \leq 1$ for all i and $\sum_i \|\phi_i\| = 1$, represents a family of physical processes *one* of which is bound to occur. (Such a family is a discrete *instrument* in the sense of [9, 12].)

In many cases, one wants to impose some further constraint on the possible dynamics of a system represented by an abstract state space A. By a *dynamical semigroup* for A, we mean a closed, convex set $\mathfrak{D}_A$ of norm-contractive positive linear mappings $\tau : A \to A$, closed under composition and containing the identity mapping Id_A. We understand $\mathfrak{D}_A$ as representing the set of all physically possible processes on A. (Here, "physically" refers to the use of this framework for abstractly formulating possible physical theories; the framework could also have other applications, so the terminology "operationally possible" might be more accurate. With this caveat, however, we will stick with "physically.")

A state space equipped with a distinguished dynamical semigroup, we call a *dynamical model*. Note that any abstract state space can be regarded as a dynamical model if we take $\mathfrak{D}_A$ to be the set of all norm-contractive positive linear mappings $A \to A$. *In the balance of this paper, we take it as a standing assumption, relaxed only where explicitly noted, that this is the case.*

Self-Duality and Weak Self-Duality In both classical and quantum settings, A carries a natural inner product with respect to which there is a *canonical* order-isomorphism $A \simeq A^*$. Indeed, in both classical and quantum cases, the positive cone is *self dual*, in that

$$A_+ = A^+ := \{\alpha \in V | \forall \beta \in V_+ \langle \alpha, \beta \rangle \geq 0\}.$$

This property is a very special one, not shared by most abstract state spaces. For an example, let A be the abstract state space of the "square bit" of Example 2.1, with its square space Ω of normalized states. As noted above, for each side of Ω, there is an effect taking the value 1 along that side, with the effects corresponding to opposite sides summing to 1. These effects in fact generate the extremal rays of the dual cone, which thus also has a square cross-section. Hence, the cones A_+ and A_+^* are isomorphic. Nevertheless, A_+ is not self-dual, as A^+ is the image of A_+ under a rotation by $\pi/4$.

In this paper, we shall call a finite-dimensional ordered space *weakly self-dual* iff, as in the example above, there exists an order isomorphism (that is, a bijective, positive linear mapping with positive inverse) $\phi : V \simeq V^*$. This is a far less stringent condition than self-duality.

A classical result of Koecher [16] and Vinberg [26] shows that any finite-dimensional self-dual cone that is *homogeneous*, in the sense that any interior point can be mapped to any other by an affine symmetry (automorphism) of the cone, and *irreducible* in the sense that the cone is not a direct sum of simpler cones, is either the cone of positive self-adjoint elements of some full matrix ∗-algebra over the reals, complexes or quaternions, or is the cone generated by a ball-shaped base, or is the set of positive self-adjoint 3×3 matrices over the octonions.

Thus, self-duality, plus irreducibility and homogeneity, brings us within hailing distance of Hilbert space quantum mechanics. One might hope to motivate these conditions in operational terms. In this paper, we make some progress in this direction by identifying *weak* self-duality of a system as a necessary condition for a composite of three copies of the system to support conclusive (probabilistic) teleportation, and a condition not much stronger than homogeneity on the space of normalized states of the system to be teleported, as sufficient for the existence of a tripartite model permitting deterministic teleportation.

3. Composite Systems

In order to discuss teleportation protocols, it is important to consider composite systems having, at a minimum, three components: one belonging to the sender ("Alice"), another to the receiver ("Bob"), and a third, accessible to the sender but entangled with the receiver, to serve as a channel across which the sender's state can be teleported. In this section, we review the account of bipartite state spaces given in [5], and extend it to cover systems having three or more components. In doing so, we identify a non-trivial condition on such composites, which we term *regularity*, that will play an important role in our discussion of teleportation protocols in the sequel.

In order to maintain the flow of discussion, the proofs of several results from this section have been placed in a brief appendix.

3.1. Bipartite Systems. It will be convenient, in what follows, to identify the algebraic tensor product, $A \otimes B$, of two vector spaces A and B with the space of all bilinear forms on $A^* \times B^*$. In particular, if $\alpha \in A$ and $\beta \in B$, we identify the pure tensor $\alpha \otimes \beta$ with the bilinear form defined by

$$(\alpha \otimes \beta)(a, b) = a(\alpha)b(\beta)$$

for all $a \in A^*$ and $b \in B^*$. If A and B are *ordered* vector spaces, we call a form $\omega \in A \otimes B$ *positive* iff $\omega(a, b) \geq 0$ for all positive functionals $a \in A^*$ and $b \in B^*$. Note that if $\alpha \in A$ and $\beta \in B$ are positive, then $\alpha \otimes \beta$ is a positive form. Note, too, that the set of positive forms is a cone in $A \otimes B$.

DEFINITION 3.1. *The* maximal tensor product *of ordered vector spaces A and B, denoted $A \otimes_{max} B$, is $A \otimes B$, equipped with the cone of all positive forms. Their* minimal tensor product, *denoted $A \otimes_{min} B$, is $A \otimes B$ equipped with the cone of all positive linear combinations of pure tensors.*

The maximal and minimal tensor products are exactly the injective and projective tensor products discussed by Wittstock in [**25**]; see also [**11, 19**]. It is not difficult to show that, in our present finite-dimensional setting, $(A \otimes_{max} B)^* = A^* \otimes_{min} B^*$ and $(A \otimes_{min} B)^* = A^* \otimes_{max} B^*$.

If (A, u_A) and (B, u_B) are abstract state spaces representing two physical systems, then subject to a plausible no-signaling condition and a "local observability" assumption guaranteeing that the correlations between local observables determine the global state, the largest sensible model for a bipartite system having physically separated components modeled by A and B is $A \otimes_{max} B$, with order unit given by $u^{AB} = u_A \otimes u_B$. (For a more detailed development of this point, see [**24**] or [**5**].)

Accordingly, we model a (non-signaling, locally observable) *composite system* with components (A, u_A) and (B, u_B) by the algebraic tensor product of A and B, ordered by *any* cone lying between the maximal and minimal tensor cones, and with order unit $u_{AB} = u_A \otimes u_B$. We shall write AB, generically, for such a state space, denoting the convex set $u_{AB}^{-1}(1)$ of normalized states by Ω_{AB}.

It will be important, below, to remember that all states, in whatever cone we use, can be represented as linear combinations of pure product states, as these span $A \otimes B$. Unless the sets of normalized states for A or B are simplices – that is, unless one system at least is classical – the minimal and maximal tensor products are quite different, with the latter containing many more normalized states than the former. These additional states we term *entangled*; states in $A \otimes_{min} B$, we term *separable*.

Marginal and Conditional States Every state ω in a bipartite system AB has natural *marginal states* $\omega^A \in A$ and $\omega^B \in B$, given respectively by

$$\omega^A(a) = \omega(a \otimes u_B) \text{ and } \omega^B(b) = \omega(u_A \otimes b)$$

for all $a \in A^*$ and $b \in B^*$. We also have un-normalized conditional states, given by
$$\omega_a^B(b) = \omega(a,b) = \omega_b^A(a)$$
and their normalized versions,
$$\widetilde{\omega}_a^B(b) = \frac{\omega(a,b)}{\omega^A(a)} \text{ and } \widetilde{\omega}_b^A(a) = \frac{\omega(a,b)}{\omega^B(b)}$$
if the marginal states are non-zero, and set equal to 0 otherwise, so that the expected identities $\omega(a,b) = \omega_b^A(a)\omega^B(b) = \omega^A(a)\omega_a^B(b)$ hold. Using these, it is not difficult to show that, just as in quantum theory, the marginals of an entangled state are necessarily mixed, while those of an unentangled pure state are necessarily pure.

EXAMPLE 3.2. As an example of these ideas, consider again the "square bit" of Example 2.1, i.e., the abstract state space associated with two measurements $E = \{x, x'\}$ and $F = \{y, y'\}$, with $\Omega \simeq [0,1]^2$. Let A denote $X = \{x, x', y, y'\}$, regarded as a subset of $V(\Omega)^*$. A normalized state $\omega \in V(\Omega) \otimes_{max} V(\Omega)$ determines, and is determined by, a joint probability assignment $X \times X \to [0,1]$, given by $\omega(i,j) = (i \otimes j)(\omega)$ for all $i, j \in X$. Regarding this as a block matrix, with 2×2 blocks indexed by E and F, we see that ω sums to unity in each block, and also satisfies the no-signaling principle: $\omega(x,\cdot) + \omega(x',\cdot) = \omega(y,\cdot) + \omega(y',\cdot)$, and similarly in the second variable. Conversely, it can be shown [**6, 24**] that any such array corresponds to an element of $V(\Omega) \otimes_{max} V(\Omega)$. A famous example is the "PR Box" state [**20**]

	x	x'	y	y'
x	1/2	0	1/2	0
x'	0	1/2	0	1/2
y	1/2	0	1/2	0
y'	0	1/2	0	1/2

For this state, the CHSH parameter $S = \omega(x,y) + \omega(x,y') + \omega(x',y) - \omega(x',y')$ achieves the a priori maximum value of 4, violating not only Bell's inequality, but also the Tsirel'son bound constraining quantum states.

Dynamically Admissible Composites It is reasonable to suppose that, if $\tau_A \in \mathfrak{D}_A$ and $\tau_B \in \mathfrak{D}_B$ are physically admissible processes on A and B, respectively, then, for any state ω on a composite system AB,
$$(\tau_A \otimes \tau_B)(\omega) : a, b \mapsto \omega(\tau_A^* a, \tau_B^* b)$$
is a state of AB. When this is the case, let us say that the composite system AB is *dynamically admissible*. Equivalently, AB is dynamically admissible iff for all $\tau_A \in \mathfrak{D}_A, \tau_B \in \mathfrak{D}_B$, AB_+ is stable under $\tau_A \otimes \tau_B$ acting on $A \otimes B$. Note that both minimal and maximal tensor products are stable under any pure tensor of positive operators, so these are dynamically admissible regardless of the dynamics.

Where $\mathfrak{D}_A$ and $\mathfrak{D}_B$ – as per our standing assumption – comprise *all* norm-contractive positive mappings $A \to A$ and $B \to B$, respectively, AB is dynamically admissible iff its positive cone AB_+ is stable under $\tau_1 \otimes \tau_2$ for *all* positive mappings $\tau_1 : A \to A$ and $\tau_2 : B \to B$. Although the minimal and maximal tensor products $A \otimes_{min} B$ and $A \otimes_{max} B$ both enjoy this property, it is highly non-trivial. Indeed, if $A = \mathcal{B}_h(\mathbf{H})$ and $B = \mathcal{B}_h(\mathbf{K})$, the spaces of self-adjoint operators on Hilbert spaces $\mathbf{H}$ and $\mathbf{K}$, and $AB = \mathcal{B}_h(\mathbf{H} \otimes \mathbf{K})$, the usual quantum-mechanical composite state space, then the cone AB_+ is stable only under products of *completely* positive mappings. However, this difficulty is easily met: one need only define a composite of two dynamical models $(A, \mathfrak{D}_A)$ and $(B, \mathfrak{D}_B)$ to be a model $(AB, \mathfrak{D}_{AB})$ where AB is a dynamically admissible composite of A and B, and $\mathfrak{D}_{AB}$ is a semigroup of norm-contractive positive mappings $AB \to AB$ containing all products $\tau_A \otimes \tau_B$ where $\tau_A \in \mathfrak{D}_B$ and $\tau_B \in \mathfrak{D}_B$. In the balance of this paper, results will be formulated for composites of state spaces, rather than of dynamical models; however, these can easily be modified to accommodate the latter.

Bipartite states and effects as operators Elements of the tensor product $A \otimes B$ and of its dual $(A \otimes B)^*$ can be regarded as operators $A^* \to B$ and $A \to B^*$, respectively. Indeed, every $f \in (A \otimes B)^*$ induces a linear mapping $\widehat{f} : A \to B^*$, uniquely defined by the condition that
$$\widehat{f}(\alpha)(\beta) = f(\alpha \otimes \beta).$$
The mapping $f \mapsto \widehat{f}$ is a linear isomorphism. Note also that, if f is positive, then so is $\widehat{f}$ (though not conversely, unless we use the maximal tensor product). Similarly, any $\omega \in A \otimes B$ induces a linear mapping $\widehat{\omega} : A^* \to B$, uniquely defined by the condition that
$$\widehat{\omega}(f)(g) = (f \otimes g)(\omega)$$
for all $f, g \in V^*$. Again, the mapping $\omega \mapsto \widehat{\omega}$ is a linear isomorphism. Also, since elements of the maximal tensor product $A \otimes_{max} B$ are precisely those corresponding to positive bilinear forms, $\widehat{\omega}$ will be a positive operator, regardless of which tensor product we use. In the special case in which ω is a pure tensor, say $\omega = \beta \otimes \gamma$, we have
$$\widehat{(\beta \otimes \gamma)}(f) = f(\beta)\gamma.$$
In the sequel, we shall write $\widehat{AB}$ for the set of operators $\widehat{\omega}$ corresponding to $\omega \in AB$, ordered by the cone of operators $\widehat{\omega}$ with $\omega \in AB_+$. For example, $A \widehat{\otimes_{max}} B$ is simply the space $\mathcal{L}(A, B)$, ordered by the cone of positive operators.

Note that the operator $\widehat{\omega}$ corresponding to a *normalized* state in AB has the property that $\widehat{\omega}(u)(u) = 1$, i.e., $\widehat{\omega}(u)$ is a state. Conversely, given a positive linear mapping $\phi : A^* \to B$ with the property that $\phi(u_A)$ is a state, the bilinear form $\omega(a, b) := \phi(a)(b)$ defines an element of the maximal tensor product, with $\phi = \widehat{\omega}$. It is useful to note ([**11**], Equation 16) that any positive operator $\phi : A^* \to B$ has operator norm (induced by the above-defined order-unit and base norms on A^* and B) given by
$$\|\phi\| = \|\phi(u)\|_B$$
where $\| \cdot \|_B$ denotes the base-norm on B; hence, bipartite states correspond exactly to positive operators of norm 1.

Similarly, if f is a bipartite effect in $A^* \otimes_{max} A^*$, then the mapping $\widehat{f} : A \to A^*$ takes any state α to the effect $\widehat{f}(\alpha)(\beta) = f(\alpha \otimes \beta)$. Evidently, this is no greater than unity on Ω, so we have $\widehat{f}(\alpha) \leq u$ for all $\alpha \in \Omega$; conversely, any such positive mapping defines a bipartite effect.

3.2. Multipartite Systems. Up to a point, the foregoing considerations readily extend to composite systems involving more than two components. Suppose $(A_1, u_1), ..., (A_n, u_n)$ are abstract state spaces. As above, call an n-linear form on

$A_1^* \times \cdots \times A_n^*$ *positive* iff it takes non-negative values on all n-tuples $f = (f_1, ..., f_n)$ of positive functionals $f_i \in A_i^*$. Given states $\alpha_i \in A_{i+}$ for $i = 1, ..., n$, the product state $\alpha_1 \otimes \cdots \otimes \alpha_n$, defined by $(\otimes_i \alpha_i)(f) = \Pi_i \alpha_i(f_i)$, is obviously positive in this sense.

DEFINITION 3.3. A *composite* of state spaces (A_i, u_i), $i = 1, ..., n$, is any space A of n-linear forms on $A_1^* \cdots A_n^*$, ordered by any cone of positive forms containing all product states, and with with order-unit given by $u = u_1 \otimes \cdots \otimes u_n$.

This is equivalent to saying that A contains all product states, and A^* contains all product effects. Examples of composites of, say, three spaces A, B and C would include $A \otimes_{max} B \otimes_{max} C$, $A \otimes_{min} B \otimes_{min} C$, and mixed composites such as $A \otimes_{min} (B \otimes_{max} C)$. Extending the terminology of the previous section, we shall call a composite A of state spaces (A_i, u_i) *dynamically admissible* iff A_+ is stable under mappings of the form $\bigotimes_i \tau_i$ where $\tau_i : A_i \to A_i$ are arbitrary positive mappings. A product of dynamical models $(A_i, \mathfrak{D}_i)$ is a dynamical model $(A, \mathfrak{D})$ where A is a dynamically admissible model of $A_1, ..., A_n$ and $\mathfrak{D}$ is a dynamical semigroup that includes all products of mappings $\tau_i \in \mathfrak{D}_i$.

Regular composites Suppose now that A is a composite of $A_1, ..., A_n$, and that $J \subseteq \{1, ..., n\}$. Given a list of positive linear functionals $f = (f_i) \in \Pi_{i \in I \setminus J} A_i^*$ and a state $\omega \in A_+$, we may define a $|J|$-linear form ω_f^J on $\Pi_{j \in J} A_j^*$ by setting

$$\omega_f^J(g) = \omega(f \otimes g),$$

where $(f \otimes g)_i$ is g_i if $i \in J$ and f_i otherwise. We refer to ω_f^J as a *partially evaluated* state. The set of such partially-evaluated states ω_f^J generates a cone in $\bigotimes_{j \in J} A_j$; together with the order unit $\otimes_{j \in J} u_j$, this defines an abstract state space A^J, which we call the *J-partial sub-system*, and which we take to represent the subsystem corresponding to the set of elementary systems A_j with $j \in J$.

In the simplest cases, we should expect that that a composite of "elementary" systems $A_1, ..., A_n$ can equally be regarded as a composite of complex sub-systems A^J obtained through an arbitrary coarse-graining of the index set $I = \{1, ..., n\}$. This suggests the following

DEFINITION 3.4. A composite A of state spaces $A_1, ..., A_n$ is *regular* iff, for all partitions $\{J_1, ..., J_k\}$ of $\{1, ..., n\}$, A is a composite, in the sense of Definition 1, of the partial systems $A^{J_1}, ..., A^{J_k}$.

Equivalently, A is a regular composite of $A_1, ..., A_n$ iff for all partitions $J_1, .., J_k$ of $\{1, .., n\}$, and for all sequences of states $\mu_k \in A^{J_k}$, the product state $\bigotimes_k \mu_k$ belongs to A, and for all sequences of effects $f_k \in (A^{J_k})^*$, the product effect $\bigotimes_k f_k$ belongs to A^*.

We regard regularity as an eminently reasonable restriction on a model of a composite physical system, at least in cases in which the components retain their separate identities (so that the systems are "separated"). As we shall see in the sequel, regularity is sufficient to ground a weak analogue of a teleportation protocol, which we call *remote evaluation*. In the balance of this section, we collect some examples of regular composites, and adduce some technical results concerning the notion of regularity.

As a matter of notational convenience, we'll write ABC for a composite of three systems A, B and C, denoting by AB, BC, and AC the three bipartite subsystems. In this case, the condition that ABC be regular amounts to requiring that

$$AB \otimes_{min} C \leq ABC \leq AB \otimes_{max} C$$

and similarly for A and BC and for AC and B. Equivalently, we require that

$$AB \otimes_{min} C \leq ABC \text{ and } (AB)^* \otimes_{min} C^* \leq (ABC)^*.$$

As an example, let us show that the mixed tensor product

$$A \otimes_{min} (B \otimes_{max} C)$$

is a regular composite of A, B and C. The only interesting coarse-grainings here are $\{\{A,B\},\{C\}\}$ and $\{\{A,C\},\{B\}\}$. To analyze the first of these, suppose that $\omega = \sum_i t_i \alpha_i \otimes \mu_i$ where $\alpha_i \in A_+$ and $\mu_i \in (B \otimes_{max} C)_+$. Then for all $c \in C^*$,

$$\omega_c^{AB} = \sum_i t_i \alpha_i \otimes \widehat{\mu}_i(c),$$

a positive linear combination of positive elements of A and B; hence, $\omega_c^{AB} \in (A \otimes_{min} B)_+$, so $AB = A \otimes_{min} B$. It follows that, if $\gamma \in C_+$, we have

$$\omega_c^{AB} \otimes \gamma \in (A \otimes_{min} B \otimes_{min} C)_+$$
$$\leq ((A \otimes_{min} B) \otimes_{max} C)_+ = (AB \otimes_{max} C)_+ .$$

A similar argument applies to the bipartition $\{\{A,C\},B\}$.

In the next section (see Corollary 4.2), we'll show that $A \otimes_{max} (B \otimes_{min} C)$ is also regular. An example of a non-regular composite is

$$(A \otimes_{min} A) \otimes_{max} (A \otimes_{min} A)$$

where A is weakly self-dual. This follows from considerations involving entanglement swapping, as discussed in section 6.

The following lemma collects a number of facts about composites and regular composites that will be used freely – and often tacitly – in the sequel. (For a proof, see the appendix.)

LEMMA 3.5. *Let A be a composite of systems $A_1, ..., A_n$. Then*
(a) *If $K \subseteq J \subseteq \{1,...,n\}$, then $(A^J)^K = A^K$.*
(b) *If A is regular, then $(A^J)_+ = \{\omega_u^J | \omega \in A_+\}$.*
(c) *If A is regular, so is A^J for every $J \subseteq \{1,...,n\}$.*

Probabilistic Theories Roughly, by a *probabilistic theory*, we mean a class $\mathcal{C}$ of probabilistic models – that is, abstract state-spaces – closed under some construction or constructions whereby systems can be composed. Examples would include the class of all classical systems (i.e., systems with simplicial state spaces), the class of all quantum systems with the usual quantum-mechanical state space, the class obtained by forming the maximal tensor products of quantum systems, the convexified version of Spekkens' "toy theory" [23], etc. In principle, this idea might be given a precise category-theoretic formulation (something we expect to pursue in a subsequent paper); here, we content ourselves with a more informal treatment.

Consider a class $\mathcal{C}$ of state spaces equipped with a specific coupling $A, B \mapsto A \odot B$, where $A \odot B$ is a composite of A and B. We shall call $\odot$ *associative* if for all $A, B, C \in \mathcal{C}$, $A \odot (B \odot C) \simeq (A \odot B) \odot C$ under the obvious association mapping (defined on product states by $\alpha \otimes (\beta \otimes \gamma) \mapsto (\alpha \otimes \beta) \otimes \gamma$. The straightforward but tedious proof of the following can be found in the appendix:

PROPOSITION 3.6. *If $\odot$ is associative, then for all $A_1, ..., A_n \in \mathcal{C}$, $A_1 \odot \cdots \odot A_n$ is a regular composite of $A_1, ..., A_n$.*

It follows that composites constructed using only the maximal, or only the minimal, tensor product are regular, as are composite quantum systems. For later purposes, if $\mathcal{C}$ is a class of abstract state spaces closed under an associative coupling $\odot$ preserving isomorphism, we shall call the pair $(\mathcal{C}, \odot)$ a *monoidal theory*. (By *preserving isomorphism*, we mean that if $A \simeq B$ and $C \simeq D$, then $(A \odot C) \simeq (B \odot D)$.) It is by no means obvious that every sensible theory must be monoidal, however – for instance, we may wish to consider theories in which one can form tripartite systems of the form $A \otimes_{min} (B \otimes_{max} C)$, in which there is maximal entanglement between B and C, but no entanglement at all between A and either B or C. There is certainly precedent for such mixed tensorial constructions, e.g., in Hardy's causaloid framework for quantum gravity [**13**]. On the other hand, as spelled out in Section 6, considerations involving entanglement swapping place some nontrivial restrictions on non-monoidal theories.

Remark: In the interest of clarity, it will sometimes be helpful in the sequel to adorn an element of a factor in a tensor product with a superscript indicating to which factor it belongs, writing, for instance, $\alpha \otimes \beta$ or $\alpha^A \otimes \beta^B$ for product states in $A \otimes B$, or f^{AB} for an arbitrary bipartite effect in $(A \otimes B)^*$. On occasion, both ornamented and unornamented forms – e.g., ω and ω^{AB} – may occur in the same calculation; when they do, they refer to the same object.

4. Conclusive Teleportation

Suppose ABC is a composite of state spaces A, B and C. If f is an effect on AB and ω is a state in BC, then we have positive linear mappings $\widehat{f} : A \to B^*$ and $\widehat{\omega} : B^* \to C$. Their composite, $\widehat{\omega} \circ \widehat{f}$, is a positive operator $A \to C$. If ABC is a *regular* composite, we also have, for any state $\alpha \in A$ and any effect $c \in C^*$, that $\alpha \otimes \omega$ is a state in ABC and $f \otimes c$ is an effect in $(ABC)^*$. We now make a technically trivial but crucial observation:

LEMMA 4.1. *With notation as above, the un-normalized conditional state of $\alpha \otimes \omega$ given an effect $f \in AB$ is*
$$(\alpha^A \otimes \omega^{BC})_f^C = \widehat{\omega}(\widehat{f}(\alpha)).$$

Proof: As pure tensors generate BC, it is sufficient to check this in the case that $\omega = \beta \otimes \gamma$. Then, for any $b \in B^*$, $\widehat{\omega}(b) = \beta(b)\gamma$ (using, here, our convention of identifying a state space with its double dual). Note also that $f(\alpha \otimes \beta) = \beta(\widehat{f}(\alpha))$. Hence, for any $c \in C^*$, $(f \otimes c)(\alpha \otimes \omega) = f(\alpha \otimes \beta)\gamma(c) = \beta(\widehat{f}(\alpha))\gamma(c) = \widehat{\omega}(\widehat{f}(\alpha))(c)$. $\square$

COROLLARY 4.2. *For any state spaces A, B and C,*
  (i) *There is a canonical embedding*
$$A \otimes_{min} (B \otimes_{max} C) \leq (A \otimes_{min} B) \otimes_{max} C.$$

(ii) The composite $(A \otimes_{min} B) \otimes_{max} C$ is regular.

Proof: By Lemma 4.1, any product state $\alpha \otimes \omega$ with $\alpha \in A_+$ and $\omega \in (B \otimes_{max} C)_+$ yields a positive bilinear form on $(A \otimes_{min} B)^* \times C^*$, namely, $(\alpha \otimes \omega)(f, c) = c(\widehat{\omega}(\widehat{f}(\alpha)))$. Hence, we have a natural positive linear mapping $A \otimes_{min} (B \otimes_{max} C) \to ((A \otimes_{min} B)^* \otimes_{min} C^*)^*$; the last is isomorphic to $(A \otimes_{min} B) \otimes_{max} C$. This establishes (i).

To show that $(A \otimes_{min} B) \otimes_{max} C$ is regular, we first observe that $BC = B \otimes_{max} C$. Indeed, let $\mu \in B \otimes_{max} C$, and let $\widehat{\mu}$ be the associated positive operator $B^* \to C$. Let α be some fixed state in A. Given $f \in (A \otimes_{min} B)^* \simeq \mathcal{L}_+(A, B^*)$ and $c \in C^*$, set
$$\omega(f, c) = \widehat{\mu}(\widehat{f}(\alpha))(c):$$
this is bilinear in f and in c, and positive where both f and c are positive, and so, defines an element $\omega \in (A \otimes_{min} B) \otimes_{max} C$. We now observe that the reduced state $\omega_{u_A}^{BC}$, evaluated on a pair of effects $(b, c) \in B^* \times C^*$, yields
$$\begin{aligned}
\omega_{u_A}^{BC}(b, c) &= \omega(u_A, b, c) \\
&= \widehat{\mu}(\widehat{(u_A \otimes b)}(\alpha))(c) \\
&= \widehat{\mu}(b)(c) = \mu(b, c).
\end{aligned}$$

Thus, $\omega_{u_A}^{BC} = \mu$. This shows that $B \otimes_{max} C \leq BC$; the reverse inclusion is trivial, so $BC \simeq B \otimes_{max} C$, as claimed. We now have, by part (i), that
$$\begin{aligned}
A \otimes_{min} (BC) &= A \otimes_{min} (B \otimes_{max} C) \\
&\leq (A \otimes_{min} B) \otimes_{max} C = ABC.
\end{aligned}$$

Obviously, we have $ABC \leq A \otimes_{max} (B \otimes_{max} C) = A \otimes_{max} BC$. The corresponding result for the coarse-graining $\{\{AC\}, \{B\}\}$ follows similarly (or by symmetry), and that for $\{\{A\}, \{B, C\}\}$ is trivial, so so ABC is regular. □

We can interpret Lemma 4.1 in information-processing terms as follows. Suppose two parties, Alice and Bob, have access to systems A and B, respectively. Suppose, moreover, that Alice's system consists of two subsystems, A_1 and A_2, with A_1 in an unknown state α. If the total Alice-Bob system is represented by a regular composite $AB = A_1 A_2 B$, then if f is an effect on A and ω is a known state on $A_2 B$, we may prepare $A_1 A_2 B$ in the joint state $\alpha \otimes \omega$: if Alice performs a measurement on $A = A_1 A_2$ having f as a possible outcome, then, conditional upon securing this outcome, the conditional state of B is, up to normalization, $\widehat{\omega}(\widehat{f}(\alpha))$. Thus, we may say that Alice has evaluated a known mapping, namely $\widehat{\omega} \circ \widehat{f}$, on an unknown input α, simply by securing f as a measurement outcome. In the sequel, we refer to this protocol as *remote evaluation*.

This is obviously reminiscent of a teleportation protocol. Indeed, conclusive teleportation can be regarded as the special case of remote evaluation in which the mapping $\widehat{\omega} \circ \widehat{f}$ is invertible. Suppose that $\eta : A_1 \simeq B$ is a fixed isomorphism between Alice's system A_1 and Bob's system B (allowing us to say what we mean by saying a state of B is the same as a state of A_1). Suppose, further, that the unknown state α is recoverable from the *normalized* conditional state $\widetilde{(\alpha \otimes \omega)}_f^B$ by means of a physically admissible process τ, depending on f but not on α: upon securing a measurement outcome corresponding to f, Alice can then instruct Bob to make the correction τ; once this is done, she is certain that the conditional state of Bob's system B – whatever it is – is identical (up to η) to the original, but unknown, state α.

In fact, we can distinguish two situations: one in which the correction operation τ is certain to succeed, and another in which it may fail, but in which this failure will be apparent to Bob. In the latter case, if Alice as well as Bob is to be certain that the state has been transferred, the teleportation protocol requires an additional step: Alice must wait for Bob to report the success of the correction. We shall refer to these as *strong* and *weak* conclusive teleportation, respectively. Notice that the standard (one-outcome post-selected) quantum teleportation protocol is an instance of a strong teleportation protocol.

We make this language precise as follows. To avoid needless repetition, here and in the balance of this paper $A_1 A_2 B$ denotes a regular composite of state spaces A_1, A_2 and B with A_1 isomorphic to B by a fixed isomorphism $\eta : A_1 \simeq B$; and f is an effect on $A_1 A_2$ and ω is a state in $A_2 B$.

DEFINITION 4.3. We say that the pair (f, ω) is a *conclusive teleportation protocol* on $A_1 A_2 B$ iff there exists a norm-contractive linear mapping $\tau : B \to B$, called a *correction*, such that, for every normalized state $\alpha \in \Omega_{A_1}$,

$$\tau(\widetilde{(\alpha \otimes \omega)}_f^B) = t_\alpha \eta(\alpha)$$

for some constant $t_\alpha > 0$. If τ can be so chosen that $t_\alpha = 1$ for all α, we say that the protocol (f, ω) is *strong*.

By Lemma 4.1, the conditional state $\widetilde{(\alpha \otimes \omega)}_f^B$ can be expressed as $\widehat{\omega}(\widehat{f}(\alpha))/u(\widehat{\omega}(\widehat{f}(\alpha)))$. Let

$$\mu := \widehat{\omega} \circ \widehat{f} : A_1 \to B,$$

noting that this is a norm-contractive positive mapping. Then (f, ω) is a teleportation protocol iff there exists a norm-contractive positive mapping $\tau : B \to B$ with

$$\tau(\mu(\alpha)) = t_\alpha \|\mu(\alpha)\| \eta(\alpha)$$

for all $\alpha \in \Omega_A$. Notice that $t_\alpha = u(\tau(\widetilde{\mu(\alpha)}))$, i.e., t_α is the probability that the correction τ *succeeds* in the conditional state $\widetilde{\mu(\alpha)}$). Accordingly, a strong protocol is one for which there exists a correction that is certain to succeed.

Theorem 4.5, below, gives a complete characterization of conclusive teleportation protocols, strong or otherwise, in terms of the mapping $\mu = \widehat{\omega} \circ \widehat{f}$. We require an easy preliminary

LEMMA 4.4. *Let A and B be any abstract state spaces. Let $\phi, \psi : A \to B$ be any two linear mappings with ψ injective. If, for every $\alpha \in \Omega_A$, there is a constant $k(\alpha)$ such that $\phi(\alpha) = k(\alpha)\psi(\alpha)$, then in fact $k(\alpha) \equiv k$, a constant not depending on α.*

Proof: Let α and β be distinct, and hence, linearly independent, elements of Ω_A, and consider $\gamma = (\alpha + \beta)/2$. Then we have

$$\phi(\gamma) = k(\gamma)\psi(\gamma) = (k(\gamma)/2)(\phi(\alpha) + \phi(\beta))$$

and also

$$\phi(\gamma) = (\phi(\alpha) + \phi(\beta))/2 = (k(\alpha)\psi(\alpha) + k(\beta)\psi(\beta))/2.$$

Thus,
$$(k(\alpha) - k(\gamma))\psi(\alpha) + (k(\beta) - k(\gamma))\psi(\beta) = 0.$$
Since ψ is injective, $\psi(\alpha)$ and $\psi(\beta)$ are linearly independent in B; hence, $k(\alpha) - k(\gamma) = k(\beta) - k(\gamma) = 0$, whence $k(\alpha) = k(\beta)$. □

Recall that an *order-isomorphism* between abstract state spaces is a positive linear bijection with a positive inverse, while an *isomorphism* also preserves normalization.

THEOREM 4.5. *Let $\mu := \widehat{\omega} \circ \widehat{f} : A_1 \to B$. Then*

(a) (f, ω) *is a conclusive teleportation protocol iff μ is an order isomorphism; in this case $\tau = s(\eta \circ \mu^{-1})$ where $s \leq 1/\|\mu^{-1}\| \leq 1$.*

(b) (f, ω) *is a strong teleportation protocol iff μ is proportional to an isomorphism; in this case, the correction τ is a symmetry of B.*

Proof:

(a) Suppose first that (f, ω) is a teleportation protocol. Then there exists a positive, norm-contractive mapping $\tau : B \to B$ such that, for all $\alpha \in \Omega_{A_1}$,
$$\tau(\mu(\alpha)) = t_\alpha \|\mu(\alpha)\| \eta(\alpha)$$
for some constant $t_\alpha > 0$. As η is injective, Lemma 4.4 implies that
$$t_\alpha \|\mu(\alpha)\| \equiv s,$$
a constant independent of α. Note that, as τ is norm-contractive, $s < 1$. Since Ω_{A_1} spans A_1, we have $\tau \circ \mu = s\eta$. It follows that $\tau : B \to B$ is a surjective linear mapping. As we are working in finite dimensions, this implies that τ is invertible; we have
$$\tau^{-1} = \mu \circ \frac{1}{s}\eta^{-1},$$
which is positive. Thus, τ is an order-isomorphism. It follows $\mu = \tau^{-1} \circ s\eta$ is also an order-isomorphism.

For the converse, suppose that μ is an order-isomorphism. Then $\eta \circ \mu^{-1}$ is also an order-isomorphism. Let
$$\tau := s(\eta \circ \mu^{-1})$$
where $s < 1/\|\mu^{-1}\|$. As $\|\eta\| = 1$, we have $\|\tau\| \leq s\|\eta\|\|\mu^{-1}\| < 1$, so τ is norm-contractive. Now $\tau \circ \mu = s\eta$. For all α, let $t_\alpha = s/\|\mu(\alpha)\|$ (noting that $\|\mu(\alpha)\| > 0$, since μ is injective), so that
$$\tau(\mu(\alpha)) = s\eta(\alpha) = t_\alpha \|\mu(\alpha)\| \eta(\alpha).$$

(b) Suppose first that $\mu = k\phi$ for some isomorphism $\phi : A_1 \to B$. Then $\phi : A_1 \to B$ and some positive constant k. Let $\tau = \eta \circ \phi^{-1}$: then $\tau(\mu(\alpha)) = k\eta(\alpha)$ for all α. Since $k = \|\mu(\alpha)\|$ for all α, we have a strong teleportation protocol.

For the converse, suppose (f, ω) is a strong teleportation protocol. Thus, there exists a norm-contractive positive mapping $\tau : B \to B$ such that, for all $\alpha \in \Omega_{A_1}$,
$$\tau(\mu(\alpha)) = \|\mu(\alpha)\| \eta(\alpha).$$
We claim that τ is a symmetry. To see this, let $\Gamma = \mu(\Omega_{A_1}) := \{\mu(\alpha) | \alpha \in \Omega_{A_1}\}$, and set $\widetilde{\Gamma} = \{\widetilde{\gamma} | \gamma \in \Gamma\}$; note that this set is a convex subset of Ω_B.[1] Now,

---

[1] To spell this out, suppose $\gamma_1, \gamma_2 \in \Gamma$ with normalized versions $\widetilde{\gamma}_1 = t_1\gamma_1, \widetilde{\gamma}_2 = t_2\gamma_2 \in \widetilde{\Gamma}$. Now consider a convex combination
$$\alpha = p\widetilde{\gamma}_1 + q\widetilde{\gamma}_2$$

$\tau(\tilde{\mu}(\alpha)) = \eta(\alpha) \in \Omega_B$, so τ effects an affine bijection of Γ onto Ω_B. It follows that the affine span of Γ equals that of Ω, whence, that τ preserves the affine span of the latter – which is exactly the hyperplane $u_B^{-1}(1)$. As τ is positive, it also preserves the positive cone B_+, whence, τ preserves $B_+ \cap u^{-1}(1) = \Omega_B$. Thus, τ is a symmetry, as claimed. It remains to show that μ is proportional to an isomorphism. But as we have
$$\tau(\mu(\alpha)) = \|\mu(\alpha)\|\eta(\alpha),$$
we also have
$$\mu(\alpha) = \|\mu(\alpha)\|\tau^{-1}(\eta(\alpha))$$
for all $\alpha \in \Omega_{A_1}$. Invoking Lemma 4.4, we see that $\|\mu(\alpha)\| \equiv k$, a constant independent of α – whence, $\mu = k\tau^{-1} \circ \eta$. $\square$

Remarks: (1) For Bob to be able to apply the correction mapping τ, the latter must belong to the dynamical semigroup $\mathfrak{D}_B$. Given our simplifying assumption that $\mathfrak{D}_B$ comprises all norm-contractive positive mappings on B, this is automatic, but in a treatment using more general dynamical models, it would need to be assumed as part of the definition of a teleportation protocol.

(2) If (f,ω) is a teleportation protocol on $A_1 A_2 B$, then we can regard it also as a teleportation protocol on $A_1 \otimes_{min} (A_2 \otimes_{max} B)$, as the latter is regular, f is an effect on $A_1 \otimes_{min} A_2$, and ω is a state in $A_2 \otimes_{max} B$. Thus, all teleportation protocols involving regular composites of A_1, A_2 and B live, so to speak, in $A_2 \otimes_{min} (A_2 \otimes_{max} B)$.

One can regard non-strong conclusive teleportation protocols as *inherently inefficient*. The question arises, whether an inefficient protocol can always be replaced with one that is perfectly efficient. We show that this is always possible when the composite is dynamically admissible. (Recall that under our standing assumption, a composite is dynamically admissible iff its positive cone is closed under products of positive mappings on the factors.)

COROLLARY 4.6. *Suppose $A_1 A_2 B$ is dynamically admissible. If (f,ω) is a conclusive teleportation protocol with correction τ, then let $\omega' \in A_2 B$ be the state defined, for all $a \in A_2^*$ and $b \in B^*$, by*
$$\omega'(a,b) = \omega\widetilde{(a,\tau(b))}.$$

Then (f,ω') is a strong conclusive teleportation protocol, requiring no correction.

Proof: Since (f,ω) is a conclusive teleportation protocol, there exists a positive mapping $\tau : B \to B$ such that
$$\tau \circ \widehat{\omega} \circ \widehat{f} = s\eta$$

---

where $p, q \geq 0$ with $p + q = 1$. Then
$$\alpha = pt_1\gamma_1 + qt_2\gamma_2.$$

Let
$$\gamma = \frac{pt_1}{pt_1 + qt_2}\gamma_1 + \frac{pt_2}{pt_1 + qt_2}\gamma_2 \in \Gamma$$
and note that
$$\alpha = (pt_1 + qt_2)\gamma = \tilde{\gamma} \in \tilde{\Gamma}.$$

for some constant s. Since A_1A_2B is dynamically admissible, $\omega' \in A_2B$. It is easily verified that $\widehat{\omega}' = (\tau \circ \omega)/\|\tau \circ \omega\|$; hence,

$$\widehat{\omega}' \circ \widehat{f} = \frac{s}{\|\tau \circ \omega\|}\eta.$$

Thus, $\widehat{\mu}' := \widehat{\omega}' \circ \widehat{f}$ is proportional to a symmetry, so $(f, \widehat{\omega})$ is a strong conclusive teleportation protocol, by Theorem 4.5. Moreover, as the symmetry in question is η itself, no correction is required. □

It follows from Theorem 4.5 that if a bipartite state ω on A_2B and a bipartite state f on A_1A_2 supply a conclusive teleportation protocol, then the positive linear mappings $\widehat{f}$ and $\widehat{\omega}$ are respectively injective and surjective. We can be somewhat more precise about the geometry of the situation. Let us say that a *compression* on an ordered space V is a positive mapping $P : V \to V$ such that $P^2 = P$. Equivalently, P's range, $P(V)$, is an ordered subspace of V, and $P(\alpha) = \alpha$ for all $\alpha \in P(V)$. As an example, let K be a cube, and let F be a face thereof; the obvious affine surjection $K \to F$ extends to a compression $V(K) \to V(F)$.

Suppose now that (f, ω) is a conclusive teleportation protocol on A_1A_2B with an order-isomorphic correction $\tau : B \to B$, so that

$$\tau \circ \widehat{\omega} \circ \widehat{f} = s\eta$$

for some constant $s > 0$. Then $\widehat{f} : A_1 \to A_2^*$ is an order-embedding, and and $\widehat{\omega} : A_2^* \to B$ is a positive surjection. Let

$$P := \widehat{f} \circ \eta^{-1} \circ \tau \circ \widehat{\omega} : A_2^* \to A_2^* :$$

an easy computation shows that P is a compression in the above-defined sense, with range equal to the image of $\widehat{f}$.

Conversely, suppose we are given an effect f such that $\widehat{f} : A_1 \to A_2^*$ takes A_1 order-isomorphically onto the range of a compression $P : A_2^* \to A_2^*$. Let $\widehat{f}^+ :$ Ran$(P) \to A_1$ be the inverse of $\widehat{f}$'s co-restriction to Ran(P), and let $\alpha_o = \widehat{f}^+(u_{A_2})$, i.e, the unique element of A_{1+} such that $\widehat{f}(\alpha_o) = P(u_{A_2})$. Define

$$\widehat{\omega}' := \frac{1}{\|\alpha_o\|} \eta \circ \widehat{f}^+ \circ P.$$

Then $\widehat{\omega}'(u_{A_2}) = \eta(\alpha)/\|\alpha_o\| \in \Omega_B$ (since η is an isomorphism, hence norm-preserving), whence, $\widehat{\omega}'$ corresponds to a normalized state ω' in $A_2 \otimes_{max} B$. The pair (f, ω') gives us a *strong* – and correction-free – teleportation protocol on $A_1 \otimes_{min} (A_2 \otimes_{max} B)$. If A_1A_2B is dynamically admissible, then $\omega' \in A_2B$, and indeed, is precisely the state ω' defined in Corollary 4.6.

Summarizing:

THEOREM 4.7. *Let A_1, A_2 and $B \simeq A_1$ be abstract state spaces. A regular composite A_1A_2B supports a conclusive teleportation protocol iff there exists an effect f on A_1A_2, a state ω in A_2B, and a compression $P : A_2^* \to A_2^*$ such that $\widehat{f}$, co-restricted to Ran(P), is an order-isomorphism $A_1 \simeq$ Ran(P) and $\widehat{\omega}$, restricted to Ran(P), is an order-isomorphism Ran$(P) \simeq B$.*

COROLLARY 4.8. *$A_1 \otimes_{min} (A_2 \otimes_{max} A_1)$ supports conclusive teleportation with $\eta(\alpha) = \alpha$ for all states $\alpha \in A_1$ iff $A_1 \leq A_2^*$ is the range of a compression $P : A_2^* \to A_2^*$.*

Proof: Suppose first that we have a compression $P : A_2^* \to A_1^*$: regarding P as a positive surjection $\pi : A_2^* \to A_1$, and letting $\iota : A_1 \to A_2^*$ be the positive inclusion mapping, we have $\pi \in \widehat{(A_2 \otimes_{max} A_1)}$ and $\iota \in (A_1 \widehat{\otimes_{min}} A_2)^*$. As $(\pi \circ \iota)(\alpha) = \alpha$ for all $\alpha \in A_1$, Theorem tells us that $A_1 A_2 A_1 = A_1 \otimes_{min} (A_2 \otimes_{max} A_1)$ supports conclusive teleportation.

Conversely, if $A_1 \otimes_{min} (A_2 \otimes_{max} A_1)$ supports conclusive teleportation, then by Corollary 4.6, there exist positive operators $\widehat{\omega} : A_2^* \to A_1$ and $\widehat{f} : A_1 \to A_2^*$ with $\widehat{\omega} \circ \widehat{f} : A_1 \to A_1$ an isomorphism, in which case $P := \widehat{f} \circ \widehat{\omega}$ is a compression. $\square$

We also have

COROLLARY 4.9. *Let $A_1 A_2 B$ be a regular composite of three pairwise isomorphic, weakly self-dual state spaces. If $A_2 B$ contains a state ω with $\widehat{\omega} : A_2^* \simeq B$, then $A_1 A_2 B$ supports conclusive teleportation. In particular, $A_1 \otimes_{min} (A_2 \otimes_{max} B)$ supports conclusive teleportation.*

Remark: As observed above, the standing assumption that for a system A, its dynamical semigroup $\mathfrak{D}_A$ is the set of all positive maps on A, strongly restricts the nature of dynamically admissible tensor products, and is, for example, incompatible with the usual quantum tensor product. However, our definitions and results concerning teleportation are easily adapted to the setting of regular composites of arbitrary dynamical models: as noted above, the definition of a teleportation protocol in that setting requires that the correction mapping τ^B on B belong to the dynamical semigroup $\mathfrak{D}_B$; with this modification, one has one has obvious analogues of Theorems 4.5 and 4.7, and of Corollary 4.6.

5. Deterministic Teleportation

As in the previous section, $A_1 A_2 B$ is a regular composite of three state spaces A_1, A_2 and B, with B isomorphic to A_1. In order for $A_1 A_2 B$ to support a *deterministic* teleportation protocol, we require a bipartite state $\omega \in A_2 B$ and an *observable* $\{f_1, ..., f_n\}$ on $A_1 A_2$ such that for every state α in A_1 and for each i, the state α is recoverable from the conditional state of $\alpha \otimes \omega$ given outcome (effect) f_i.

DEFINITION 5.1. Let $A_1 A_2 B$ be a regular composite of A_1, A_2 and B with $B \simeq A_1$ via a fixed isomorphism $\eta : A_1 \to B$. If ω is a state in $A_2 B$ and $E = (f_1, ..., f_n)$ is an observable on $A = A_1 A_2$, we shall say that the pair (E, ω) realizes a *deterministic teleportation protocol* iff, for each effect $f_i \in E$, the pair (f_i, ω) realizes a strong conclusive teleportation protocol.

The idea is that, upon measuring E and obtaining outcome f_i, Alice instructs Bob to apply a suitable correction τ_i; the conditional state of B is then $\eta(\alpha)$. Note that, by Theorem 4.5, the correction τ_i must be a symmetry of B.

At present, it is not clear to us exactly what conditions on the pair A_1, A_2 will be necessary in order to secure a deterministic teleportation protocol. However, Theorem 5.2 below provides a wealth of examples of systems which, while weakly self-dual, are neither classical nor quantum, but can nevertheless by combined so as to support a deterministic teleportation protocol. In particular, self-duality is not necessary for deterministic teleportation.

In what follows, let A be an abstract state space carrying an action of a finite group G that preserves the state space Ω. Note that there is a canonical dual action of G on A^* given by
$$(ga)(\alpha) = a(g^{-1}\alpha)$$
for all $g \in G$, $a \in A^*$, and $\alpha \in A$. Note, too, that the order-unit $u = u_A$ is invariant under this action, i.e., $gu = u$ for all $g \in G$. A state ω is called *G-equivariant* if for all $g \in G$ and all effects $a \in A^*$ we have

(5.1) $$g\widehat{\omega}(a) = \widehat{\omega}(ga) .$$

THEOREM 5.2. *Let A be weakly self-dual, and suppose G is a finite group acting on A, in such a way that (i) G acts transitively on the extreme points of Ω, and (ii) there exists a G-equivariant isomorphism $A^* \simeq A$. Then $A \otimes_{min} (A \otimes_{max} A)$ supports a deterministic teleportation protocol.*

Theorem 5.2 yields a wealth of examples of state spaces that, while neither classical nor quantum, do support a deterministic teleportation protocol. For an example, consider the state space of Example 2.1 obtained by taking Ω to be a unit square in $\mathbb{R}^3$, displaced one unit from the origin; A_+ is the cone generated by this square base. As observed earlier, with respect to the usual inner product, A^* can be represented as $\mathbb{R}^3$ with cone obtained by rotating A_+ by $\pi/4$. This gives us an order-isomorphism $A^* \to A$ that is equivariant with respect to the the natural action of $\mathbb{Z}_4$ on Ω. As this is transitive on the vertices of the latter, Theorem 5.2 tells us that $A \otimes_{min} (A \otimes_{max} A)$ will support a deterministic teleportation protocol. Similar considerations show that the same conclusion holds whenever Ω_A is any regular polygon.

For the proof of Theorem 5.2, we need an easy lemma.

LEMMA 5.3. *Let A and G be as in Theorem 5.2. Then there exists a unique invariant normalized state $\omega_o \in \Omega_A$.*

Proof: Notice, first, that there is certainly at least one fixed state, namely $(1/|G|)\omega_o = \sum_{g \in G} g\alpha_o$, where α_o is any one extreme state. To see that there can be no more than one such state, let Γ denote the set of G-fixed points of Ω. Observe that Γ is an affine section of Ω; hence, if Γ contains more than a single point, it contains an affine line, which must intersect the topological boundary of Ω. Let α be a fixed state belonging to this boundary: equivalently, α is fixed, and belongs to a proper face of Ω. Let F be the smallest face containing α: for each $g \in G$, gF is again a face containing α, so $F \subseteq gF$. In other words, F is invariant. But since F is a proper face and G acts transitively on Ω's extreme points, this is impossible. □

Proof of Theorem 5.2: Let A, G and ω_o be as above. By assumption, there is an equivariant order-isomorphism $\phi : A^* \to A$; normalizing if necessary, we can assume that $\phi = \widehat{\omega}$ for some bipartite state on AB. We claim that $\widehat{\omega}(u) = \omega_o$. Indeed, for all $g \in G$, we have
$$g\widehat{\omega}(u) = \widehat{\omega}(gu) = \widehat{\omega}(u).$$
Thus, $\widehat{\omega}(u)$ is G-invariant; but there is only one invariant state, namely ω_o.

Now, for all $g \in G$, let $f_g \in (A \otimes_{max} A)^*$ correspond to the operator
$$\widehat{f}_g = \frac{1}{|G|}\widehat{\omega}^{-1} \circ g.$$

We claim that $E = \{f_g\}$ is an observable, and (E, ω_o) realizes a strong deterministic teleportation protocol. To see this, note that for every $\alpha \in A$, $\frac{1}{|G|} \sum_{g \in G} g\alpha$ is a G-invariant state, and hence, by Lemma 5.3, equals ω_o. Thus,

$$\sum_{g \in G} f_g(\alpha) = \sum_{g \in G} \frac{1}{|G|} \widehat{\omega}^{-1}(g\alpha)$$
$$= \widehat{\omega}^{-1}\left(\frac{1}{|G|} \sum_{g \in G} g\alpha\right)$$
$$= \widehat{\omega}^{-1}(\omega_o) = u$$

(appealing, in the last step, to the fact that $\widehat{\omega}(u) = \omega_o$). So $\sum_{g \in G} f_g = u$, i.e., $g \mapsto f_g$ is an observable. Moreover,

$$\widehat{\omega}(\widehat{f_g}(\alpha)) = \widehat{\omega}(\widehat{\omega}^{-1}(g\alpha)) = g\alpha.$$

Thus, $\widehat{\omega} \circ \widehat{f_g}$ acts as the group element $g \in G$ – and hence, in particular, has a norm-preserving inverse. $\square$

Remarks: If the group G is compact, we can replace the discrete observable $\{f_g | g \in G\}$ in Theorem 5.2 by the continuous G-valued density $g \mapsto f_g := \int_G \omega^{-1} \circ g \, d\mu(g)$, where μ is the normalized Haar measure on G. While it is far from clear that we should want to regard this as a "continuously indexed observable" in any literal sense, it may be that discrete, coarse-grained versions of the effect-valued measure $B \mapsto \int_{g \in B} f_g d\mu(g)$ (B ranging over Borel subsets of G) can each underwrite some form of approximate teleportation protocol, of which a deterministic protocol is in some sense the limiting case. We defer exploration of this possibility to a future paper.

6. Entanglement Swapping

Consider a scenario in which Alice and Bob each possess one wing of two non-local, bipartite systems, say $S_1 = A_1 B_1$ and $S_2 = A_2 B_2$. We may model this situation by supposing that the total system, S, is a composite of the four components A_1, A_2, B_1 and B_2. We then have, in addition to the two non-local marginal systems S_1 and S_2, two local systems, $A = A_1 A_2$ and $B = B_1 B_2$ corresponding to Alice and Bob, respectively.

Suppose now that f is an effect on $A = A_1 A_2$ and μ and ω are states in $S_1 = A_1 B_1$ and $S_2 = A_2 B_2$, respectively. We have corresponding positive operators $\widehat{f} : A_1 \to A_2^*$, $\widehat{\omega} : A_2^* \to B_2$, and $\widehat{\mu}^* : B_1^* \to A_1$ (the dual of $\widehat{\mu} : A_1^* \to B_1$). Composing, we obtain a positive operator $\widehat{\omega} \circ \widehat{f} \circ \widehat{\mu}^* : B_1^* \to B_2$, corresponding to a sub-normalized state in $B_1 \otimes_{max} B_2$. The question arises, does this belong to the marginal state space $B = B_1 B_2$? Equivalently, can we implement the mapping in question by (un-normalized) conditionalization on the outcome of a measurement on A?

If S is a *regular* composite of A_1, A_2, B_1 and B_2, the answer is yes: $\mu \otimes \omega$ is then a legitimate state on $S = AB$, whence, for all $f \in A^*$, the partially evaluated state $(\mu \otimes \omega)_B(f) = (\mu \otimes \omega)(f \otimes -)$ lies in B. Now notice the following analogue of Lemma 4.1 (proved in the same way, i.e,. by checking it on pure tensors):

LEMMA 6.1. *With notation as above,*
$$(f^A \otimes g^B)(\mu^{S_1} \otimes \omega^{S_2}) = g^B(\widehat{\omega} \circ \widehat{f} \circ \widehat{\mu}^*).$$

It follows that
$$\widehat{\omega} \circ \widehat{f} \circ \widehat{\mu}^* = (\mu \otimes \omega)_B(f) \in B,$$

as claimed. This is analogous to the remote evaluation protocol of Section 3: conditional upon Alice securing a measurement outcome corresponding to f^A, the conditional state of Bob's system $B = B_1 \otimes B_2$ corresponds to the operator $\widehat{\omega} \circ \widehat{f^A} \circ \widehat{\mu}^*$. We might call this *state-pivoting*, as one can easily verify that the marginal state of B_1 is undisturbed.

Where the operation $\widehat{\omega} \circ \widehat{f}$ can be reversed, this protocol can be used to transfer the state μ from subsystem S_1 to subsystem B, as in conventional entanglement-swapping. Indeed, suppose that (i) $A_1 = B_2$, (ii) there exists a conclusive teleportation protocol for the tripartite system $A_1 A_2 B_2 - \omega$ in S_2 and an effect f in A^* such that $\widehat{\omega} \circ \widehat{f}$ is proportional to the identity operator on A_1. Then, for any $\mu \in S_2$, Lemma 6.1 tells us that:
$$(\mu \otimes \omega)_f^B = \mu.$$

That is, conditional on the occurrence of f in some measurement by Alice on system A, the state of Bob's system B is μ. In this situation, we may say that μ has been teleported from S_1 *through* ω to B.

The same considerations also allow us to convert an effect f on A into a subnormalized state on B. Indeed, if S_1 and S_2 contain states η_1 and η_2, respectively, corresponding to order-isomorphisms $\widehat{\eta}_i : B_i^* \simeq A_i$ for $i = 1, 2$, then the mapping $\widehat{f} \mapsto \widehat{\eta}_1 \circ \widehat{f} \circ \widehat{\eta}_2^*$ gives us an order-preserving linear injection from A^* to B. Pursuing this a bit further, let $(\mathcal{C}, \odot)$ be a monoidal theory, as defined in Section 2. Let us say that a state-space $A \in \mathcal{C}$ is $\mathcal{C}$-*self dual* iff there exists a state $\eta \in A \odot A$ with $\widehat{\eta} : A^* \to A$ an isomorphism, and $\widehat{\eta}^{-1} : A \to A^*$ corresponding to an effect in $(A \odot A)^*$. It follows from the above, with $A_1 = A_2$ and $B_1 = B_2$, that if A and B are $\mathcal{C}$-self dual, then so is $A \odot B$.

Four-part disharmonies The entanglement-swapping protocol described above can be applied negatively, to show that certain four-part composites aren't regular.

EXAMPLE 6.2. Consider any four non-classical state spaces A_1, A_2, B_1 and B_2 with $B_2 \simeq A_1$. If A_1, A_2 and B_2 support a conclusive teleportation protocol (in particular, if all three are isomorphic and weakly self-dual) then the composite
$$S := (A_1 \otimes_{min} A_2) \otimes_{max} (B_1 \otimes_{min} B_2)$$
cannot be regular. Indeed, arguing as in the proof of Corollary 1, we see that the reduced system $B := B_1 B_2$ is precisely $B_1 \otimes_{min} B_2 \simeq B_1 \simeq A_1 \otimes_{min} B_1$, while $S_1 := A_1 B_1$ is $A_1 \otimes_{max} B_1$. Since A_1 and B_1 are non-classical, we can find an entangled state $\omega \in A_1 \otimes_{max} B_1$. If the composite were regular, we could apply the entanglement-swapping protocol of Lemma 6.1 to pivot ω to an entangled state on $B = B_1 \otimes_{min} B_2$ – which is absurd, as the latter contains no entangled states.

A similar disharmony obtains between the maximal and the usual tensor products of quantum systems [7]. Consider a situation in which two quantum-mechanical systems, represented by state spaces A and B, are coupled by means of the maximal tensor product to form $A \otimes_{max} B$. Suppose also that A and B are themselves composite systems, say $A = A_1 \otimes A_2$ and $B = B_1 \otimes B_2$, where $\otimes$ is the usual quantum-mechanical tensor product. Then an application of Lemma 6.1 shows that if ω is a maximally entangled state on $A_2 \otimes B_2$ and $\rho \in A_1 \otimes_{max} B_1$ is what we might call an *ultra-entangled* state of $A_1 \otimes B_1$ – that is, a state of the maximal tensor product not belonging to $A_1 B_1$ – then conditional on a suitable maximally entangled outcome for a measurement on A, one finds that ρ has apparently been teleported through ω, and now resides in $B_1 \otimes B_2$ – which is absurd, as the latter is an ordinary composite quantum system hosting no ultra-entangled states.

7. Conclusions and Prospectus

We have established necessary and sufficient conditions for a composite of three probabilistic models to admit a conclusive teleportation protocol. We have also provided a class of examples illustrating that deterministic teleportation can be supported by weakly self-dual probabilistic models that are far from being either classical or quantum-mechanical. Along the way, we have developed tools for manipulating regular composites that are likely to be useful in any systematic study of categories of probabilistic models, and particularly categories equipped with more than a single tensor product.

It remains an open problem to find non-trivial necessary and sufficient conditions for a deterministic teleportation protocol to exist. Theorem 5.2 is a step in this direction; however, one would like a sharp criterion for the existence of a G-equivariant isomorphism $A^* \simeq A$, where G is a finite or, more generally, compact group acting transitively on the extreme points of Ω_A.

Looking further ahead, one would like to consider in detail the categorical structure of probabilistic theories subject to precise axioms governing remote evaluation, teleportation, etc., making contact with the rapidly developing theory of information processing in compact-closed categories [1, 2, 3, 21, 22].

Acknowledgements Significant parts of this work were done at the following conferences, retreats and workshops during 2007: (i) New Directions in the Foundations of Physics, College Park, MD (HB, JB, ML, AW); (ii) Philosophical and Formal Foundations of Modern Physics, Les Treilles, (HB); (iii) Operational Theories as Foils to Quantum Theory, Cambridge, supported by the Foundational Questions Institute (FQXi) and SECOQC (HB, JB, ML, AW); (iv) Operational Approaches to Quantum Theory, Paris (HB, AW). We wish to thank the organizers of these events, Jeffrey Bub and Rob Rynasiewicz (i), Alexei Grinbaum (ii, iv), Michel Bitbol (ii), and Tony Short and Rob Spekkens (iii), for the invaluable opportunities they provided for us to work on this project.

References

1. S. Abramsky and B. Coecke, A categorical semantics of quantum protocols, in *Proceedings of the 19th Annual IEEE Symposium on Logic in Computer Science: LICS 2004*, pp. 415–425, IEEE Computer Society (2004); also quant-ph/0402130v5 (2004, revised 2007)
2. S. Abramsky and B. Coecke, Abstract Physical Traces, Theory and Applications of Categories **14** (2005), 111-124.

3. J. Baez. Quantum quandaries: a category-theoretic perspective. `quant-ph/0404040`, 2004.
4. 2 J. Barrett, Information processing in general probabilistic theories, Phys. Rev. A. **75** (2007) 032304-
5. H. Barnum, J. Barrett, M. Leifer and A. Wilce, Cloning and Broadcasting in Generic Probabilistic Models, quant-ph/061129 (2006)
6. H. Barnum, J. Barrett, M. Leifer and A. Wilce, A general no-cloning theorem, Phys. Rev. Lett. **99** 240501 (2007).
7. H. Barnum, C. Fuchs, J. Renes and A. Wilce, Influence-free states on coupled quantum-mechanical systems, quant-ph/0507108 (2005)
8. C.H. Bennett, G. Brassard, C. Crépeau, R. Jozsa, A. Peres and W.K. Wootters, Teleporting an unknown quantum state via dual classical and Einstein-Podolsky-Rosen channels, Physical Review Letters, Vol. 70 (1993), 1895 1899.
9. E. B. Davies and J. T. Lewis, An operational approach to quantum probability, Comm. Math. Phys. **17** (1970) 239-260
10. C. M. Edwards, The operational approach to quantum probability I, Comm. Math. Phys. **17** (1971), 207-230.
11. A. J. Ellis, Linear operators in partially ordered normed vector spaces, J. London Math. Soc. **41** (1966) 323-332.
12. A. Holevo, Radon-Nikodym derivatives of quantum instruments J. Math. Phys. **39** (1998) 1373-
13. L. Hardy, A framework for probabilistic theories with non-fixed causal structure, J. Phys. A. **40** (2007) 3081
14. L. Hardy, Disentangling nonlocality and teleportation quant-ph/9906123 (1999)
15. M. Kläy, D. J. Foulis, and C. H. Randall, Tensor products and probability weights, Int. J. Theor. Phys. **26** (1987), 199-219.
16. Koecher, Die geoodätischen von Positivitaätsbereichen, Math. Annalen **135** (1958) 192-202.
17. G. Ludwig, *An Axiomatic Basis of Quantum Mechanics 1, 2*, Springer-Verlag, 1985, 1987.
18. G. Mackey, *Mathematical Foundations of Quantum Mechanics*, Benjamin, 1963.
19. I. Namioka and R. Phelps, Tensor products of compact convex sets, Pacific J. Math. **9** (1969), 469-480.
20. S. Popescu and D. Rohrlich, Quantum nonlocality as an axiom, Foundations of Physics **24** (1994), 379-385.
21. P. Selinger. Towards a semantics for higher-order quantum computation. In Proceedings of the 2nd International Workshop on Quantum Programming Languages, Turku Finland, pages 127–143. Turku Center for Computer Science, 2004. Publication No. 33.
22. P. Selinger. Dagger compact closed categories. *Electronic Notes in Theoretical Computer Science*, 170:139–163, 2007. Proceedings of the 3rd International Workshop on Quantum Programming Languages (QPL 2005), Chicago.
23. R. Spekkens, Evidence for the epistemic view of quantum states: a toy theory, Phys. Rev. A. **75** (2007) 032110
24. A. Wilce, Tensor products in generalized measure theory, Int. J. Theor. Phys. **31** (1992), 1915-1928.
25. G. Wittstock, Ordered normed tensor products, in H. Neumann and H. Hartkamper (eds.), Foundations of quantum mechanics and ordered linear spaces, Springer Lecture Notes in Physics, 1974.
26. E. B. Vinberg, Homogeneous cones, Dokl. Acad. Nauk. SSSR **141** (1960) 270-273; English trans. Soviet Math. Dokl. **2** (1961) 1416-1619.

Appendix: proofs from section 3

Proof of Lemma 3.5 (a) Let $\mu \in (A^J)_+$ be a positive linear combination $\mu = \sum_p t_p(\omega_p)^J_{a^p}$ of reduced states, where for all p, $\omega_p \in A$ and $a^p = (a^p_i) \in \Pi_{i \in I \setminus J} A^*_i$. Then for any $b = (b_j) \in \Pi_{j \in J \setminus K}$, we have $\mu^K_b = \sum_p t_p(\omega_p)^J_{a^p}(b) = \sum_p (\omega_p)^K_{a^p \otimes b} \in A^K$. It follows that $((A^J)^K)_+ \subseteq (A^K)_+$. For the converse, let $\omega \in A_+$: for any $a = (a_i) \in \Pi_{i \in I} A_i$, we have $a = b \otimes c$ where $b = (b_j) \in \Pi_{j \in J \setminus K} A_j$ and $c = (c_k) \in \Pi_{k \in K} A_k$. Thus, $\omega_a^K = (\omega_b^J)_c^K \in ((A^J)^K)_+$.

For (b), suppose $\omega \in A$ and $a = (a_i)$ in $\Pi_{i \notin I} A_i^*$. Pick any $c = (b_j) \in \Pi_{j \in J} A_j^*$: we can set
$$\alpha = \omega_a^J \in A_J \text{ and } \beta = \omega_b^{I-J} \in A_{I-J}.$$
If A is regular, we then have $\alpha \otimes \beta \in A$, whence,
$$\alpha = (\alpha \otimes \gamma)_{u_{I-J}}^J.$$
Part (c) follows from (a) and (b). □

Proof of Proposition 3.6 Let $A = \bigodot_{i \in I} A_i$. We first show that, for any set $J \subseteq I$, $A^J = \bigodot_{j \in J} A_j$). By assumption, we have
$$A \simeq (\bigodot_{j \in J} A_j) \odot (\bigodot_{k \in I \setminus J} A_k) \geq (\bigodot_{j \in J} A_j) \otimes_{min} (\bigodot_{k \in I \setminus J} A_k).$$
It follows that, for every $\mu \in \bigodot_{j \in J} A_j$, and for any $\nu \in \bigodot_{k \in I \setminus J} A_k$, $\mu \otimes \nu \in A$; hence,
$$\mu = (\mu \otimes \nu)_{\otimes_{k \in I \setminus J} u_k}^J \in A_J.$$
Thus, $\bigodot_{j \in J} A_j \leq A^J$.

For the reverse inclusion, note that we also have
$$(\bigodot_{j \in J} A_j) \odot (\bigodot_{k \in I \setminus J} A_k) \leq (\bigodot_{j \in J} A_j) \otimes_{max} (\bigodot_{k \in I \setminus J} A_k);$$
hence, for any $\omega \in A$ and any $f \in (\bigodot_{k \in K} A^k)^*$—in particular, for any $f = (f_k)_{k \in I \setminus J}$—we have $\omega_f^J \in \bigodot_{j \in J} A_j$. The rest of the proof now proceeds easily. If $J_1, ..., J_m$ is a partition of I, then we have $A = \bigodot_{p=1}^m (\bigodot_{j \in J_p} A_p) = \bigodot_{p=1}^m A^{J_p}$. Since $\odot$ is a coupling, this last is a composite of A^{J_p}, $p = 1, ..., m$. □

CCS-3: Information Sciences, MS B256, Los Alamos National Laboratory, Los Alamos, NM 87545 USA
Current address: Department of Physics and Astronomy, University of New Mexico
E-mail address: `hnbarnum@aol.com, hbarnum@unm.edu`

Perimeter Institute for Theoretical Physics, 31 Caroline Street N, Waterloo, Ontario N2L 2Y5, Canada
Current address: Department of Mathematics, Royal Holloway, University of London Egham, Surrey TW20 0EX, United Kingdom
E-mail address: `jon.barrett@rhul.ac.uk`

Institute for Quantum Computing, University of Waterloo, Waterloo, Ontario, Canada
Current address: Department of Physics and Astronomy, University College London, London WC1E 6BT, England
E-mail address: `matt@mattleifer.info`

Department of Mathematical Sciences, Susquehanna University, Selinsgrove, PA 17870 USA
E-mail address: `wilce@susqu.edu`

Fixed Points in Epistemic Game Theory

Adam Brandenburger, Amanda Friedenberg, and H. Jerome Keisler

ABSTRACT. The epistemic conditions of "rationality and common belief of rationality" and "rationality and common assumption of rationality" in a game are characterized by the solution concepts of a "best-response set" (Pearce [**26**, 1984]) and a "self-admissible set" (Brandenburger, Friedenberg, and Keisler [**13**, 2008]), respectively. We characterize each solution concept as the set of fixed points of a map on the lattice of rectangular subsets of the product of the strategy sets. Of note is that both maps we use are non-monotone.

1. Introduction

Topological fixed-point arguments have a long tradition in game theory. von Neumann's [**28**, 1928] proof of his famous Minimax Theorem made use of a topological fixed-point argument. Nash's existence proof for his equilibrium concept is also a topological fixed-point argument—using Brouwer's Theorem ([**14**, 1910]) in [**23**, 1950] and [**25**, 1951] and Kakutani's Theorem ([**17**, 1941]) in [**24**, 1950]. Subsequently, the Brouwer and Kakutani theorems became the standard tools in existence arguments in game theory.

Order-theoretic fixed-point arguments also have a role in game theory. These arguments are prominent in epistemic game theory (EGT). The purpose of this note is to explain the role of order-theoretic fixed-point arguments in EGT.

This note focuses on two solution concepts that arise from the EGT approach: best-response sets and self-admissible sets. The treatment is in finite games. (The definitions of these concepts will be laid out later.) We will characterize these solution concepts as arising from fixed points of certain non-monotone maps.

There is an important stream of papers that treats iterated dominance concepts in general infinite games. See, e.g., Apt [**2**, 2007a], [**3**, 2007b], [**4**, 2007c], and Apt and Zvesper [**5**, 2007]. The main focus in this stream of papers is on the case of monotonicity, but non-monotonic maps also arise.

2000 *Mathematics Subject Classification.* Primary 91A10, 91A26; Secondary 47H10.

We thank Samson Abramsky, Krzysztof Apt, the editors, and a referee for very helpful comments, and Ariel Ropek and Andrei Savochkin for research assistance. Financial support from the Stern School of Business, the Olin School of Business, and the W.P. Carey School of Business is gratefully acknowledged. Some of this material appeared in our "Admissibility in Games" (December 2004 version), which was superseded by our "Admissibility in Games," *Econometrica*, 76, 2008, 307-352.

Mathematical logic, theoretical computer science, and lattice theory are other areas where order-theoretic fixed point methods are commonly employed. See Moschovakis [**22**, 1974], Libkin [**18**, 2010], Abramsky and Jung [**1**, 1994], and Davey and Priestley [**15**, 2002] for standard presentations. The order-theoretic maps most often used there are monotonic, but non-monotonic maps also arise. A notable instance of the non-monotone case is Martin's [**19**, 2000] theory of measurements on domains. It would be very interesting to investigate whether what is known in other areas about the non-monotone case could be applied to EGT. We leave this to future work.

2. Epistemic Game Theory

We now give a very brief sketch of the EGT approach. The classical description of the strategic situation consists of a game matrix or game tree. The idea is that players may face uncertainty about how others play the game. Under the EGT approach, this uncertainty is part of the description of the strategic situation. That is, the description consists of a game (matrix or tree) and beliefs about the play of the game. These "beliefs about the play of the game" are, in fact, the so-called "hierarchies about the play of the game," i.e., what each player thinks about "the strategies other players select," what each player thinks about "what others think about 'the strategies others select,'" etc.[1]

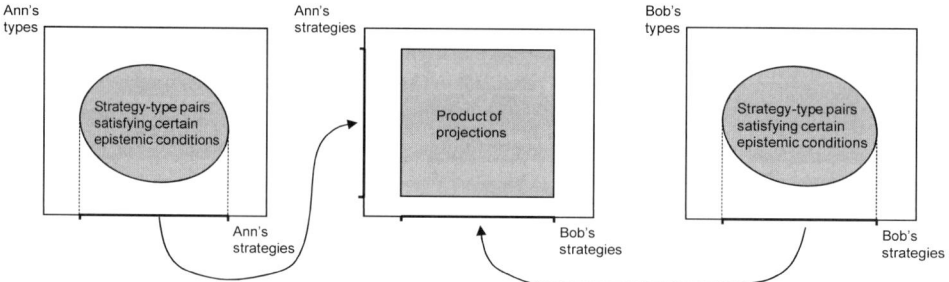

Figure 2.1

Harsanyi [**16**, 1967] introduced a **type structure** model to describe these hierarchies. Formally, this adds **types** for each player, where the type describes what that player thinks, think other players think, ... The Appendix provides a formal treatment of this step, in a baseline case. (See Siniscalchi [**27**, 2008] for a survey of the relevant literature.)

Figure 2.1 depicts the set-up in the two-player case. The starting point is the strategy-type pairs for Ann and Bob, given by the left-hand and right-hand panels. We then impose epistemic conditions. These correspond to the shaded sets in the left-hand and right-hand panels. The idea is that the epistemic conditions correspond to "strategic reasoning." For example, we might consider those strategy-type pairs for Ann that satisfy the condition that Ann is **rational**—i.e., that her strategy is optimal for her given what, according to her type, she thinks Bob will play.

---

[1]This discussion presumes that the structure of the game (e.g., payoff functions) is "transparent" among the players. If it is not, then the description consists of the game and beliefs about both the structure of the game and the play of the game.

(Ann's remaining strategy-type pairs satisfy the condition that she is irrational. Below, we will be more precise about what "thinks" means.) A basic epistemic condition is that Ann is rational, thinks Bob is rational, and so on. In this case, we take the shaded set in the left-hand panel to be the strategy-type pairs for Ann that satisfy this condition. Likewise, we take the shaded set in the right-hand panel to be the strategy-type pairs for Bob that are rational, think Ann is rational, and so on.

The **characterization** question is whether we can identify the strategies that can be played under such epistemic conditions, by looking only at the game (matrix or tree). That is, can we identify the projections of the shaded sets into the strategy sets, as depicted in the middle panel, without reference to the type structure model? There are several such characterization results in EGT. They differ according to how the terms "rationality" and "thinks" are formalized. The different formalizations, in turn, reflect different concepts of "strategic reasoning" (i.e., different epistemic conditions) that we the analysts can impose on games.

The remainder of this section formalizes the epistemic descriptions and the epistemic conditions. Section 4 returns to the question of characterization.

Fix a two-player finite strategic-form game $\langle S^a, S^b, \pi^a, \pi^b \rangle$, where S^a and S^b are finite strategy sets for Ann and Bob, and $\pi^a : S^a \times S^b \to \mathbb{R}$ and $\pi^b : S^a \times S^b \to \mathbb{R}$ are their payoff functions.[2]

Begin with the most basic formalization. Append Polish spaces T^a and T^b of types for Ann and Bob. Here, a type t^a for Ann is associated with a probability measure on the Borel subsets of $S^b \times T^b$. A strategy-type pair $(s^a, t^a) \in S^a \times T^a$ for Ann is **rational** if s^a maximizes her expected payoff, under the marginal on S^b of the probability measure associated with t^a. In this case, "thinks" means "belief." Say Ann **believes** $E^b \subseteq S^b \times T^b$ if E^b is Borel and the probability measure associated with type t^a assigns probability 1 to E^b. These and subsequent definitions have counterparts with Ann and Bob interchanged.

A second formalization follows Brandenburger, Friedenberg, and Keisler [13, 2008]. Again, append Polish spaces T^a and T^b of types for Ann and Bob. Here, a type t^a for Ann is associated with a lexicographic probability system on the Borel subsets of $S^b \times T^b$. A lexicographic probability system (Blume, Brandenburger, and Dekel [10, 1991]) is a finite sequence of mutually singular probability measures. It is to be thought of as a sequence of hypotheses—a primary hypothesis, a secondary hypothesis, ...,—held by Ann about Bob's strategy and type. A strategy-type pair $(s^a, t^a) \in S^a \times T^a$ for Ann is (**lexicographically**) **rational** if s^a lexicographically maximizes her sequence of expected payoffs, calculated under the marginals on S^b of the sequence of probability measures associated with t^a. In this case, "thinks" means "assumption." If T^b is finite, say Ann **assumes** E^b if each point in E^b receives positive probability under an earlier probability measure in the sequence than does any point not in E^b. Alternatively put, the event E^b is "infinitely more likely" than the event not-E^b. See [13, 2008, Section 5] for a general treatment (i.e., in the case of infinite T^b).

For the purposes of this note, we will need one key property of both "belief" and "assumption." Return to the general term "thinks," in order to subsume both cases. Define a thinking operator C^a from the family of Borel subsets of $S^b \times T^b$

---

[2]We restrict attention to two-player games for notational simplicity, but our analysis immediately extends to games with three or more players.

to T^a by
$$C^a(E^b) = \{t^a \in T^a : t^a \text{ thinks } E^b \text{ is true}\}.$$

AXIOM 2.1 (Conjunction). *Fix a type* $t^a \in T^a$ *and Borel sets* $E_1^b, E_2^b, \ldots$ *in* $S^b \times T^b$. *Suppose, for each* m, *that* $t^a \in C^a(E_m^b)$. *Then* $t^a \in C^a(\bigcap_m E_m^b)$.

In words, this says that if Ann thinks that each event E_m^b is true, then she thinks the joint event $\bigcap_m E_m^b$ is true. It is immediate from the rules of probability that "belief" satisfies this conjunction property. For the case of "assumption," see ([**13**, 2008, Property 6.3]).

Given Borel sets $E^a \subseteq S^a \times T^a$ and $E^b \subseteq S^b \times T^b$, define $E_1^a = E^a$, $E_1^b = E^b$, and for $m \geq 1$,
$$E_{m+1}^a = E_m^a \cap [S^a \times C^a(E_m^b)].$$
(For this to be well-defined, the sets E_m^a and E_m^b must be Borel.)

DEFINITION 2.1. *The event that* $E^a \times E^b$ **is true and commonly thought** *is*
$$\bigcap_{m=1}^{\infty} E_m^a \times \bigcap_{m=1}^{\infty} E_m^b.$$

In the situations we study in this note, $E^a \times E^b$ is the event that Ann and Bob are rational. Thus, the event $\bigcap_{m=1}^{\infty} E_m^a \times \bigcap_{m=1}^{\infty} E_m^b$ is either the event that there is **rationality and common belief of rationality** (**RCBR**) or the event that there is **rationality and common assumption of rationality** (**RCAR**).

3. Epistemic Fixed Points

Given two Polish spaces P^a, P^b, let $\mathcal{B}(P^a, P^b)$ be the set of all rectangles $E^a \times E^b$ where E^a is a Borel subset of P^a and E^b is a Borel subset of P^b. Proposition 3.2 below will show that each of the events RCBR and RCAR is a fixed point of a mapping Γ from $\mathcal{B}(S^a \times T^a, S^b \times T^b)$ to itself. This suggests that we should be able to describe the strategies playable under RCBR or RCAR via fixed points of mappings from $\mathcal{B}(S^a, S^b)$ to itself. Section 5 will show that this is indeed the case. In sum: This section is about epistemic fixed points, i.e., fixed points in $\mathcal{B}(S^a \times T^a, S^b \times T^b)$. Section 5 will be about fixed points in the game matrix, i.e., fixed points in the smaller space $\mathcal{B}(S^a, S^b)$ where the type structure plays no part.

Define a mapping Γ from $\mathcal{B}(S^a \times T^a, S^b \times T^b)$ to itself, as follows. Given $E^a \times E^b \in \mathcal{B}(S^a \times T^a, S^b \times T^b)$, set
$$\Gamma(E^a \times E^b) = (E^a \times E^b) \cap ([S^a \times C^a(E^b)] \times [S^b \times C^b(E^a)]).$$
Note that the mapping Γ depends on C^a and C^b. In words, Γ maps an event $E^a \times E^b$ to the event that $E^a \times E^b$ is true, and Ann and Bob think their respective components of $E^a \times E^b$ are true. The next lemma is immediate:

LEMMA 3.1. *The event* $E^a \times E^b$ *is a fixed point of* Γ, *i.e.* $\Gamma(E^a \times E^b) = E^a \times E^b$, *if and only if*
$$\begin{aligned} E^a &\subseteq S^a \times C^a(E^b), \\ E^b &\subseteq S^b \times C^b(E^a). \end{aligned}$$

When C^a stands for belief, Lemma 3.1 says the fixed points of Γ are the so-called "self-evident events." (The concept of a self-evident event is due to Monderer-Samet [**21**, 1989].) Lemma 3.1 is used in the proofs of the following two propositions.

PROPOSITION 3.2. *Fix $E^a \times E^b \in \mathcal{B}(S^a \times T^a, S^b \times T^b)$. The event*
$$\bigcap_{m=1}^{\infty} E_m^a \times \bigcap_{m=1}^{\infty} E_m^b$$
is a fixed point of Γ.

PROOF. Using the definitions,
$$\bigcap_{m=1}^{\infty} E_m^a = E_1^a \cap \bigcap_{m=1}^{\infty}[S^a \times C^a(E_m^b)] \subseteq \bigcap_{m=1}^{\infty}[S^a \times C^a(E_m^b)] =$$
$$S^a \times \bigcap_{m=1}^{\infty} C^a(E_m^b) \subseteq S^a \times C^a(\bigcap_{m=1}^{\infty} E_m^b),$$
where the last inclusion relies on conjunction (Axiom 2.1). □

Proposition 3.2 says that the event RCBR (resp. RCAR) is a fixed point of the map Γ when C^a, C^b correspond to belief (resp. assumption). We also have:

PROPOSITION 3.3. *Suppose $E^a \times E^b \in \mathcal{B}(S^a \times T^a, S^b \times T^b)$ is a fixed point of Γ. Then $E_m^a = E^a$ and $E_m^b = E^b$ for all m.*

PROOF. This is immediate for $m = 1$, so suppose it is true for m. We have
$$E_{m+1}^a = E_m^a \cap [S^a \times C^a(E_m^b)] = E^a \cap [S^a \times C^a(E^b)],$$
using the induction hypothesis. But since $E^a \times E^b$ is a fixed point,
$$E^a \cap [S^a \times C^a(E^b)] = E^a,$$
and so $E_{m+1}^a = E^a$, as required. □

An important reference on fixed points on epistemic structures is Barwise [**7**, 1988].

4. Best-Response Sets and Self-Admissible Sets

We now turn to fixed-point characterizations of the strategies playable under RCBR and RCAR. We undertake these characterizations in two steps. First, we review the existing strategic characterizations of RCBR and RCAR. These are couched in terms of best-response sets (Pearce [**26**, 1984]) and self-admissible sets (Brandenburger, Friedenberg, and Keisler [**13**, 2008]). Then, we point to fixed-point characterizations of these sets.

Throughout, we treat RCBR and RCAR in parallel rather than in sequence. The reason is that certain mathematical techniques that are typically associated with self-admissible sets (and so RCAR) will be useful in our fixed-point characterization of best-response sets.[3]

We begin with some preliminary definitions and lemmas. Given a finite set Ω, let $\mathcal{M}(\Omega)$ denote the set of all probability measures on Ω. Write $\operatorname{Supp}\sigma$ for the support of $\sigma \in \mathcal{M}(X)$.

The definitions to come all have counterparts with a and b reversed. Extend π^a to $\mathcal{M}(S^a) \times \mathcal{M}(S^b)$ by taking $\pi^a(\sigma^a, \sigma^b)$, for $\sigma^a \in \mathcal{M}(S^a)$, $\sigma^b \in \mathcal{M}(S^b)$, to be the expectation of π^a under $\sigma^a \otimes \sigma^b$:
$$\pi^a(\sigma^a, \sigma^b) = \sum_{(s^a, s^b) \in S^a \times S^b} \sigma^a(s^a) \sigma^b(s^b) \pi^a(s^a, s^b).$$

---

[3] We are grateful to a referee for asking us to emphasize this point.

DEFINITION 4.1. *Fix $X \times Y \subseteq S^a \times S^b$ with $Y \neq \emptyset$, and $\sigma^b \in \mathcal{M}(S^b)$. A strategy $s^a \in X$ is σ^b-**justifiable with respect to** $X \times Y$ if $\sigma^b(Y) = 1$ and $\pi^a(s^a, \sigma^b) \geq \pi^a(r^a, \sigma^b)$ for every $r^a \in X$. Say s^a is **justifiable with respect to** $X \times Y$ if s^a is σ^b-justifiable with respect to $X \times Y$ for some σ^b.*

DEFINITION 4.2. *Fix $Q^a \times Q^b \subseteq S^a \times S^b$. The set $Q^a \times Q^b$ is a **best-response set (BRS)** if for each $s^a \in Q^a$ there is a $\sigma^b \in \mathcal{M}(S^b)$ such that:*

(i) *s^a is σ^b-justifiable with respect to $S^a \times Q^b$;*
(ii) *if r^a is also σ^b-justifiable with respect to $S^a \times Q^b$, then $r^a \in Q^a$;*

and likewise for each $s^b \in Q^b$.

The original definition of a BRS is due to Pearce [**26**, 1984]. Definition 4.2 is from Battigalli and Friedenberg [**8**, 2011]. It differs from Pearce's definition in two ways: players choose only pure (not mixed) strategies, and condition (ii) is new.[4] Condition (ii) is important in the epistemic characterization; see the statement at the beginning of Section 5.

DEFINITION 4.3. *Fix $X \times Y \subseteq S^a \times S^b$ with $Y \neq \emptyset$. A strategy $s^a \in X$ is **strongly dominated with respect to** $X \times Y$ if there is a $\sigma^a \in \mathcal{M}(S^a)$, with $\sigma^a(X) = 1$, such that $\pi^a(\sigma^a, s^b) > \pi^a(s^a, s^b)$ for every $s^b \in Y$. Otherwise, say s^a is **undominated with respect to** $X \times Y$.*

The following equivalence is standard. (Necessity is immediate. Sufficiency is proved via the supporting hyperplace theorem.)

LEMMA 4.4. *Fix $X \times Y \subseteq S^a \times S^b$ with $Y \neq \emptyset$. A strategy $s^a \in X$ is justifiable with respect to $X \times Y$ if and only if it is undominated with respect to $X \times Y$.*

The next definition picks out a notable BRS.

DEFINITION 4.5. *Set $S_0^i = S^i$ for $i = a, b$, and define inductively*

$$S_{m+1}^i = \{s^i \in S_m^i : s^i \text{ is undominated with respect to } S_m^a \times S_m^b\}$$

*A strategy $s^i \in S_m^i$ is called m-**undominated**. A strategy $s^i \in \bigcap_{m=0}^{\infty} S_m^i$ is called **iteratively undominated (IU)**.*

By finiteness, there is a (first) number M such that $\bigcap_{m=0}^{\infty} S_m^i = S_M^i \neq \emptyset$ for $i = a, b$. It is easy to check that the IU set is a BRS and every BRS is contained in the IU set.

We now repeat this development for self-admissible sets.

DEFINITION 4.6. *Fix $X \times Y \subseteq S^a \times S^b$ with $Y \neq \emptyset$. A strategy $s^a \in X$ is **weakly dominated with respect to** $X \times Y$ if there is a $\sigma^a \in \mathcal{M}(S^a)$, with $\sigma^a(X) = 1$, such that $\pi^a(\sigma^a, s^b) \geq \pi^a(s^a, s^b)$ for every $s^b \in Y$, and $\pi^a(\sigma^a, s^b) > \pi^a(s^a, s^b)$ for some $s^b \in Y$. Otherwise, say s^a is **admissible with respect to** $X \times Y$.*

DEFINITION 4.7. *Fix $Y \subseteq S^b$ with $Y \neq \emptyset$. Say r^a **supports** s^a **with respect to** Y if there is a $\sigma^a \in \mathcal{M}(S^a)$ with $r^a \in \operatorname{Supp} \sigma^a$ and $\pi^a(\sigma^a, s^b) = \pi^a(s^a, s^b)$ for all $s^b \in Y^b$. If r^a supports s^a with respect to S^b, say simply r^a **supports** s^a.*

DEFINITION 4.8 (Brandenburger, Friedenberg, and Keisler [**13**, 2008]). *Fix $Q^a \times Q^b \subseteq S^a \times S^b$. The set $Q^a \times Q^b$ is a **self-admissible set (SAS)** if:*

---

[4]David Pearce (private communication) told one of us that he was aware of this condition, but to keep things simple did not include it in his definition in [**26**, 1984].

(i) each $s^a \in Q^a$ is admissible with respect to $S^a \times S^b$;
(ii) each $s^a \in Q^a$ is admissible with respect to $S^a \times Q^b$;
(iii) if $r^a \in S^a$ supports some $s^a \in Q^a$, then $r^a \in Q^a$;

and likewise for each $s^b \in Q^b$.

The next equivalence is a special case of a classic result in convex analysis due to Arrow, Barankin, and Blackwell [**6**, 1953].

LEMMA 4.9. *Fix $X \times Y \subseteq S^a \times S^b$ with $Y \neq \emptyset$. A strategy $s^a \in X$ is admissible with respect to $X \times Y$ if and only if there is a $\sigma^b \in \mathcal{M}(S^b)$, with $\text{Supp}\, \sigma^b = Y$, such that s^a is σ^b-justifiable with respect to $X \times Y$.*

The next lemma rewrites Definition 4.8 in a way that brings out the comparison with BRS's.

LEMMA 4.10. *A set $Q^a \times Q^b$ is an SAS if and only if:*
(i) *for each $s^a \in Q^a$, there is a $\sigma^b \in \mathcal{M}(S^b)$, with $\text{Supp}\, \sigma^b = S^b$, such that:*
- s^a *is σ^b-justifiable with respect to $S^a \times S^b$,*
- *if $r^a \in S^a$ is also σ^b-justifiable with respect to $S^a \times S^b$, then $r^a \in Q^a$;*

(ii) *for each $s^a \in Q^a$, there is a $\rho^b \in \mathcal{M}(S^b)$, with $\text{Supp}\, \rho^b = Q^b$, such that s^a is ρ^b-justifiable with respect to $S^a \times Q^b$;*

and likewise for each $s^b \in Q^b$.

To show that an SAS satisfies the conditions in this Lemma, use Lemma D.4 in Brandenburger, Friedenberg, and Keisler [**13**, 2008]. For the converse, use Lemma D.2 in Brandenburger, Friedenberg, and Keisler [**13**, 2008].

The next definition picks out a notable SAS.

DEFINITION 4.11. *Set $\overline{S}_0^i = S^i$ for $i = a, b$, and define inductively*
$$\overline{S}_{m+1}^i = \{s^i \in \overline{S}_m^i : s^i \text{ is admissible with respect to } \overline{S}_m^a \times \overline{S}_m^b\}.$$
*A strategy $s^i \in \overline{S}_m^i$ is called m-**admissible**. A strategy $s^i \in \bigcap_{m=0}^\infty \overline{S}_m^i$ is called **iteratively admissible (IA)**.*

By finiteness, there is a (first) number N such that $\bigcap_{m=0}^\infty \overline{S}_m^i = \overline{S}_N^i \neq \emptyset$ for $i = a, b$. The IA set is an SAS (Brandenburger and Friedenberg [**12**, 2010, Proposition 5.1]). But, unlike the case with IU and BRS's, it need not be the case that every SAS is contained in the IA set. See Example 5.9 to come.

5. Fixed-Point Characterizations

BRS's characterize the epistemic condition of RCBR: Fix a game $\langle S^a, S^b, \pi^a, \pi^b \rangle$ and an associated type structure, where each type is mapped to a (single) probability measure. Define "believes" and "rationality" as before. Then, the projection to $S^a \times S^b$ of the RCBR event (which lies in $S^a \times T^a \times S^b \times T^b$) constitutes a BRS of the game. Conversely, every BRS of $\langle S^a, S^b, \pi^a, \pi^b \rangle$ arises in this way, for a suitable choice of type structure. This follows from Battigalli and Friedenberg [**8**, 2011, Theorem 5.1].

Likewise, SAS's characterize the epistemic conditions of RCAR: Fix a game $\langle S^a, S^b, \pi^a, \pi^b \rangle$ and an associated type structure, where each type is now mapped to a lexicographic probability system. Define "assumes" and "(lexicographic) rationality" as before. Then, the projection to $S^a \times S^b$ of the RCAR event constitutes an SAS of the game. Conversely, every SAS of $\langle S^a, S^b, \pi^a, \pi^b \rangle$ arises in this way, for a suitable choice of type structure. This is Theorem 8.1 in Brandenburger, Friedenberg, and Keisler [13, 2008].

So, to deliver on our promised fixed-point characterizations of RCBR and RCAR in terms of strategies played, it remains to provide fixed-point characterizations of BRS's and SAS's.

Fix $X \times Y \subseteq S^a \times S^b$ with $Y \neq \emptyset$. The next two lemmas are standard.

LEMMA 5.1. *A strategy $s^a \in X$ is σ^b-justifiable with respect to $X \times Y$ if and only if s^a is admissible with respect to $X \times \mathrm{Supp}\,\sigma^b$.*

LEMMA 5.2. *Fix $s^a \in X$ and $\sigma^b \in \mathcal{M}(S^b)$ such that $\pi^a(s^a, \sigma^b) \geq \pi^a(q^a, \sigma^b)$ for all $q^a \in X$. If r^a supports s^a, then $\pi^a(r^a, \sigma^b) \geq \pi^a(q^a, \sigma^b)$ for all $q^a \in X$.*

LEMMA 5.3. *Suppose $s^a \in S^a$ is undominated (resp. admissible) with respect to $X \times Y$. If r^a supports s^a, then r^a is undominated (resp. admissible) with respect to $X \times Y$.*

PROOF. By Lemma 4.4 (resp. Lemma 4.9), there is a $\sigma^b \in \mathcal{M}(S^b)$, with $\sigma^b(Y) = 1$ (resp. $\mathrm{Supp}\,\sigma^b = Y^b$) such that $\pi^a(s^a, \sigma^b) \geq \pi^a(q^a, \sigma^b)$ for all $q^a \in X$. By Lemma 5.2, r^a is then undominated (resp. admissible) with respect to $X \times Y$. □

Given $\sigma^b \in \mathcal{M}(S^b)$, write $\mathcal{J}(\sigma^b)$ for the set of strategies $s^a \in S^a$ that are σ^b-justifiable.

DEFINITION 5.4. *Say that σ^b **minimally justifies** s^a **with respect to** $X \times Y$ if s^a is σ^b-justifiable with respect to $X \times Y$ and, for each $\rho^b \in \mathcal{M}(S^b)$ such that s^a is ρ^b-justifiable with respect to $X \times Y$, we have $\mathcal{J}(\sigma^b) \subseteq \mathcal{J}(\rho^b)$.*

LEMMA 5.5. *If s^a is justifiable with respect to $S^a \times Y$, there is a σ^b that minimally justifies s^a with respect to $S^a \times Y$.*

PROOF. Suppose s^a is justifiable with respect to $S^a \times Y$. Then, by Lemma 5.1, there is $\emptyset \neq Z_k \subseteq Y$ such that s^a is admissible with respect to $S^a \times Z_k$. Let Z be the union of all such Z_k. Then, s^a is admissible with respect to $S^a \times Z$. To see this, suppose not, i.e., suppose there is $\sigma^a \in \mathcal{M}(S^a)$ with $\pi^a(\sigma^a, s^b) \geq \pi^a(s^a, s^b)$ for every $s^b \in Z$, and $\pi^a(\sigma^a, s^b) > \pi^a(s^a, s^b)$ for some $s^b \in Z$. Then, we can find some $Z_k \subseteq Z$ such that $\pi^a(\sigma^a, s^b) \geq \pi^a(s^a, s^b)$ for every $s^b \in Z_k$, and $\pi^a(\sigma^a, s^b) > \pi^a(s^a, s^b)$ for some $s^b \in Z_k$. This contradicts the fact that s^a is admissible with respect to each $S^a \times Z_k$.

We have established that s^a is admissible with respect to $S^a \times Z$. By Lemma D.4 in Brandenburger, Friedenberg, and Keisler [13, 2008], there is then a $\sigma^b \in \mathcal{M}(S^b)$, with $\mathrm{Supp}\,\sigma^b = Z$, such that $\mathcal{J}(\sigma^b)$ is the set of strategies that support s^a with respect to $S^a \times Z$. We will show that σ^b minimally justifies s^a with respect to $S^a \times Y$.

Fix $\rho^b \in \mathcal{M}(S^b)$ that justifies s^a with respect to $S^a \times Y$. We show that $\mathcal{J}(\sigma^b) \subseteq \mathcal{J}(\rho^b)$. Fix $r^a \in \mathcal{J}(\sigma^b)$. Then r^a supports s^a with respect to $S^a \times Z$. Note that $\mathrm{Supp}\,\rho^b \subseteq Z$, so that r^a also supports s^a with respect to $S^a \times \mathrm{Supp}\,\rho^b$. It follows from Lemma 5.2 that $r^a \in \mathcal{J}(\rho^b)$ as required. □

Now consider the complete lattice $\Lambda = \mathcal{B}(S^a, S^b)$. (The join of two subsets is the component-by-component union. The meet is the intersection.) We will define a map $\Phi : \Lambda \to \Lambda$ so that the fixed points of Φ are the BRS's. Specifically, for $Q^a \times Q^b \in \mathcal{B}(S^a, S^b)$, set $(s^a, s^b) \in \Phi(Q^a \times Q^b)$ if either: (a) $s^a \in Q^a$ and s^a is justifiable with respect to $S^a \times Q^b$, or (b) $s^a \in \mathcal{J}(\sigma^b)$ for some σ^b that minimally justifies some $r^a \in Q^a$ with respect to $S^a \times Q^b$. (The analogous conditions must hold for s^b.)

PROPOSITION 5.6. *If $Q^a \times Q^b$ is a BRS, then it is a fixed point of Φ, i.e., $\Phi(Q^a \times Q^b) = Q^a \times Q^b$. Conversely, if $Q^a \times Q^b$ is a fixed point of Φ, then it is a BRS.*

PROOF. Fix a BRS $Q^a \times Q^b$. We will show that it is a fixed point of Φ. First, fix $(s^a, s^b) \in \Phi(Q^a \times Q^b)$. We will show that $(s^a, s^b) \in Q^a \times Q^b$. Indeed, suppose that $s^a \notin Q^a$. Then there is an $r^a \in Q^a$ such that r^a is justifiable with respect to $S^a \times Q^b$ and, for any σ^b such that r^a is σ^b-justifiable with respect to $S^a \times Q^b$, $s^a \in \mathcal{J}(\sigma^b)$. It follows from condition (ii) of the definition of a BRS that $s^a \in Q^a$, a contradiction. Likewise, we reach a contradiction if we suppose that $s^b \notin Q^b$, so we conclude that $\Phi(Q^a \times Q^b) \subseteq Q^a \times Q^b$. Next, fix $(s^a, s^b) \in Q^a \times Q^b$. By condition (i) of the definition of a BRS and condition (a) of the definition of Φ, we get $(s^a, s^b) \in \Phi(Q^a \times Q^b)$, establishing that $Q^a \times Q^b \subseteq \Phi(Q^a \times Q^b)$.

For the converse, suppose $Q^a \times Q^b = \Phi(Q^a \times Q^b)$. Fix $s^a \in Q^a$ and note that, by condition (a) of the definition of Φ, there is a $\sigma^b \in \mathcal{M}(S^b)$ that justifies s^a with respect to $S^a \times Q^b$. By Lemma 5.5, we can choose σ^b to minimally justify s^a with respect to $S^a \times Q^b$. Then, by condition (b) of the definition of Φ, we get that s^a satisfies conditions (i)-(ii) of a BRS (using the measure σ^b). We can make the same argument for each $s^b \in Q^b$. This establishes that $Q^a \times Q^b$ is a BRS. □

EXAMPLE 5.7. *The map Φ is not monotone (increasing). Consider the game in Figure 5.1. We have $\Phi(\{(U, L)\}) = \{U, D\} \times \{L, R\}$ but $\Phi(\{U\} \times \{L, R\}) = \{U\} \times \{L, R\}$.*

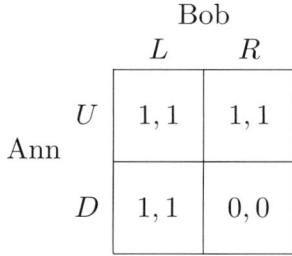

Figure 5.1

Next, we define a map $\Psi : \Lambda \to \Lambda$ so that the fixed points of Ψ are the SAS's. Specifically, set $(s^a, s^b) \in \Psi(Q^a \times Q^b)$ if either: (a) $s^a \in Q^a$ and satisfies conditions (i)-(ii) of the definition of an SAS; or (b) s^a supports an $r^a \in Q^a$ that satisfies these conditions. (The analogous conditions must hold for s^b.)

Here is the analog to Proposition 5.6:

PROPOSITION 5.8. *If $Q^a \times Q^b$ is an SAS, then it is a fixed point of Ψ. Conversely, if $Q^a \times Q^b$ is a fixed point of Ψ, then it is an SAS.*

PROOF. Fix an SAS $Q^a \times Q^b$. If $s^a \in Q^a$, then s^a satisfies condition (a) for Ψ. Likewise for s^b. Thus $Q^a \times Q^b \subseteq \Psi(Q^a \times Q^b)$. Next, fix $(s^a, s^b) \in \Psi(Q^a \times Q^b)$. We need to show that $s^a \in Q^a$. If $s^a \notin Q^a$ then s^a supports r^a for some $r^a \in Q^a$. But then condition (iii) of an SAS implies $s^a \in Q^a$, a contradiction.

For the converse, fix $(s^a, s^b) \in Q^a \times Q^b = \Psi(Q^a \times Q^b)$. If s^a satisfies condition (a) for Ψ, then it satisfies conditions (i) and (ii) of an SAS. Next suppose s^a fails condition (a) for Ψ, i.e., is inadmissible with respect to $S^a \times S^b$, or $S^a \times Q^b$, or both. But then s^a must satisfy condition (b) for Ψ, i.e. s^a supports r^a for some $r^a \in Q^a$ satisfying conditions (i) and (ii) of an SAS. By Lemma 5.3, s^a is then admissible with respect to both $S^a \times S^b$ and $S^a \times Q^b$, a contradiction. Finally, suppose q^a supports s^a. We just saw that s^a satisfies condition (a) for Ψ, so q^a satisfies condition (b) for Ψ. Thus $q^a \in \text{proj}_{S^a} \Psi(Q^a \times Q^b) = Q^a$. This establishes condition (iii) of an SAS. □

EXAMPLE 5.9. *Like* Φ, *the map* Ψ *is non-monotone. Consider the game in Figure 5.2. We have* $\Psi(\{U\} \times \{L, R\}) = \{U\} \times \{L, R\}$ *but* $\Psi(\{U, D\} \times \{L, R\}) = \{(U, R)\}$. *Note also that the fixed points of* Ψ *are* $\{U\} \times \{L, R\}$, $\{(U, R)\}$, *and* $\{(M, L)\}$. *The SAS* $\{(M, L)\}$ *is the IA set. We see that, different from BRS vs. IU, the SAS's need not be contained in the IA set.*

	Bob L	Bob R
Ann U	2, 2	2, 2
Ann M	3, 1	0, 0
Ann D	0, 0	1, 3

Figure 5.2

Appendix A. From Types to Hierarchies

In this appendix, we focus on the basic formalization, where a type of Ann is associated with a probability measure on the Borel subsets on the strategies and types of Bob. We show how, in this case, types naturally induce hierarchies of beliefs.[5] For this construction, it may not be immediately clear how to generalize from the two-player case, so we prefer to treat the n-player case explicitly.

Given a Polish space Ω, write $\mathcal{B}(\Omega)$ for the Borel σ-algebra on Ω. Also, extend our earlier notation to write $\mathcal{M}(\Omega)$ for the space of all Borel probability measures on Ω, where $\mathcal{M}(\Omega)$ is endowed with the topology of weak convergence (and so is again Polish). Given sets $X^1, \ldots, X^n$, write $X^{-i} = \prod_{j \neq i} X^j$.

---

[5]This presentation is repeated from Brandenburger and Friedenberg [**11**, 2008, Section 8], which itself closely follows Mertens-Zamir [**20**, 1985, Section 2] and Battigalli-Siniscalchi [**9**, 2002, Section 3].

Fix an n-player strategic-form game $\langle S^1, \ldots, S^n; \pi^1, \ldots, \pi^n \rangle$, where S^i is the finite set of strategies for player i and $\pi^i : S \to \mathbb{R}$ is i's payoff function. An $(S^1, \ldots, S^n)$-**based type structure** is a structure

$$\langle S^1, \ldots, S^n; T^1, \ldots, T^n; \lambda^1, \ldots, \lambda^n \rangle,$$

where each T^i is a Polish space and each $\lambda^i : T^i \to \mathcal{M}(S^{-i} \times T^{-i})$ is continuous. Members of T^i are called **types** for player i.

Associated with each type t^i for each player i in a type structure is a hierarchy of beliefs about the strategies chosen. To see this, inductively define sets Y_m^i, by setting $Y_1^i = S^{-i}$ and

$$Y_{m+1}^i = Y_m^i \times \prod_{j \neq i} \mathcal{M}(Y_m^j).$$

Define continuous maps $\rho_m^i : S^{-i} \times T^{-i} \to Y_m^i$ inductively by

$$\rho_1^i(s^{-i}, t^{-i}) = s^{-i},$$
$$\rho_{m+1}^i(s^{-i}, t^{-i}) = (\rho_m^i(s^{-i}, t^{-i}), (\delta_m^j(t^j))_{j \neq i}),$$

where $\delta_m^j = \underline{\rho}_{-m}^j \circ \lambda^j$ and, for each $\mu \in \mathcal{M}(S^{-j} \times T^{-j})$, $\underline{\rho}_{-m}^j(\mu)$ is the image measure under ρ_m^j.

Standard arguments (see [**11**, 2008, Appendix B] for details) show that these maps are indeed continuous, and so are well-defined. Define a continuous map $\delta^i : T^i \to \prod_{m=1}^{\infty} \mathcal{M}(Y_m^i)$ by $\delta^i(t^i) = (\delta_1^i(t^i), \delta_2^i(t^i), \ldots)$. (Again, see [**11**, 2008, Appendix B] for details.) Then $\delta^i(t^i)$ is the hierarchy of beliefs about strategies induced by type t^i.

References

[1] Abramsky, S., and A. Jung, "Domain Theory," in Abramsky, S., D. Gabbay, and T. Maibaum (eds.), *Handbook of Logic in Computer Science*, III, Oxford University Press, 1994, 1–168.

[2] Apt, K., "Epistemic Analysis of Strategic Games with Arbitrary Strategy Sets," in Samet, D. (ed.), *Proceedings of the 11th Conference on Theoretical Aspects of Rationality and Knowledge (TARK XI)*, Presses univ. de Louvain, 2007, 22–38.

[3] Apt, K., "Relative Strength of Strategy Elimination Procedures," *Economics Bulletin*, 3, 2007, 1–9.

[4] Apt, K., "The Many Faces of Rationalizability," *The B.E. Journal of Theoretical Economics (Topics)*, 7, 2007, Article 18.

[5] Apt, K., and J. Zvesper, "Common Beliefs and Public Announcements in Strategic Games with Arbitrary Strategy Sets," 2007. Available at http://arxiv.org/abs/0710.3536.

[6] Arrow, K., E. Barankin, and D. Blackwell, "Admissible Points of Convex Sets," in Kuhn, H., and A. Tucker (eds.), *Contributions to the Theory of Games*, Volume II, Princeton University Press, 1953, 87–91.

[7] Barwise, J., "Three Views of Common Knowledge," in Vardi, M., (ed.), *Proceedings of the Second Conference on Theoretical Aspects of Reasoning about Knowledge*, Morgan Kaufmann, 1988, 365–379.

[8] Battigalli, P., and A. Friedenberg, "Forward Induction Reasoning Revisited," 2011. Forthcoming in *Theoretical Economics*.

[9] Battigalli, P., and M. Siniscalchi, "Strong Belief and Forward-Induction Reasoning," *Journal of Economic Theory*, 106, 2002, 356–391.

[10] Blume, L., A. Brandenburger, and E. Dekel, "Lexicographic Probabilities and Choice under Uncertainty," *Econometrica*, 59, 1991, 61–79.

[11] Brandenburger, A., and A. Friedenberg, "Intrinsic Correlation in Games," *Journal of Economic Theory*, 141, 2008, 28–67.

[12] Brandenburger, A., and A. Friedenberg, "Self-Admissible Sets," *Journal of Economic Theory*, 145, 2010, 785–811.

[13] Brandenburger, A., A. Friedenberg, and H.J. Keisler, "Admissibility in Games," *Econometrica*, 76, 2008, 307–352.
[14] Brouwer, L., "Über eineindeutige stetige Transformationen von Flächen in Sich," *Mathematische Annalen*, 67, 1910, 176–180.
[15] Davey, B., and H. Priestley, *Introduction to Lattices and Order*, 2nd edition, Cambridge University Press, 2002.
[16] Harsanyi, J.C., "Games with Incomplete Information Played by 'Bayesian' Players, I-III," *Management Science*, 1967, 159–182.
[17] Kakutani, S., "A Generalization of Brouwer's Fixed Point Theorem," *Duke Mathematical Journal*, 8, 1941, 457–459.
[18] Libkin, L. *Elements of Finite Model Theory*, Ttxts in Theoretical Computer Science, 2010.
[19] Martin, K., "The Measurement Process in Domain Theory," in *Proceedings of the 27th International Colloquium on Automata, Languages, and Programming*, July 2000, 116–126.
[20] Mertens, J.-F., and S. Zamir, "Formulation of Bayesian Analysis for Games with Incomplete Information," *International Journal of Game Theory*, 14, 1985, 1–29.
[21] Monderer, D., and D. Samet, "Approximating Common Knowledge with Common Beliefs," *Games and Economic Behavior*, 1, 1989, 170–190.
[22] Moschovakis, Y. *Elementary Induction on Abstract Structures*, North-Holland Elsevier, 1974.
[23] Nash, J., "Non-Cooperative Games," doctoral dissertation, Princeton University, 1950.
[24] Nash, J., "Equilibrium Points in n-Person Games," *Proceedings of the National Academy of Sciences*, 36, 1950, 48–49.
[25] Nash, J., "Non-Cooperative Games," *Annals of Mathematics*, 54, 1951, 286–295.
[26] Pearce, D., "Rational Strategic Behavior and the Problem of Perfection," *Econometrica*, 52, 1984, 1029–1050.
[27] Siniscalchi, M., "Epistemic Game Theory: Beliefs and Types," in Durlauf, S., and L. Blume (eds.), *The New Palgrave Dictionary of Economics*, 2nd edition, Palgrave Macmillan, 2008.
[28] Von Neumann, J., "Zur Theorie der Gesellschaftsspiele," *Mathematische Annalen*, 100, 1928, 295–320.

NYU STERN SCHOOL OF BUSINESS, NEW YORK UNIVERSITY, NEW YORK, NY 10012
E-mail address: adam.brandenburger@stern.nyu.edu
URL: www.stern.nyu.edu/~abranden

W.P. CAREY SCHOOL OF BUSINESS, ARIZONA STATE UNIVERSITY, TEMPE, AZ 85287
E-mail address: amanda.friedenberg@asu.edu
URL: www.public.asu.edu/~afrieden

DEPARTMENT OF MATHEMATICS, UNIVERSITY OF WISCONSIN-MADISON, MADISON, WI 53706
E-mail address: keisler@math.wisc.edu
URL: www.math.wisc.edu/~keisler

Spekkens's toy theory as a category of processes

Bob Coecke and Bill Edwards

ABSTRACT. We provide two mathematical descriptions of Spekkens's toy qubit theory, an inductively one in terms of a small set of generators, as well as an explicit closed form description. It is a subcategory **MSpek** of the category of finite sets, relations and the cartesian product. States of maximal knowledge form a subcategory **Spek**. This establishes the consistency of the toy theory, which has previously only been constructed for at most four systems. Our model also shows that the theory is closed under both parallel and sequential composition of operations (= symmetric monoidal structure), that it obeys map-state duality (= compact closure), and that states and effects are in bijective correspondence (= dagger structure). From the perspective of categorical quantum mechanics, this provides an interesting alternative model which enables us to describe many quantum phenomena in a discrete manner, and to which mathematical concepts such as basis structures, and complementarity thereof, still apply. Hence, the framework of categorical quantum mechanics has delivered on its promise to encompass theories other than quantum theory.

1. Introduction

In 2007 Rob Spekkens proposed a toy theory [20] with the aim of showing that many of the characteristic features of quantum mechanics could result from a restriction on our knowledge of the state of an essentially classical system. The theory describes a simple type of system which mimics many of the features of a quantum qubit. The success of the toy theory in replicating characteristic quantum behaviour is, in one sense, quite puzzling, since the mathematical structures employed by the two theories are quite different. Quantum mechanics represents states of systems by vectors in a Hilbert space, while processes undergone by systems are represented by linear maps. In contrast, the toy theory represents states by subsets and processes by relations. This 'incomparability' means that the mathematical origins of the similarities (and differences) between the two theories are not easy to pinpoint.

2000 *Mathematics Subject Classification.* 81P10, 18B10, 18D35.

This work was supported by EPSRC Advanced Research Fellowship EP/D072786/1 and by EU-FP6-FET STREP QICS.

This work was supported by EPSRC PhD Plus funding at Oxford University Computing Laboratory and an FQXi Large Grant.

In this paper we consider the toy theory from a new perspective - by looking at the 'algebra' of how the processes of the theory combine. The mathematical structure formed by the composition of processes in any physical theory is a *symmetric monoidal category* [14], with the objects of the category representing systems, and the morphisms representing processes undergone by these systems - we term this the *process category* of the theory. Work initiated by Abramsky and Coecke [2], and continued by many other authors [17, 5, 9] has investigated the structure of the process category of quantum mechanics. Mathematical structures of this category have been identified which correspond directly to key physical features of the theory. In principle, any theory whose processes form a category with these mathematical features should exhibit these quantum-like physical features. It will be gratifying then to note that the process category of the toy theory shares many of these features, thus 'accounting for' the similarities which this theory has with quantum mechanics.

The main aim of this paper is to characterise the process category of the toy theory. This turns out to be less straightforward than one might have hoped! In the case of quantum mechanics it is straightforward to see that the states of a system correspond exactly with the vectors of the Hilbert space describing that system, and that processes correspond exactly with linear maps; thus we can immediately conclude that the process category of quantum mechanics is **Hilb**, whose objects are Hilbert spaces and whose morphisms are linear maps. Such a quick and easy statement of the process category for the toy theory is not possible. This is due largely to the way that the valid states and processes of the theory are defined: this is via an inductive procedure where the valid states and transformations for a collection of $n-1$ systems must be known before those for a collection of n systems can be deduced.

There is also a certain degree of ambiguity in the way in which the theory was originally stated, and, since the toy theory was never fully constructed (until now), its consistency was not even guaranteed. Another feature which was not addressed in the theory is *compositional closure*, that is: *If we compose two valid processes, either in parallel or sequentially (when types match), do we again obtain a valid process?* Obviously, this is a natural operational requirement, and as indicated above, this notion of compositional closure is the operational cornerstone to modelling physical processes in a symmetric monoidal category. In this paper, it is exactly the compositional closure assumption which allows us to formulate the toy theory in terms of no more than a few generators, with clear operational meanings.

The structure of the paper is as follows. We begin with a very brief summary of the toy bit theory, and a discussion of the ambiguities in its definition. We then define a categories **Spek** and **MSpek** which we claim is the process category for the toy theory. This is a sub-category of **Rel**, whose objects are sets and whose morphisms are relations. The next section is devoted to showing the general form of the relations which constitute the morphisms of **MSpek**. We then go on to argue that these relations are exactly those which describe the processes of the toy theory, thus demonstrating that **MSpek** is the process category of the toy theory. Finally we note some of the key categorical features of **MSpek** and link these to the characteristically quantum behaviour exhibited by the toy bit theory.

The category **Spek** was first proposed by us in [6], and we provide some explicit pointers to useful information therein throughout this text.

2. The toy theory

For full details of Spekkens's toy bit theory, the reader is referred to the original paper [20]. Here we provide a very brief summary of the key points. There is just one type of elementary system in the theory, which can exist in one of four states. We will denote the set of these four states, the *ontic state space* by IV = $\{1, 2, 3, 4\}$. Alternatively we can depict it graphically as:

(2.1)

The ontic states are to be distinguished from the *epistemic* states, which describe the extent of our knowledge about which ontic state the system occupies. For example we might know that the system is either in ontic state 1 or 2. We would depict such an epistemic state in the following fashion:

(2.2)

The epistemic state is intended to be the analogue of the quantum state. Note that an epistemic state is simply a subset of the ontic state space. Given a composite system of n elementary systems, its ontic state space is simply the Cartesian product of the ontic state spaces of the composite systems, IV^n. Thus such a system has 4^n ontic states.

The key premise of the theory is that our knowledge of the ontic state is restricted in a specific way so that only certain epistemic states are allowed. We refer the reader to the original paper for the full statement of this epistemic restriction, which Spekkens refers to as the *knowledge balance principle*. It suffices here to say that, given an n-component system, an epistemic state is a $2^n, 2^{n+1}, \ldots$ or 2^{2n} element subset of the ontic state space, the 2^n case being the situation of maximal knowledge and the 2^{2n} case being the situation of total ignorance.

For a single system this means that there are six epistemic states of maximal knowledge:

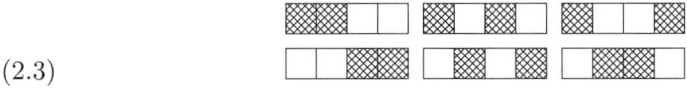

(2.3)

and one of non-maximal knowledge:

(2.4)

A consequence of the knowledge balance principle is that any measurement on a system inevitably results in a probabilistic disturbance of that system's ontic state. A measurement in the toy theory essentially corresponds to asking which of a collection of subsets of the ontic state space contains the ontic state. An example would be a measurement on a single system corresponding to the question "is the ontic state in the subset $\{1, 2\}$ or the subset $\{3, 4\}$?". If the initial epistemic state of the system was the subset $\{1, 3\}$ then answering this question would allow us to pin down the ontic state precisely, to 1 if we get the first answer or 3 if we get the second, in violation of the knowledge balance principle. Thus we hypothesise that if, for example, we get the first answer to our question, then the ontic state undergoes a random disturbance, either remaining in state 1 or moving into state 2, with equal probability. The epistemic state following the measurement then would be $\{1, 2\}$. The effect of such a probabilistic disturbance on the *epistemic* state is best modelled by viewing it as a *relation* on the ontic state space, defined by

$1 \sim \{1,2\}, 2 \sim \{1,2\}, 3 \sim \emptyset, 4 \sim \emptyset$, where the notation denotes that the element 1 in the domain relates to both the elements 1 and 2 in the codomain, the element 2 in the domain, while the elements 3 and 4 in the domain relate to no elements in the codomain. Below we will omit specification of those elements in the domain that relate to no elements in the codomain. Thus in general the transformations of the theory are described by *relations*.

In his original paper Spekkens derives the states and transformations allowed in the theory, apparently appealing only to the following three principles:

(1) The epistemic state of any system must satisfy the knowledge balance principle globally i.e. it should be a $2^n, 2^{n+1}, \ldots$ or 2^{2n} element subset of the ontic state space.

(2) When considering a composite system, the 'marginal' epistemic state of any subsystem should also satisfy the knowledge balance principle. By the marginal epistemic state we mean the following. Suppose the whole system has n components, and we are interested in an m-component subsystem. The epistemic state of the whole system will be some set of n-tuples: to get the marginal state on the m-component system we simply delete the $n - m$ entries from each tuple which correspond to the subsystems which are not of interest to us.

(3) Applying a valid transformation to a valid state should result in a valid state.

The second principle is used, for example, to rule this out as an epistemic state for a two-component system:

(2.5)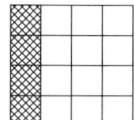

The third principle is used extensively. It is invoked to show that the transformations on a single element system constitute permutations on the set of ontic states, since any other function on IV would lead to some valid epistemic state being transformed into an invalid state.

Another illuminating example of the third principle in action is the elimination of this state as a valid epistemic state:

(2.6)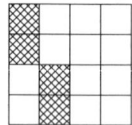

This state would be allowed by the first and second principles, but upon making a certain measurement on one of the systems, the resulting measurement disturbance would transform it into this state:

(2.7)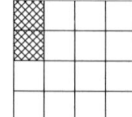

which clearly fails to satisfy the first principle.

The mode of definition of the theory, via the three principles stated earlier, raises some interesting issues. Firstly, this approach seems to be necessarily iterative. Compare it, for example, to how we would describe the form of valid states in quantum mechanics - in one line we can say that they are the normalised vectors of the system's state space. Secondly, do these rules actually uniquely define the toy theory? There does seem to be a problem with the third rule. When we use it to rule out the state in diagram 2.6, our argument is that we already know what the valid measurement disturbance transformations on a single elementary system are, and when we apply one of them to this state we obtain a state which clearly violates the knowledge balance principle. However, it is not clear that we could not have made the alternative choice - that this state should be valid, and therefore that the transformation which we had previously thought was valid could no longer be considered as such. It seems that considerations other than the three rules above come into deciding which should be the valid states of the theory, but it is nowhere clearly stated exactly what they are.

The second point touched slightly on our final issue: is the theory as Spekkens presents it consistent? He derives valid states and transformations for systems with up to three elementary components. However, can we be sure that these states and transformations, when combined in more complex situations involving four or more elementary systems, won't yield a state which clearly violates the knowledge balance principle? Currently there seems to be no such proof of consistency. In fact, in the process of re-expressing the theory in categorical terms, we will develop such a proof.

3. Process categories of quantum-like theories

We briefly review the key structural features of the process categories of quantum-like theories. A physicist-friendly tutorial of the category theoretic preliminaries is in [8]. A survey on the recent applications of these to quantum theory and quantum information is in [4].

DEFINITION 3.1. A *symmetric monoidal category* $(\mathcal{C}, I, -\otimes -)$ is a category equipped with the following extra structure: a bifunctor $-\otimes - : \mathcal{C} \times \mathcal{C} \to \mathcal{C}$; a *unit object* I; and four natural isomorphisms, left and right unit: $\lambda_A : A \cong I \otimes A$, $\rho_A : A \cong A \otimes I$, associative: $\alpha_{A,B,C} : (A \otimes B) \otimes C \cong A \otimes (B \otimes C)$ and commutative: $\sigma_{A,B} : A \otimes B \cong B \otimes A$. Furthermore these objects and natural isomorphisms obey a series of *coherence conditions* [14].

As argued in [8], the process category of any physical theory is a SMC. The bifunctor $-\otimes-$ is interpreted as adjoining two systems to make a larger compound system. **Hilb** becomes a SMC with the tensor product as the bifunctor, and $\mathbb{C}$ as the unit object. **Rel** is a SMC with the Cartesian product as the bifunctor and the single element set as the unit object.

There is a very useful graphical language for describing SMCs, due to Joyal and Street [12], which we will make extensive use of. It traces back to Penrose's earlier diagrammatic notation for abstract tensors [16].

In this language we represent a morphism $f : A \to B$ by a box:

$$\begin{array}{c} A \boxed{\ f\ } B \end{array}$$

(3.1)

$g \circ f$, the composition of morphisms $f : A \to B$ and $g : B \to C$ is depicted as:

(3.2)
$$\begin{array}{c} A \quad \boxed{f} \quad B \quad \boxed{g} \quad C \end{array}$$

The identity morphism 1_A is actually just written as a straight line — this makes sense if you imagine composing it with another morphism. Turning to the symmetric monoidal structure, a morphism $f : A \otimes B \to C \otimes D$ is depicted:

(3.3)
$$A \otimes B \ \boxed{f} \ C \otimes D \quad = \quad \begin{array}{c} A \\ B \end{array} \boxed{f} \begin{array}{c} C \\ D \end{array}$$

and if $f : A \to B$ and $g : C \to D$ then $f \otimes g$ is depicted as:

(3.4)
$$A \otimes C \ \boxed{f \otimes g} \ B \otimes D \quad = \quad \begin{array}{c} A \ \boxed{f} \ B \\ C \ \boxed{g} \ D \end{array}$$

The identity object I is not actually depicted in the graphical language. Morphisms $\psi : I \to A$ and $\pi : A \to I$ are written as:

(3.5)
$$\triangleleft\!\psi\ \ A \qquad\qquad A\ \ \pi\!\triangleright$$

The associativity and left and right unit natural isomorphisms are also implicit in the language. The symmetry natural isomorphism is depicted as:

(3.6)
$$\begin{array}{c} A \diagdown\!\diagup B \\ B \diagup\!\diagdown A \end{array}$$

In fact the graphical language is more than just a useful tool; it enables one to derive all equational statements that follow from the axioms of a SMC:

THEOREM 3.2 ([12, 18]). *Two morphisms in a symmetric monoidal category can be shown to be equal using the axioms of a SMC iff the diagrams corresponding to these morphisms in the graphical language are isomorphic, where by diagram isomorphism we mean that the boxes and wires of the first are in bijective correspondence with the second, preserving the connections between boxes and wires.*

The categories described by the graphical language are in fact *strict SMCs*, that is, those for which λ_A, ρ_A and $\alpha_{A,B,C}$ are equalities. Mac Lane's Strictification Theorem [14, p.257], which establishes an equivalence of any SMC with a strict one, enables one to apply the diagrammatic notation to any SMC.

We will often refer to morphisms of type $\psi : I \to A$ as *states*, since in a process category they represent the preparation of a system A in a given state. In **Hilb** such morphisms are in bijection with the vectors of the Hilbert space A; in **Rel** they correspond to subsets of the set A.

The process categories of *quantum-like* theories possess a range of additional structures.

DEFINITION 3.3. A *dagger category* is a category equipped with a contravariant involutive identity-on-objects functor $(-)^\dagger$. A *dagger symmetric monoidal category* (†-SMC) is a symmetric monoidal category with a dagger functor such that: $(A \otimes B)^\dagger = A^\dagger \otimes B^\dagger$, $\lambda_A^{-1} = \lambda_A^\dagger$, $\rho_A^{-1} = \rho_A^\dagger$, $\sigma_{A,B}^{-1} = \sigma_{A,B}^\dagger$, and $\alpha_{A,B,C}^{-1} = \alpha_{A,B,C}^\dagger$.

Hilb is a †-SMC with the *adjoint* playing the role of the dagger functor. **Rel** is a †-SMC with *relational converse* as the dagger functor.

DEFINITION 3.4 ([**13**]). In a SMC $\mathcal{C}$ a *compact structure* on an object A is a tuple
$$\{A, A^*, \eta_A : I \to A^* \otimes A, \epsilon_A : A \otimes A^* \to I\},$$
where A^* is a dual object to A which may or may not be equal to A, and η_A and ϵ_A satisfy the conditions:

(3.7)
$$\begin{array}{ccc}
A & \xrightarrow{\rho_A} A \otimes I \xrightarrow{1_A \otimes \eta_A} & A \otimes (A^* \otimes A) \\
\downarrow 1_A & & \downarrow \alpha_{A,A^*,A} \\
A & \xleftarrow{\lambda_A^{-1}} I \otimes A \xleftarrow{\epsilon_A \otimes 1_A} & (A \otimes A^*) \otimes A
\end{array}$$

and the dual diagram for A^*. A *compact closed* category $\mathcal{C}$ is a SMC in which all $A \in \mathrm{Ob}(\mathcal{C})$ have compact structures.

Hilb is a compact closed category, where for each object A the *Bell-state* ket $\sum_i |i\rangle \otimes |i\rangle$ (with $\{|i\rangle\}$ a basis for the Hilbert space A) familiar from quantum mechanics provides the morphism η_A, and the corresponding bra acts as ϵ_A. Perhaps unsurprisingly, given this example, Abramsky and Coecke in [**2**] showed that this structure underlies the capacity of any quantum-like theory to exhibit information processing protocols such as *teleportation* and *entanglement swapping*. Furthermore Abramsky has shown [**1**] that any theory whose process category is compact closed will obey a generalised version of the no-cloning theorem.

For the compact structures of interest in this paper A^* is always equal to A – see proposition 3.9 below.[1] Hence, from here on we won't distinguish these two objects.

This simplifies the graphical language, which can be extended by introducing special elements to represent the morphisms of the compact structure η_A and ϵ_A:

(3.8)

Equation 3.7 and its dual are then depicted as:

(3.9)

This extension of the graphical language now renders it completely equivalent to the axioms of a compact closed category; for a detailed discussion we refer to [**18**].

THEOREM 3.5. *Two morphisms in a compact closed category can be shown to be equal using the axioms of compact closure iff the diagrams corresponding to these morphisms in the graphical language are isotopic.*

---

[1] An analysis of the coherence conditions for these *self-dual* compact structures is in [**19**].

Compact closure is of particular importance to us since in any compact closed category there will be *map-state duality*: a bijection between the hom-sets $\mathcal{C}(I, A \otimes B)$ and $\mathcal{C}(A, B)$ (in fact both these will further be in bijection with $\mathcal{C}(A \otimes B, I)$):

(3.10)

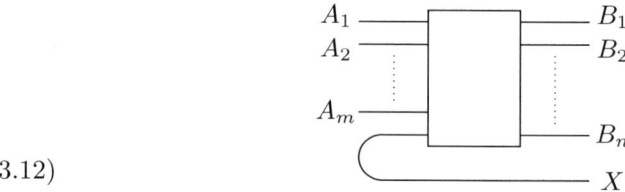

and

(3.11)

That this is a bijection follows from equation 3.9. If we have a morphism with larger composite domain and codomain the number of hom-sets in bijection increases dramatically. For example the morphisms of the hom-sets $\mathcal{C}(A_1 \otimes \cdots \otimes A_m \otimes X, B_1 \otimes \cdots \otimes B_n)$ and $\mathcal{C}(A_1 \otimes \cdots \otimes A_m, B_1 \otimes \cdots \otimes B_n \otimes X)$ are in bijection: explicitly the conversion between morphisms from the two sets can be depicted as:

(3.12)

Clearly manoeuvres like this can convert any 'input' line into an 'output' line, by using the unit and co-unit morphisms to 'bend lines around'.

DEFINITION 3.6. A *diagram equivalence class*[2] (DEC) is a set of morphisms in a compact closed category which can be inter-converted by composition with the units and co-units of the factors of their domains and codomains.

The final structure of interest is the *basis structure*, which can be seen as the 'dagger-variant' of Carboni and Walters's Frobenius algebras [3]. This structure is discussed at length in [5].

DEFINITION 3.7. In a †-SMC a *basis structure* Δ on an object A is a commutative isometric dagger Frobenius comonoid $(A, \delta : A \to A \otimes A, \epsilon : A \to I)$. For more details on this definition see section 4 of [5] where basis structures are referred to as 'observable structures'. We represent the morphisms δ and ϵ graphically as:

(3.13)

Basis structures are so named because in **Hilb** they are in bijection with orthonormal bases, via the correspondence:

(3.14) $\quad \delta : \mathcal{H} \to \mathcal{H} \otimes \mathcal{H} :: |i\rangle \mapsto |i\rangle \otimes |i\rangle \qquad \epsilon : \mathcal{H} \to \mathbb{C} :: |i\rangle \mapsto 1$

---

[2]The terminology is inspired by the fact that the diagrams of all members of a class are essentially the same, only differing in the orientations of their input and output arrows.

where $\{|i\rangle\}_{i=1,\ldots,n}$ is an orthonormal basis for $\mathcal{H}$. This is proved in [10].

PROPOSITION 3.8 (See e.g. [5]). *Two basis structures $(A, \delta_A, \epsilon_A)$ and $(B, \delta_B, \epsilon_B)$ induce a third basis structure $(A \otimes B, \delta_{A \otimes B}, \epsilon_{A \otimes B})$, with:*

$$\text{(3.15)} \qquad \delta_{A \otimes B} = (1_A \otimes \sigma_{A,B} \otimes 1_B) \circ (\delta_A \otimes \delta_B) \qquad \epsilon_{A \otimes B} = \epsilon_A \otimes \epsilon_B,$$

PROPOSITION 3.9 (See e.g. [9]). *Any basis structure induces a self-dual dagger compact structure, with $A = A^*$ and $\eta_A = \delta \circ \epsilon^\dagger$.*

Basis structures are identified in [5] as a key structure underlying a variety of quantum-like features, for example the existence of incompatible observables, and information processing tasks such as the quantum fourier transform.

4. The categories Spek and MSpek

FIGURE 1. Category-theoretic graffiti on an Oxford bridge.

In the section we give the definitions of two key categories. The first definition is a convenient stepping stone to the second:

DEFINITION 4.1. The category **Spek** is a subcategory of **Rel**, defined inductively, as follows:
- The objects of **Spek** are the single-element set $\text{I} = \{*\}$, the four element set $\text{IV} := \{1, 2, 3, 4\}$, and its n-fold Cartesian products IV^n.
- The morphisms of **Spek** are all those relations generated by composition, Cartesian product and relational converse from the following generating relations:
  (1) All permutations $\{\sigma_i : \text{IV} \to \text{IV}\}$ of the four element set, represented diagrammatically by:

(4.1)
$$—\boxed{\sigma_i}—$$

There are 24 such permutations and they form a group, S_4.

(2) A relation $\delta_{\mathbf{Spek}} : \mathrm{IV} \to \mathrm{IV} \times \mathrm{IV}$ defined by:

$$1 \sim \{(1,1),(2,2)\}$$
$$2 \sim \{(1,2),(2,1)\}$$
$$3 \sim \{(3,3),(4,4)\}$$
$$4 \sim \{(3,4),(4,3)\};$$

represented diagrammatically by:

(4.2)

(3) a relation $\epsilon_{\mathbf{Spek}} : \mathrm{IV} \to \mathrm{I}$ defined by $\{1,3\} \sim *$ and represented diagrammatically by:

(4.3)

(4) the relevant unit, associativity and symmetry natural isomorphisms.

PROPOSITION 4.2. [6] $(\mathrm{IV}, \delta_{Spek}, \epsilon_{Spek})$ *is a basis structure in* **FRel** *and hence also in* **Spek**.

This category turns out to be the process category for the fragment of the toy theory containing only epistemic states of maximal knowledge. In particular, as discussed in detail in [6], the interaction of the basis structure $(\mathrm{IV}, \delta_{\mathbf{Spek}}, \epsilon_{\mathbf{Spek}})$ and the permutations of S_4 results in three basis structures, analogous to the Z-, the X- and the Y-bases of a qubit.

Furthermore, as also discussed in [6]§4, **Spek** also contains operations which in quantum theory correspond with projection operators, while these are not included in the toy theory. Indeed, compact closure of **Spek** implies *map-state duality*: operations and bipartite states are in bijective correspondence.[3]

The process category for the full toy theory has a similar definition:

DEFINITION 4.3. *The category* **MSpek** *is a sub-category of* **Rel**, *with the same objects as* **Spek**. *Its morphisms are all those relations generated by composition, Cartesian product and relational converse from the generators of* **Spek** *plus an additional generator:*

$$\bot_{\mathbf{MSpek}} : \mathrm{I} \to \mathrm{IV} :: \{*\} \sim \{1,2,3,4\}$$

By construction, both **Spek** and **MSpek** inherit dagger symmetric monoidal structure from **Rel**, with Cartesian product being the monoidal product, and relational converse acting as the dagger functor.

PROPOSITION 4.4. ***Spek** and **MSpek** are both compact closed.*

PROOF. By propositions 4.2 and 3.8 it follows that each object in these categories has a basis structure, and hence by proposition 3.9 they both are compact closed. □

The primary task of the remainder of this paper is to demonstrate that **MSpek** is indeed the process category of Spekkens's toy theory.

---

[3]We see this as an improvement, and in [21] also Spekkens expressed the desire for theories to have this property, as well as having a dagger-like structure, in the sense that states and effects should be in bijective correspondence.

5. The general form of the morphisms of Spek and MSpek

We first make some preliminary observations.

DEFINITION 5.1. A **Spek** *diagram* is any valid diagram in the graphical language introduced in section 3 which can be formed by linking together the diagrams of the **Spek** generators, as described in definition 4.1.

There is clearly a bijection between the possible compositions of **Spek** generators, and **Spek** diagrams. A **Spek** diagram with m inputs and n outputs represents a morphism of type $IV^m \to IV^n$: a relation between sets IV^m and IV^n. The number of relations between two finite sets A and B is clearly finite itself: it is the power set of $A \times B$. Thus the hom-set $\mathbf{FRel}(A,B)$ is finite. Since $\mathbf{Spek}(IV^m, IV^n) \subseteq \mathbf{FRel}(IV^m, IV^n)$ we can be sure that the hom-sets of **Spek** are finite. On the other hand, there is clearly an infinite number of **Spek** diagrams which have m inputs and n inputs - we can add more and more internal loops to the diagrams. Thus many diagrams represent the same morphism. However the morphisms of **Spek** are, by definition, all those relations resulting from arbitrary compositions of the generating relations, i.e. any relation that corresponds to one of the infinity of **Spek** diagrams. Hence any proof about the form of the morphisms in **Spek** is going to have to be a result about the relations corresponding to each possible **Spek** diagram, even though in general many diagrams correspond to a single morphism.

If we know the relation corresponding to one diagram in one of **Spek**'s diagram equivalence classes (recall definition 3.6), then it is straightforward to determine the relations corresponding to all of the other diagrams.

LEMMA 5.2. *Given a **Spek** diagram and corresponding relation:*

(5.1) $$R : X_1 \times X_2 \times \cdots \times X_m \to Y_1 \times Y_2 \times \cdots \times Y_n$$

then the relation corresponding to the following diagram:

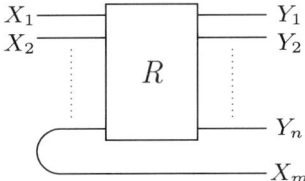

(5.2)

is given by, for all $x_i \in X_i$:

(5.3) $$(x_1, \ldots, x_{m-1}) \sim \left\{ (y_1, \ldots, y_n, x_m) \; \middle| \; \begin{array}{c} y_i \in Y_i \\ x_m \in X_m \\ (y_1, \ldots, y_n) \in R(x_1, \ldots, x_m) \end{array} \right\}$$

where $R(x_1, \ldots, x_m)$ is the subset of $Y_1 \times \cdots \times Y_n$ which is related by R to $(x_1, \ldots, x_m)$.

Every diagram equivalence class in **Spek** has at least one diagram of type $I \to IV^n$, representing a state, where we make every external line an output. Relations of this type can be viewed as subsets of the set IV^n and it will be convenient for us to concentrate on characterising these morphisms. Via lemma 5.2 any results on the general form of states will translate into results on the general form of all morphisms. In what follows we will therefore make no distinction between the inputs and outputs of a **Spek**-diagram: a diagram with m inputs and n outputs will simply be referred to as a $(m+n)$-legged diagram.

Our proof will involve building up **Spek** diagrams by connecting together the generating morphisms. Here we show what various diagram manipulations mean in concrete terms for the corresponding relations. Henceforth, remembering that we only have to consider states, we will assume that the relation corresponding to any n-legged diagram is of type $\mathrm{I} \to \mathrm{IV}^n$.

First we introduce some terminology. The *composite* of an m-tuple $(x_1,\ldots,x_m)$ and an n-tuple $(y_1,\ldots,y_n)$ is the $(m+n)$-tuple $(x_1,\ldots,x_n,y_1,\ldots,y_n)$ from II^{m+n}. By the ith-*remnant* of an n-tuple we mean the $(n-1)$-tuple obtained by deleting its ith component. By the i,j^{th}-*remnant* of an n-tuple (where $i > j$) we mean the $(n-2)$-tuple obtained by deleting the jth component of its ith-remnant (or equivalently, deleting the $(i-1)^{th}$ component of its jth-remnant).

EXAMPLE 5.3. *Consider linking two diagrams, the first representing the relation* $R : \mathrm{I} \to X_1 \times \cdots \times X_m$ *the second representing the relation* $S : \mathrm{I} \to Y_1 \times \cdots \times Y_n$ *via a permutation P, to form a new diagram as shown:*

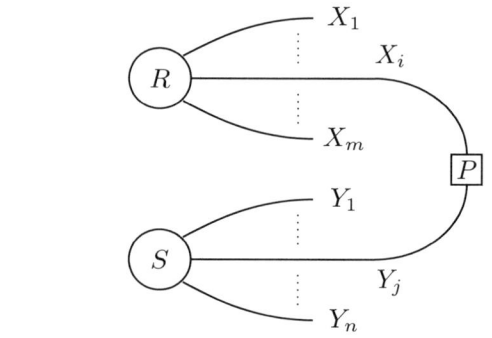

(5.4)

The relation corresponding to this diagram is given by
(5.5)
$$* \sim \left\{ (x_1,\ldots,x_{i-1},x_{i+1},\ldots,x_m,y_1,\ldots,y_{j-1},y_{j+1},\ldots,y_n) \;\middle|\; \begin{array}{c} (x_1,\ldots,x_m) \in R(*) \\ (y_1,\ldots,y_n) \in S(*) \\ x_i = P(y_j) \end{array} \right\}$$

Or, in less formal language, for every pair of a tuple from R and a tuple from S obeying the condition $x_i = P(y_j)$, we form composite of the ith remnant of the tuple from R, and the jth remnant of the tuple from S.

EXAMPLE 5.4. *Given a diagram representing the relation* $R : \mathrm{I} \to X_1 \times \cdots \times X_m$, *consider forming a new diagram by linking the ith and j^{th} legs of the original* *diagram via a permutation P.*

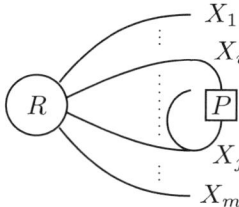

(5.6)

The relation corresponding to this diagram is given by:

(5.7) $\quad * \sim \left\{ (x_1, \ldots, x_{i-1}, x_{i+1}, \ldots, x_{j-1}, x_{j+1}, \ldots, x_m) \;\middle|\; \begin{array}{c} (x_1, \ldots, x_m) \in R(*) \\ x_i = P(x_j) \end{array} \right\}$

Or, in less formal language, we take the i,j^{th}-remnant of every tuple for which $x_i = P(x_j)$.

EXAMPLE 5.5. *Consider linking two diagrams, the first representing the relation $R : \mathrm{I} \to X_1 \times \cdots \times X_n$ the second representing the relation $S : \mathrm{I} \to X_i$ via a permutation P, to form a new diagram as shown:*

(5.8)

The relation corresponding to this diagram is given by:

(5.9) $\quad * \sim \left\{ (x_1, \ldots, x_{i-1}, x_{i+1}, \ldots, x_n) \;\middle|\; \begin{array}{c} (x_1, \ldots, x_n) \in R(*) \\ x_i \in P(S(*)) \end{array} \right\}$

Or, in less formal language, we take the ith remnant of every tuple for which $x_i \in P(S(*))$.

5.1. Structure of the construction. It is convenient to single out a particular sub-group of the S_4 permutation sub-group. This consists of the four permutations which don't mix between the sets $\{1,2\}$ and $\{3,4\}$: $(1)(2)(3)(4)$ (the identity), $(12)(3)(4)$, $(1)(2)(34)$ and $(12)(34)$. We term these the *phased* permutations. All other permutations are termed *unphased*. We single out this sub-group because the relations corresponding to **Spek** diagrams generated from $\delta_{\mathbf{Spek}}$, $\epsilon_{\mathbf{Spek}}$, and the four phased permutations have a significantly simpler general form. We term such diagrams *phased diagrams*. All other diagrams are termed *unphased diagrams*. A morphism which corresponds to a phased diagram is termed a *phased morphism*, all other morphisms being *unphased morphisms*.

It is straightforward to see that any unphased **Spek** diagram can be viewed as a collection of phased sub-diagrams linked together via unphased permutations. We refer to these sub-diagrams as *zones*. Furthermore, note that any permutation in S_4 can be written as a product of phased permutations and the unphased permutation $(1)(3)(24)$ (which we denote by Σ). Since a single phased permutation constitutes a phased zone with two legs, we can in fact view an unphased diagram as a collection of phased zones linked together by the permutation Σ:

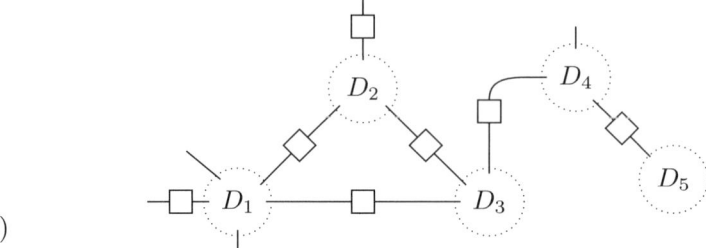

(5.10)

Here D_i represents a phased sub-diagram and a square box represents an unphased permutations. Note that such a diagram is not necessarily planar, i.e. it may involve crossing wires. We distinguish between *external zones* which have external legs (e.g. D_1, D_2 and D_5 in diagram 5.10), and *internal zones* all of whose legs are connected to other legs within the diagram (e.g. D_3 and D_4). To the external legs we associate an enumeration, such that legs from the same external zone appear consecutively.

In the first stage of the proof we determine the general form of the relation corresponding to a *phased diagram*. This stage itself splits into two phases: first we determine the general form of the morphisms of a new category **HalfSpek**, and then we show how to use this result to prove our main result in **Spek**. In the second stage we draw on the results of the first to determine the general form of the relation corresponding to *any* **Spek** diagram.

5.2. The general form of the morphisms of HalfSpek. We build up to the full theorem via a simplified case. For this we need a new category.

DEFINITION 5.6. The category **HalfSpek** is a subcategory of **FRel**. It is defined inductively, as follows:
- The objects of **HalfSpek** are the single-element set $\mathbb{I} = \{*\}$, the two element set $\mathbb{II} := \{0, 1\}$, and its n-fold Cartesian products $\mathbb{II}^n$.
- The morphisms of **HalfSpek** are all those relations generated by composition, Cartesian product and relational converse from the following generating relations:
  (1) All permutations $\{\sigma_i : \mathbb{II} \to \mathbb{II}\}$ of the two element set. There are 2 such permutations, the identity and the operation σ which swaps the elements of $\mathbb{II}$. Together they form the group Z_2.
  (2) A relation $\delta_{\textbf{Half}} : \mathbb{II} \to \mathbb{II} \times \mathbb{II}$ defined by:
  $$0 \sim \{(0,0), (1,1)\} \quad 1 \sim \{(0,1), (1,0)\} \,;$$
  (3) a relation $\epsilon_{\textbf{Half}} : \mathbb{II} \to \mathbb{I} :: 0 \sim *$

PROPOSITION 5.7. [6] $(\mathbb{II}, \delta_{Half}, \epsilon_{Half})$ *is a basis structure in* **FRel** *and hence also in* **HalfSpek**.

REMARK 5.8. *The existence of this basis structure came as a surprise to the authors. Naively one might think (as many working in the area of categorical quantum mechanics initially did) that on a set X in* **FRel** *there is a single basis structure with δ given by $x \sim (x, x)$ for all $x \in X$. The 'basis vectors' (or copyable points in the language of [5]) are then the elements of this set. But this is not the case. There are many 'non-well-pointed' basis structures such as $(\mathbb{II}, \delta_{Half}, \epsilon_{Half})$ for which the* *number of copyable points is less than the number of elements of the set. In related work, Pavlovic has classified all basis structures in **FRel** and Evans et al. have identified the pairs of complementary basis structures (in the sense of* [**5**]*) among these* [**15, 11**].

Next we determine the general form of the relations which constitute the morphisms of **HalfSpek**, to which the considerations made at the beginning of Section 5 also apply.

We say that an element of $\mathbb{II}^n$ has *odd* parity if it has an odd number of '1' elements, and that it has *even* parity if it has an even number of '1' elements. We will use P to represent a particular parity, odd or even, and P' will represent the opposite parity. Whether an odd-parity n-tuple has an odd or even number of '0' elements clearly depends on whether n itself is odd or even. We could have chosen either 0 or 1 to play the role of labelling the parity; we have chosen 1 since it will turn out to be more convenient later on.

THEOREM 5.9. *The relation in* **HalfSpek** *corresponding to an n-legged* **HalfSpek**-*diagram is a subset of $\mathbb{II}^n$, consisting of all 2^{n-1} n-tuples of a certain parity, which depending on the diagram may either be even or odd: if the product of all the permutations appearing in the diagram is the identity, then parity is even, and if it is σ, then the parity is odd.*

PROOF. We prove this result by induction on the number of generators k required to construct the diagram. Remember that we need only consider those diagrams whose corresponding relations are states. There is just one possible base case ($k = 1$), a diagram composed purely of the generator $\epsilon^\dagger_{\mathbf{Half}}$ for which $n = 1$: the corresponding state consists of the single 1-tuple (0), which is indeed the unique 1-tuple of even parity. Now consider a diagram D built from k generators with a corresponding state ψ consisting all 2^{n-1} n-tuples of parity P. It is easily seen that composing D with either $\epsilon^\dagger_{\mathbf{Half}}$, $\delta_{\mathbf{Half}}$ or $\delta^\dagger_{\mathbf{Half}}$ respectively yields a diagram whose corresponding state consists of all 2^{n-2} $n-1$-tuples, all 2^n $n+1$-tuples, and all 2^{n-2} $n-1$-tuples of parity P; and that composing with σ yields all 2^{n-1} n-tuples of parity P'. Finally consider producing a disconnected diagram by laying the $\epsilon^\dagger_{\mathbf{Half}}$ diagram alongside D: it is easily seen that the corresponding state consists of 2^n $n+1$-tuples of parity P. □

5.3. The general form of phased morphisms in Spek. We want to apply our results on **HalfSpek** to the category of real interest, **Spek**. To do this we first need to digress to discuss some structural features of relations. The category **Rel** has another symmetric monoidal structure, namely the *disjoint union* or *direct sum*, denoted by $\sqcup$. Concretely, if we can partition a set A into m subsets A_i, then we have $A = \sqcup_i A_i$, and recalling that a relation $R : A \to B$ is a subset of $A \times B$, we can decompose R into mn components of the form $R_{i,j} : A_i \to B_j$, such that $R = \sqcup_{i,j} R_{i,j}$. The relations $R_{i,j} : A_i \to B_j$ are termed the *components* of R *with respect to partitions* $A = \sqcup_i A_i$, $B = \sqcup_j B_j$. In category theoretic terms, this is *biproduct*, and there is a *distributive law* with respect to the Cartesian product:

$$(5.11) \qquad A \times (\sqcup_i B_i) = \sqcup_i (A \times B_i) \qquad R \times (\sqcup_i T_i) = \sqcup_i (R \times T_i)$$

for sets A, B_i and relations R, T_i. For $A = \sqcup_i A_i$, $B = \sqcup_j B_j$, $C = \sqcup_k C_k$ and $D = \sqcup_l D_l$, and relations $R : A \to B$, $S : B \to C$ and $T : C \to D$, we then have:

$$(5.12) \quad (S \circ R)_{i,k} = \bigsqcup_j S_{j,k} \circ R_{i,j} \quad (R \times T)_{i,j,k,l} = T_{k,l} \times R_{i,j} \quad R^c_{j,i} = (R_{i,j})^c$$

For $A = A_1 \sqcup A_2$ by distributivity we have:

$$(5.13) \quad A_1^m \sqcup (A_1^{m-1} \times A_2) \sqcup (A_1^{m-2} \times A_2 \times A_1) \sqcup \cdots \sqcup (A_1^{m-2} \times A_2^2) \sqcup \cdots \sqcup A_2^m$$

If for $A = A_1 \sqcup A_2$ and $R : A^m \to A^n$ the only non-empty components are $R_1 : A_1^m \to A_1^n$ and $R_2 : A_2^m \to A_2^n$ we call it *parallel*. Given parallel relations $R : A^m \to A^n$, $S : A^n \to A^p$ and $T : A^p \to A^q$ with respect to the partition $A = A_1 \sqcup A_2$, the relations:

$$(5.14) \quad S \circ R : A^m \to A^p \quad T \times R : A^{m+p} \to A^{n+q} \quad R^c : A^n \to A^m$$

are all also easily seen to be parallel with respect to the same partition of A.

We can use these insights to make a connection between **HalfSpek** and **Spek**.

PROPOSITION 5.10. *The generators of the phased morphisms of* **Spek**, *i.e.* δ_{Spek}, ϵ_{Spek} *and the phased permutations on* IV, *are all parallel with respect to the following partition of* IV $= \{1,2\} \sqcup \{3,4\}$. *We conclude that all phased morphisms of* **Spek** *are also parallel with respect to this partition. We refer to the two components of a phased* **Spek** *morphism as its* $\{1,2\}$-*component and* $\{3,4\}$-*component.*

PROPOSITION 5.11. *The* $\{1,2\}$-*components of the generators of the phased morphisms of* **Spek** *are simply the generators of* **HalfSpek** *with the elements of* II $= \{0,1\}$ *re-labelled according to* $0 \mapsto 1$, $1 \mapsto 2$. *Similarly the* $\{3,4\}$-*components of the generators of the phased morphisms of* **Spek** *are simply the generators of* **HalfSpek** *with the elements of* II $= \{0,1\}$ *re-labelled according to* $0 \mapsto 3$, $1 \mapsto 4$.

PROPOSITION 5.12. *A state* $\psi \subset \text{IV}^n$ *corresponding to a phased* **Spek** *diagram* D *is equal to the union of two states* $\psi^{12} \subset \{1,2\}$ *and* $\psi^{34} \subset \{3,4\}$. ψ^{12} *and* ψ^{34} *are obtained by the following procedure. Form a* **HalfSpek** *diagram* D^{12} *by replacing every occurence of* δ_{Spek} *and* ϵ_{Spek} *in* D *with* δ_{Half} *and* ϵ_{Half}, *and replacing every occurence of a permutation with its* $\{1,2\}$ *component, re-labelled as a* **HalfSpek** *permutation as described in proposition 5.11. Form a second* **HalfSpek** *diagram* D^{34} *in the obvious analogous fashion using* $\{3,4\}$ *components of permutations.* ψ^{12} *and* ψ^{34} *are the states corresponding to* D^{12} *and* D^{34}, *once again under the re-labelling described in proposition 5.11.*

Note that D^{12} and D^{34} will appear identical as graphs, both to each other and to D, but the labels on some of their permutations will differ.

From proposition 5.12 and theorem 5.9 now follows:

THEOREM 5.13. *A phased morphism in* **Spek** *of type* I $\to$ IV^n *is a subset of* IV^n, *consisting of* 2^n n-*tuples, divided into two classes of equal number:*

- *The first class consists of tuples of 1s and 2s, all of either odd or even parity.*
- *The second class consists of tuples of 3s and 4s, again all of either odd or even parity.*

Note that we are adopting the convention that tuples of the first class have odd parity if they have an odd number of 2s, even parity if they have an even number of 2s. Tuples of the first class have odd parity if they have an odd number of 4s, even parity if they have an even number of 4s.

5.4. The general form of arbitrary morphisms in Spek. Recall from section 5.1 that any unphased diagram can be viewed as a collection of phased zones linked together by the permutation Σ, $(1)(3)(24)$. We also enumerated the external legs.

THEOREM 5.14. *The relation in* **Spek** *corresponding to an n-legged* **Spek**-*diagram with m zones, none of which is internal, is a subset ψ of IV^n with the following form:*

(1) *It contains 2^n n-tuples; each entry corresponding to an external leg.*
(2) *All those entries corresponding to a given zone of the diagram are termed a zone of the tuple (whether we are referring to a zone of a diagram, or zone of a tuple should be clear from the context). Each zone of the tuple has a well-defined* type *(components either all 1 or 2, or all 3 or 4) and* parity *(as defined for phased relations).*
(3) *A block B is a subset of ψ such that the ith zone of all n-tuples in B has the same parity and type. The sequence of types and parities of each zone is called the* signature *of the block. The 2^n tuples of ψ are partitioned into 2^m equally sized blocks each with a unique signature.*
(4) *Each of the 2^m blocks has a different type signature - these exhaust all possible type signatures.*
(5) *The parity signature of a block is the following simple function of its type signature:*

(5.15) $$P_i = \Psi_i(T_i) + \sum_{j \in adj(i)} (T_i + T_j)$$

Here P_i and T_i are Boolean variables representing the parity and type of the ith zone. We adopt the convention that an odd parity is represented by 0 and an even parity by 1, whilst the type $\{1,2\}$ is represented by 0 and the type $\{3,4\}$ by 1. The set $adj(i)$ consists of the zones directly adjacent to the ith zone. $\Psi_i(T_i)$ denotes the parity of the type T_i tuples in the relation corresponding to the ith zone seen as an independent phased diagram (recall theorem 5.13).

THEOREM 5.15. *The relation in* **Spek** *corresponding to an n-legged* **Spek**-*diagram with m zones, of which m' are external is either:*

- *A subset, ψ of IV^n which satisfies conditions (1) and (2) of theorem 5.14 and which is partitioned into $2^{m'}$ blocks. The signatures of these blocks are determined as follows:*
  (1) *Begin with the state corresponding to the diagram obtained by adding an external leg to every internal zone (i.e. we will have 2^m blocks, each with m zones, exhausting all possible type signatures).*
  (2) *Eliminate all blocks whose type signatures do not satisfy the following constraints, one for each internal zone:*

(5.16) $$\Psi_i(T_i) + \sum_{j \in adj(i)} (T_i + T_j) = 0$$

where i is the label of the internal zone, and j labels its adjacent zones.
(3) Finally from each block delete the zones corresponding to internal zones.
- or it is equal to the empty set, $\emptyset$. This second possibility occurs iff the constraints in equation 5.16 are inconsistent.

A simple counting argument shows that within each block, every tuple with the correct type and parity signature occurs. Theorems 5.14 and 5.15 thus completely characterise the state corresponding to any **Spek**-diagram. The input data is the shape of the diagram, which determines $adj(i)$, and the 'intrinsic parities' $\Psi_i(T_i)$ of each zone. Futhermore, note from theorem 5.13 that, as we would expect, the general form of phased morphisms is a special case of the form described above, with $m = m' = 1$.

EXAMPLE 5.16. *We now give an example of theorem 5.14 in action. Consider the following schematic**Spek**-diagram (circles simply denote zones), which has three zones, all external ($m = 3$) and five external legs ($n = 5$):*

(5.17)

where the labels by each zone denote the intrinsic parities of that zone (i.e. the parity of the type-12 tuples and the parity of the type-34 tuples, recall theorem 5.13). We conclude from theorem 5.14 that the state corresponding to this diagram will consist of $2^5 = 32$ 5-tuples each with 3 zones, and that these will be partitioned into $2^3 = 8$ equally sized blocks. Every combination of types appears exactly once amongst these blocks, and the parity signatures are easily determined from equation 5.15 as follows. First we note, from the intrinsic parities that $\Psi_1(T_1) = 0$, $\Psi_2(T_2) = T_2$ and $\Psi_3(T_3) = 1 + T_3$. We then see that equation 5.15 here is essentially three equations:

(5.18)
$$\begin{cases} P_1 = & T_2 + T_3 \\ P_2 = & T_1 + T_2 + T_3 \\ P_3 = & 1 + T_1 + T_2 + T_3 \end{cases}$$

From this we can establish the signatures of the eight blocks:

(5.19)
$(Odd, 12; Odd, 12; Even, 12)$ $(Even, 12; Even, 12; Odd, 34)$
$(Even, 12; Even, 34; Odd, 12)$ $(Odd, 12; Odd, 34; Even, 34)$
$(Odd, 34; Even, 12; Odd, 12)$ $(Even, 34; Odd, 12; Even, 34)$
$(Even, 34; Odd, 34; Even, 12)$ $(Odd, 34; Even, 34; Odd, 34)$

The first zone will have two elements, the second zone one element and the third zone two elements. Within each block every possible collection of tuples consistent with the signature will appear meaning that each block will consist of four tuples.

EXAMPLE 5.17. *We go on to give an example of theorem 5.15. Consider the following diagram: it is identical to the diagram in the previous example except that* the second zone has been internalised.

(5.20)

This diagram has two external zones $m' = 2$ and four external legs $n = 4$; thus we expect the corresponding state to consist of $2^4 = 16$ 4-tuples, each with 2 zones, partitioned into $2^2 = 4$ blocks. We now determine the signatures of the blocks applying theorem 5.15. According to step (1) we begin with the blocks from the previous example. Step (2) requires that we eliminate all blocks whose types do not satisfy the constraint $T_1 + T_2 + T_3 = 0$. This leaves:

(5.21)
$(Odd, 12; Odd, 12; Even, 12)$
$(Odd, 12; Odd, 34; Even, 34)$
$(Even, 34; Odd, 12; Even, 34)$
$(Even, 34; Odd, 34; Even, 12)$

Finally we delete the second zone from each block, leaving:

(5.22)
$(Odd, 12; Even, 12)$
$(Odd, 12; Even, 34)$
$(Even, 34; Even, 34)$
$(Even, 34; Even, 12)$

The proofs of theorems 5.14 and 5.15 proceed as inductions over the process of building up a diagram by linking together phased zones via the permutation Σ. In the course of this process it is clearly possible for an external zone to become an internal zone, as its last external leg is linked to some other zone - we refer to this as *internalising* a zone. It turns out that internalisation of a zone complicates the inductive proof. To get around this we do the induction in two stages. In the first stage we build up a diagram identical to the one we are aiming for, *except* that every zone that should be internal is given a single external leg in the following fashion. Suppose we need to link together two zones via a permutation, and this will result in the internalisation of the left hand zone:

(5.23)

We instead link the left hand zone to the permutation via a $\delta_{\textbf{Spek}}$ morphism. The $\delta_{\textbf{Spek}}$ morphism becomes part of the original zone, and provides it with an external leg:

(5.24)

In the second stage we cap off all these extra external legs with the $\epsilon_{\textbf{Spek}}$ morphism. Since $\delta_{\textbf{Spek}}$ and $\epsilon_{\textbf{Spek}}$ constitute a basis structure the result is the diagram which we are aiming for:

(5.25)

5.4.1. Diagrams without internal zones.

We now prove theorem 5.14. The proof uses an induction over the process of building up a diagram. This building up can be split into two phases: firstly we connect all of the zones which will appear in the final diagram into a 'tree-like' structure with no closed loops, secondly we close up any loops necessary to yield the desired diagram. Henceforth we will represent the signature of a tuple with m zones as $(P_1, T_1; \ldots; P_m, T_m)$ where P_i is the parity of the ith zone, and T_i is its type. Again, if P is a parity, P' indicates the opposite parity, and likewise if T is a type, T' represents the other type.

LEMMA 5.18. *Consider an n-legged non-phased diagram D_1 with m zones. Suppose the corresponding state $\psi_1 \subset \text{IV}^n$ satisfies the conditions in theorem 5.14. Now consider linking the ith leg of D_1 to the jth leg of an n'-legged phased diagram D_2 (with corresponding state ψ_2), via Σ, to create an $(n+n'-2)$-legged diagram D_3 with $m+1$ external zones. We will assume that the ith leg of D_1 lies within its kth external zone. The state $\psi_3 \subset \text{IV}^{n+n'-2}$ corresponding to D_3 also satisfies the conditions in theorem 5.14.*

PROOF. By the 1-i^{th}-remnants of ψ_1 we mean the ith-remnants of those tuples in ψ_1 with a 1 in the ith position. We define the 2-, 3-, and 4-i^{th}-remnants similarly. By proposition 5.3 the elements of ψ_3 comprise all the possible composites of the x-i^{th}-remnants of ψ_1 and the $\Sigma(x)$-j^{th}-remnants of ψ_2, where $x = 1, \ldots, 4$. It is clear that the zone structure of the tuples of ψ_1 is inherited by these composites, and that the $\Sigma(x)$-j^{th}-remnants of ψ_2 constitute an additional zone within the composites - we conventionally consider this to be the $(m+1)^{\text{th}}$ zone. Thus the tuples of ψ_3 satisfy condition (2).

Consider a block $B \subset \psi_1$, with signature $(P_1, T_1; \ldots; P_k, T_k; \ldots; P_m, T_m)$, in which for definiteness we assume that $T_k = 0$ (the argument runs entirely analogously if $T_k = 1$). The composites of the 1-i^{th}-remnants of B and the 1-j^{th}-remnants of ψ_2 all have the same signature, $(P_1, T_1; \ldots; P_k, T_k; \ldots; P_m, T_m; \Psi, T_k)$, and constitute a block $B_1 \subset \psi_3$. Likewise the composites of the 2-i^{th}-remnants of B and the 4-j^{th}-remnants of ψ_2 constitute a block $B_2 \subset \psi_3$ of signature $(P_1, T_1; \ldots; P'_k, T_k; \ldots; P_m, T_m; \Psi', T'_k)$. Thus each 'parent' block in ψ_1 gives rise to two 'progeny' blocks in ψ_3. By hypothesis, each block $B \subset \psi_1$ has a unique type signature, thus the progeny blocks derived from different parent blocks are distinct. Thus ψ_3 is partitioned into 2^{m+1} blocks, thus satisfying condition (3). It is also clear that if all possible type signatures are represented by the 2^m blocks of ψ_1 then this is also true for the 2^{m+1} blocks of ψ_3, and thus that ψ_3 satisfies condition (1).

Note that B consists of 2^{n-m} n-tuples, and will have 2^{n-m-1} 1-i^{th}-remnants and a similar number of 2-i^{th}-remnants, all of which will be distinct. Similarly ψ_2 will have $2^{n'-2}$ 1-i^{th}-remnants and a similar number of 4-i^{th}-remnants, again all distinct. Thus, both B_1 and B_2 will consist of $2^{n-m-1} \cdot 2^{n'-2} = 2^{(n+n'-2)-(m+1)}$ tuples. This holds for all blocks $B \subset \psi_1$, of which there are 2^m. Thus in total ψ_3 consists of $2^{n+n'-2}$ tuples, and so satisfies condition (4).

Finally we turn to condition (5). Recall from above that a parent block of signature $(P_1, T_1; \ldots; P_k, T_k; \ldots; P_m, T_m)$ yields two progeny blocks, of signatures $(P_1, T_1; \ldots; P_k, T_k; \ldots; P_m, T_m; \Psi, T_k)$ and $(P_1, T_1; \ldots; P'_k, T_k; \ldots; P_m, T_m; \Psi', T'_k)$. Note that those progeny blocks for which the kth zone and its new adjacent zone have different types exhibit a parity flip on the kth zone, relative to the parent block, and on the new adjacent zone, relative to its 'intrinsic parity'. No such flip occurs if the zones have the same type. Note that the term $T_i + T_j$ is equal to 0 if $T_i = T_j$ and 1 if $T_i \neq T_j$. If condition (2) holds for ψ_1 then the correct parity signature for the blocks of ψ_3 can be obtained simply by adding the term $T_k + T_{m+1}$ to the P_k and P_{m+1} equations (equation 5.15). Thus we conclude that condition (5) will hold for ψ_3 as well. □

LEMMA 5.19. *Consider an n-legged diagram D with m external zones. Suppose the corresponding state ψ satisfies the conditions in theorem 5.14. Now consider forming a new $(n-2)$-legged diagram D', with corresponding state ψ', by linking the ith leg of D (in the kth zone of D), to the jth leg (in the lth zone), via Σ. ψ' also satisfies the conditions in theorem 5.14.*

PROOF. By the x, y-i, j^{th}-*remnants* of a set of tuples we mean the i, j^{th}-remnants of all those tuples with x in the ith position and y in the jth position. From proposition 5.4, ψ' consists of the $x, \Sigma(x)$-i, j^{th}-remnants of ψ. The zone structure of ψ is clearly inherited by these remnants, thus the tuples of ψ' satisfy condition (2).

Consider a block $B \subset \psi$ with signature $(P_1, T_1; \ldots; P_k, T_k; \ldots; P_l, T_l; \ldots; P_m, T_m)$. It is straightforward to see that only one quarter of the tuples in this block have i, j^{th}-remnants which are $x, \Sigma(x)$-i, j^{th}-remnants. All the $x, \Sigma(x)$-i, j^{th}-remnants have the same signature: $(P_1, T_1; \ldots; P_k, T_k; \ldots; P_l, T_l; \ldots; P_m, T_m)$ if $T_k = T_l$ and $(P_1, T_1; \ldots; P'_k, T_k; \ldots; P'_l, T_l; \ldots; P_m, T_m)$ if $T_k \neq T_l$. Thus each 'parent' block in ψ gives rise to one 'progeny' block in ψ', with one quarter as many tuples. The type signatures of the progeny blocks are identical to those of the parent blocks; by hypothesis each parent block had a different type signature and so the progeny blocks deriving from different parent blocks are all distinct. Thus ψ' is partitioned into 2^m blocks, each containing $2^{n-m}/4 = 2^{(n-2)-m}$ tuples, and so satisfies conditions (1) and (3).

Since the type signatures of progeny and parent blocks are identical, if ψ satisfies condition (4), so will ψ'. We now turn to condition (5). Closing a loop between the kth and l^{th} zones means that they now become adjacent to each other. In the previous paragraph we saw that a progeny block for which the kth and l^{th} zones have different types exhibits a parity flip on both these zones. Using similar reasoning as in the previous lemma we conclude that, if condition (2) holds for ψ, the correct parity signature for the blocks of ψ' can be obtained simply by adding the term $T_k + T_l$ to the P_k and P_l equations (equation 5.15). Thus we conclude that condition (5) will hold for ψ' as well. □

We can now prove **Theorem 5.14**:

PROOF. By induction. The base case is a diagram consisting of a single zone, and it is clear from theorem 5.13 that this satisfies all the conditions. Any other diagram is built up via two inductive steps: linking a new phased zone onto the existing diagram and closing up internal loops within the diagram. Lemmas 5.18 and 5.19 respectively show that if a diagram satisfied the conditions prior to either of these steps, the resulting new diagram will also satisfy the conditions. □

5.4.2. *Diagrams with internal zones.* We now address the issue of internalising zones. Recall that this step involves capping off external legs with the $\epsilon_{\mathbf{Spek}}$ relation. Throughout this section D will denote a diagram with no internal zones, and D' will denote the diagram obtained by internalising some of D's zones. The corresponding states will be ψ and ψ'.

PROPOSITION 5.20. *Suppose we obtain D' by capping the kth external leg of D with the ϵ_{Spek} morphism. From lemma 5.5 we conclude that ψ' consists of the $1\text{-}k^{th}$- and $3\text{-}k^{th}$-remnants of ψ.*

Suppose that in going from D to D' we internalised the ith zone of D. From the proposition above we deduce that each block $B \subset \psi$ for which $P_i = 0$ will give rise to one progeny block $B' \subset \psi'$ with the same number of tuples as B, while all those blocks for which $P_i = 1$ will give rise to no progeny blocks. From this we can conclude that those blocks which do give rise to progeny blocks satisfy a constraint on their type signatures, derived from setting $P_i = 0$ in equation 5.15. Depending on the form of $\Psi_i(T_i)$, and whether the number of zones adjacent to the ith is odd or even, this constraint takes one of four forms:

$$(5.26) \qquad \sum_{j \in \mathrm{adj}(i)} T_j = \begin{cases} 0 \\ 1 \end{cases}$$

$$(5.27) \qquad T_i + \sum_{j \in \mathrm{adj}(i)} T_j = \begin{cases} 0 \\ 1 \end{cases}$$

We describe the constraints in (5.26) as *type-0 constraints*, and those in (5.27) as *type-1 constraints*.

Suppose D has n external legs and m external zones. Suppose that in going to D' we internalise p of its zones. *Each internalisation gives rise to a corresponding constraint.* There are now two possibilities:

(1) Not all of the constraints are consistent. In this case none of the blocks in ψ satisfy all of the constraints, and none of them will give rise to progeny blocks. Thus $\psi' = \emptyset$.
(2) All p constraints are consistent, and of these p' are linearly independent (this essentially means that $p - p'$ of the constraints can be derived from the remaining p'). Each independent constraint reduces the number of blocks which can give rise to progeny by one half. Thus only $2^{m-p'}$ of the blocks in ψ give rise to progeny blocks in ψ', and ψ' can have at most $2^{m-p'}$ blocks - this maximum is attained if all of the progeny blocks are distinct.

LEMMA 5.21. *The following are equivalent:*
(1) *The constraints are consistent and there are p' linearly independent constraints.*
(2) *The $2^{m-p'}$ blocks in ψ which can give rise to progeny blocks in ψ' are partitioned into 2^{m-p} sets, each consisting of $2^{p-p'}$ blocks which all yield identical progeny blocks. Thus in total there are 2^{m-p} distinct progeny blocks. For brevity we will describe this as $(p-p')$-fold duplication of progeny blocks.*

The proof of this lemma requires a number of preliminary definitions.

DEFINITION 5.22. The *IZ-set* is the set of zones which are internalised in going from D to D'. The *non-internalised adjacent zones (nIAZs)* of an internalised zone are all the zones adjacent to it which are not themselves members of the IZ-set. An *adjacency closure set (ACS)* is a subset of the IZ-set with the minimal number of elements such that the disjoint union of the nIAZs of each element contains each nIAZ an even number of times.

EXAMPLE 5.23. *Consider the following seven zone diagram (external legs are suppressed for clarity). The filled-in zones are those which we internalise.*

(5.28)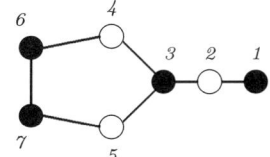

The IZ-set is $\{1,3,6,7\}$. *The nIAZs for 1 are* $\{2\}$ *for 3 are* $\{4,5\}$, *for 6 are* $\{4\}$ *and for 7 are* $\{5\}$. *Zones 3, 6 and 7 together constitute an ACS. Zone 1 is not part of any ACS.*

DEFINITION 5.24. Given a set S of zones in D, and a block $B \subset \psi$, the *S-mirror* of B, B_S is the block with the same type signature as B except on the zones in S, where the types are opposite.

PROPOSITION 5.25. *Suppose D has an ACS R. Now, so long as the blocks $B, B_R \subset \psi$ (i.e. a block and its R-mirror) both yield progeny blocks in ψ', these progeny blocks will be identical. Conversely, if any two blocks $B, B' \subset \psi$ yield identical progeny blocks in ψ', they must be mirrored with respect to some ACS in D.*

PROOF. For two blocks in ψ to give identical progeny blocks in ψ' they must have identical type and parity on every zone which is not internalised. Note from equation 5.15 that two blocks in ψ which are type mirrored on a single zone will otherwise differ only in parities on all the zones adjacent to this zone. Now consider two blocks which are type-mirrored on a set of zones R: in the case where R constitutes an ACS all the parity flips predicted by equation 5.15 cancel one another out on the zones which will still be visible in ψ'. □

PROPOSITION 5.26. *Suppose D has an ACS R whose member zones have corresponding constraints which satisfy the following condition: those zones with an odd number of adjacent zones within the IZ-set have type-1 constraints, while those zones with an even number of adjacent zones within the IZ-set have type-0 constraints. Then if a block $B \subset \psi$ satisfies the constraints and gives rise to progeny blocks in ψ', so does its R-mirror. Conversely, if a block $B \subset \psi$ and its mirror with respect to some ACS R in D both give rise to progeny blocks, the constraints corresponding to the zones of R must satisfy the condition above.*

PROOF. If a constraint contains an even number of terms relating to zones from a set R then given a block $B \subset \psi$ either (i) both B and B_R satisfy the constraint (ii) neither B nor B_R satisfy the constraint. Conversely, if both B and B_R satisfy a constraint, it must contain an even number of terms relating to zones from R. □

PROPOSITION 5.27. *Suppose D has an ACS R whose member zones have corresponding constraints which satisfy the condition in the previous proposition. Then the constraints together form a linearly dependent set. The converse is also true.*

PROOF. If R is an ACS and the condition on constraints is satisfied then each term appears in the constraints an even number of times altogether. Summing all the constraints together then results in all the terms cancelling out, yielding the single equation 0=0. This is a necessary and sufficient condition for the constraints to be linearly dependent. □

We can now prove lemma 5.21.

PROOF. From propositions 5.25 and 5.26 we conclude that for n-fold duplication to take place the IZ-set must contain n ACSs and that the constraints must satisfy the condition in proposition 5.26. The converse is also clearly true. Every linearly dependent set amongst the constraints corresponding to the internalised zones reduces the total number of linearly independent constraints by one. From proposition 5.27 we conclude that for there to be n linearly dependent sets the IZ-set must contain n ACSs and that the constraints must satisfy the condition in proposition 5.26. The converse is also clearly true. Thus we conclude that both statements in the lemma are equivalent to a third statement: the IZ-set contains $p - p'$ ACSs, and the constraints corresponding to the internalised zones satisfy the condition of proposition 5.26. □

COROLLARY 5.28. *Given a diagram D without internal zones which satisfies the conditions in theorem 5.14, a diagram D' with internal zones formed by capping off external legs of D with ϵ_{Spek} morphisms will satisfy the conditions in theorem 5.15.*

Theorem 5.15 follows as a straightforward corollary.

5.5. The general form of the morphisms of MSpek.

THEOREM 5.29. *All **MSpek** morphisms of type $I \to IV^n$ are subsets of IV^n containing $2^n, 2^{n+1}, \ldots, 2^{2n-1}$ or 2^{2n} n-tuples.*

PROOF. Any **MSpek** diagram D' can be obtained from a **Spek** diagram D simply by capping one or more legs of D with the morphism $\perp_{MSpek}$. Suppose we obtain D' by capping a single external leg of D (the ith, say) with $\perp_{MSpek}$. The state ψ' corresponding to D' consists of the ith-remnants of ψ, the state corresponding to D. Suppose D has n external legs: since D is a **Spek**-diagram ψ consists of 2^n tuples. Then, unless some of the ith-remnants of ψ are identical, ψ' will also consist of 2^n tuples, despite only having $n' = n - 1$ external legs. Furthermore it is clear that either all the ith-remnants of ψ are distinct, or ψ is partitioned into pairs, the elements of which yield identical i^{th}-remnants, meaning that the addition of each $\perp_{MSpek}$ cap either halves the number of tuples or leaves it unchanged. □

6. MSpek is the process category for the toy theory

We know that the epistemic states of the toy theory are subsets of the sets IV^n, and that the transformations on these states are relations between these sets. Thus we can see immediately that the toy theory's process category must be some sub-category of **FRel**, restricted to the objects IV^n. Furthermore we know that it cannot be the full sub-category restricted to these objects, since some subsets of IV^n clearly violate the knowledge balance principle. We will show now that (a strong candidate for)[4] the process category for the toy theory in its entirety is **MSpek**, while if we restrict the toy theory to states of maximal knowledge (consistent with the knowledge balance principle), the process category is **Spek**.

PROPOSITION 6.1. *The morphisms of the process category of the toy theory are closed under composition, Cartesian product and relational converse.*

PROOF. There is no feature of the toy theory which would put any restrictions on which operations could be composed, so we expect the states and transformations to be closed under composition. Since the Cartesian product is used by the toy theory to represent composite systems we also expect the states and transformations to be closed under Cartesian product.

Every epistemic state corresponds to an outcome for at least one measurement (measurements correspond to asking as many questions as possible from canonical sets, epistemic states correspond to the answers). Recalling the discussion of measurement in section 2, we see that given a state $\psi \subset IV^n$ the disturbance resulting from the corresponding measurement outcome can be decomposed as $\psi \circ \psi^\dagger$, where $\psi^\dagger$ is the relational converse of ψ. Thus we expect the relational converse of each state also to feature in the physical category of the theory. The toy theory state corresponding to the subset $\Psi_{\mathbf{Spek}} = \{(1,1),(2,2),(3,3),(4,4)\} \subset IV \times IV$ along with its relational converse are then easily seen to constitute a compact structure on IV. We thus have map-state duality, and it is straightforward then to show that if states are closed under relational converse, so is any morphism in the physical category. □

Note that this point sharpens our discussion about the consistency of the toy theory, in section 2. If the states and transformations which Spekkens has derived for up to three systems, under the operations of composition, Cartesian product and relational converse, yield states which violate the knowledge balance principle, then the theory as presented is inconsistent.

PROPOSITION 6.2. *All of the generating morphisms of **MSpek** are states or transformations of the toy theory, or can be derived from them by composition, Cartesian product or relational converse.*

PROOF. The only generator for which this is less than obvious is $\delta_{\mathbf{Spek}}$. This is formed by composing Spekkens's GHZ-like state (see section V of [20]) with the relational converse of the state $\Psi_{\mathbf{Spek}}$ defined in the proof above. □

PROPOSITION 6.3. *All of the states and transformations derived by Spekkens in his original paper [20] are morphisms of **MSpek**. When we restrict to states of maximal knowledge all of the states and transformations are morphisms of **Spek**.*

PROOF. By inspection of [20]. □

COROLLARY 6.4. ***MSpek** is the minimal closure under composition, Cartesian product and relational converse of the states and transformations described in [20]. **Spek** is the minimal closure under these operations of the states of [20] corresponding to maximal knowledge and the transformations which preserve them.*

---

[4]Given that the toy theory is in fact not unambiguously defined for more than three systems, there may be other extensions too. Ours is the minimal extension given compositional closure.

PROPOSITION 6.5. *All states $\psi : I \to IV^n$ of **MSpek** and **Spek** satisfy the knowledge balance principle on the system corresponding to IV^n viewed as one complete system. All those of **Spek** satisfy the principle maximally.*

PROOF. Recall that the knowledge balance principle requires that we can know the answer to at most half of a canonical question set. A system with n elementary components has 2^{2n} ontic states. A canonical set for such a system consists of $2n$ questions, each answer to a question halving the number of possibilities for the ontic state. Thus, we know the answer to m such questions ($m = 0, \ldots, n$), iff our epistemic state is a subset of IV^n with 2^{2n-m} elements. We conclude from theorem 5.29 that all states of **MSpek** satisfy the knowledge balance principle on the system as a whole. We conclude that all states of **Spek** correspond to the maximum knowledge about the system as a whole consistent with the knowledge balance principle. □

PROPOSITION 6.6. *All states $\psi : I \to IV^n$ of **MSpek** and **Spek** satisfy the knowledge balance principle on every subsystem of the system corresponding to IV^n.*

PROOF. Given an epistemic state $\psi \subset IV^n$ of a composite system with n elementary components, the 'marginal' state on some subsystem is obtained from ψ by deleting from the tuples of ψ the components corresponding to the elementary systems which are not part of the subsystem of interest. Suppose this epistemic state corresponds to a **Spek** or **MSpek** diagram, D. The elementary systems which are not part of the subsystem correspond to a certain collection of external legs of D, and, by lemma 5.5, if we cap these with the **MSpek** generator $\perp_{\mathbf{MSpek}}$, the effect on the state ψ is exactly as just described. Composing a **Spek** or **MSpek** morphism with $\perp_{\mathbf{MSpek}}$ yields some morphism of **MSpek**, which by proposition 6.5 satisfies the knowledge balance principle. □

From corollary 6.4 and propositions 6.5 and 6.6 we reach two key conclusions:
- The states and transformations derived by Spekkens in [20] for systems of up to three components are all consistent with the knowledge balance principle.
- The process category of the toy theory must, at least, contain all of the morphisms of **MSpek**.

The second conclusion begs the question, could **MSpek** be a strict sub-category of the process category of the toy theory i.e. could the toy theory contain operations not contained in **MSpek**? It is difficult to answer this question, since, as discussed at the end of section 2 it is not clear what the rigorous definition of the toy theory is, or whether there is an unambiguous way to extend it beyond three systems. Certainly, **MSpek** is the process category of a theory which coincides with Spekkens's theory up to the case of three qubits, and whose states and transformations are bound to satisfy the three rules of section 2 (the first two rules by propositions 6.5 and 6.6, and the third simply by its definition as the closure under composition of a set of generators). It is in this sense that we earlier remarked that **MSpek** is a strong candidate for the process category of the toy theory.

7. Conclusion and outlook

We achieved our goal stated in the abstract, that is, to provide a rigorous mathematical description of Spekkens' toy theory, which proves its consistency.

This was established both in terms of generators for a dagger symmetric monoidal subcategory of **FRel**, consisting of symmetries for the elementary system and a basis structure (and nothing more!), as well as in terms of an explicit description of these relations as in Theorems 5.15, 5.14 and 5.29.

This description meanwhile already has proved to be of great use, for example, in pinpointing what the essential structural difference is between the toy theory and the relevant fragment of quantum theory. In joint work with Spekkens in [**7**] we showed that the key difference between the toy theory and relevant fragment of quantum theory is the *phase group*, a group that by pure abstract nonsense can be attributed to each basis structure. In the case of the toy theory this phase group is $Z_2 \times Z_2$ while in the case of the relevant fragment of quantum theory it is Z_4. One can then show that it is this difference that causes the toy theory to be local, while the relevant fragment of quantum theory is non-local.

In this context, one may wonder wether there is a general categorical construction which would turn a 'local theory' like **Spek** into a non-local one. We also expect that the construction in this paper can be fairly straightforwardly extended beyond qubit theories, for example to qutrits [**22**].

References

1. S. Abramsky (2009) *No-cloning in categorical quantum mechanics*. In: Semantic Techniques for Quantum Computation, I. Mackie and S. Gay (eds), pages 1–28, Cambridge University Press. arXiv:0910.2401
2. S. Abramsky and B. Coecke (2004) *A categorical semantics of quantum protocols*. In: Proceedings of 19th IEEE conference on Logic in Computer Science, pages 415–425. IEEE Press. arXiv:quant-ph/0402130. Revised version (2009): *Categorical quantum mechanics*. In: Handbook of Quantum Logic and Quantum Structures, K. Engesser, D. M. Gabbay and D. Lehmann (eds), pages 261–323, Elsevier. arXiv:0808.1023
3. A. Carboni and R. F. C. Walters (1987) *Cartesian bicategories I*. Journal of Pure and Applied Algebra **49**, 11–32.
4. B. Coecke (2010) *Quantum picturalism*. Contemporary Physics **51**, 59–83. arXiv:0908.1787
5. B. Coecke and R. Duncan (2008) *Interacting quantum observables*. In: Proceedings of the 35th International Colloquium on Automata, Languages and Programming (ICALP), pages 298–310, Lecture Notes in Computer Science 5126, Springer-Verlag. Extended version (2011): *Interacting quantum observables: categorical algebra and diagrammatics*. New Journal of Physics **13**, 043016. arXiv:0906.4725
6. B. Coecke and B. Edwards (2008) *Toy quantum categories*. Electronic Notes in Theoretical Computer Science **270**, issue 1, 29–40. arXiv:0808.1037
7. B. Coecke, B. Edwards and R. W. Spekkens (2010) *Phase groups and the origin of non-locality for qubits*. Electronic Notes in Theoretical Computer Science **270**, issue 2, 15–36. arXiv:1003.5005
8. B. Coecke and E. O. Paquette (2011) *Categories for the practicing physicist*. In: New Structures for Physics, B. Coecke (ed), pages 167–271. Lecture Notes in Physics, Springer-Verlag. arXiv:0905.3010
9. B. Coecke and D. Pavlovic (2007) *Quantum measurements without sums*. In: Mathematics of Quantum Computing and Technology, G. Chen, L. Kauffman and S. Lamonaco (eds), pages 567–604. Taylor and Francis. arXiv:quant-ph/0608035.
10. B. Coecke, D. Pavlovic and J. Vicary (2008) *A new description of orthogonal bases*. To appear in Mathematical Structures in Computer Science. arXiv:0810.0812
11. J. Evans, R. Duncan, A. Lang and P. Panangaden (2009) *Classifying all mutually unbiased bases in Rel*. arXiv:0909.4453
12. A. Joyal and R. Street (1991) *The Geometry of tensor calculus* I. Advances in Mathematics **88**, 55–112.
13. G. M. Kelly and M. L. Laplaza (1980) *Coherence for compact closed categories*. Journal of Pure and Applied Algebra **19**, 193–213.

14. S. Mac Lane (2000) *Categories for the Working Mathematician* (2nd edition), Springer-Verlag.
15. D. Pavlovic (2009) *Quantum and classical structures in nondeterminstic computation.* Lecture Notes in Computer Science **5494**, page 143–157, Springer. arXiv:0812.2266
16. R. Penrose (1971) *Applications of negative dimensional tensors.* In: Combinatorial Mathematics and its Applications, D. Welsh (Ed), pages 221–244. Academic Press.
17. P. Selinger (2007) *Dagger compact closed categories and completely positive maps.* Electronic Notes in Theoretical Computer Science **170**, 139–163.
18. P. Selinger (2011) *A survey of graphical languages for monoidal categories.* In: New Structures for Physics, B. Coecke (ed), 275–337, Springer-Verlag. arXiv:0908.3347
19. P. Selinger (2010) *Autonomous categories in which $A \simeq A^*$.* Proceedings of QPL 2010.
20. R. W. Spekkens (2007) *Evidence for the epistemic view of quantum states: A toy theory.* Physical Review A **75**, 032110. arXiv:quant-ph/0401052
21. R. W. Spekkens (2007) *Axiomatization through foil theories.* Talk, July 5, University of Cambridge.
22. R. W. Spekkens (2009) *The power of epistemic restrictions in reconstructing quantum theory.* Talk, August 10, Perimeter Institute. PRISA:09080009

Oxford University, Department of Computer Science
E-mail address: coecke@cs.ox.ac.uk

Oxford University, Department of Computer Science
Current address: Perimeter Institute for Theoretical Physics
E-mail address: wae28@hotmail.com

Categorical traces from single-photon linear optics

Peter Hines and Philip Scott

ABSTRACT. We use a single-photon thought experiment, based on a modification of the Sagnac interferometer, to motivate a general construction on linear maps that has a close connection to constructions from algebraic and categorical program semantics. We analyse this general construction in terms of a category of formal power series over linear maps, and exhibit a partial categorical trace, generalising the 'particle-style' trace on Hilbert spaces [**HS10**], that has a physical realisation based on this thought-experiment.

1. Introduction

1.1. Historical background. The Sagnac interferometer is a linear-optics device whose theoretical origins [**Lo93**] predate both quantum mechanics and relativity. An experiment, described in [**Sa13i, Sa13ii**], claimed to 'demonstrate the existence of the luminiferous æther' (*"La preuve de la réalité de l'éther lumineux"*) using an interferometer that split incoming light into two counter-rotating paths around an optical loop. Sagnac's experiment is still a favorite of those who wish to disprove relativity (see *http://www.anti-relativity.com/* for examples). However, as observed by Michelson [**Br02**], the Sagnac effect cannot discriminate between (special-)relativistic and pre-relativistic theories[1].

From essentially 19th century origins, both the Sagnac interferometer and its modern incarnation as the 'Ring Laser' [**St97**] have become immensely important practical tools used in, amongst other devices, highly sensitive gyroscopes and the Sagnac effect is now a key technique in inertial and missile guidance systems (From [**St97**], *"Contrary to the supposed custom in research, the area ... has already proved its commercial and (unfortunately) its military usefulness; it is the scientific potential which has been neglected."*).

1.2. Modifying the Sagnac interferometer. In this paper, we analyse a variant of Sagnac's experiment from a quantum-mechanical perspective. We take a

---

1991 *Mathematics Subject Classification.* Primary 68Q05; Secondary 81P68.
Key words and phrases. Categorical traces, Formal Power series, Linear Optics.
The first author was supported by E.U. FP6 Grant QICS.
The second author's research is supported by an NSERC Discovery Grant, Canada.

[1] *"We will undertake this [experiment], although my conviction is strong that we shall prove only that the earth rotates on its axis, a conclusion which I think we may be said to be sure of already."* – A. Michelson, quoted in [**Br02**].

©2012 American Mathematical Society

single-photon description, and analyse the situation where the splitting of incoming light is not arbitrary (creating a quantum-mechanical 'equal superposition' of paths), but dependent on a certain quantum property (i.e. the polarisation) of the incoming light. The motivation for this is two-fold:

(1) Single-photon thought experiments in linear optics are often used as illustrations of quantum computation and information. By encoding quantum bits (qubits) on either the (superposition of) paths taken by an individual photon, or its polarisation, it is possible to implement a universal set of quantum-computational gates [**CAK05**], and hence all current quantum algorithms[2] and protocols. We give explicit examples in Appendix A.

(2) In the Sagnac interferometer, the replacement of the standard beamsplitter by a polarising beamsplitter (see Section 2.2 for these devices) gives a form of 'conditional looping'. Whether or not a photon enters the optical loop depends on its polarisation; similarly, the exiting of a photon from the optical loop is also dependent on its its polarisation.

Thus, our modified Sagnac Interferometer gives a paradigmatic example of 'quantum conditional iteration'. By analysing our thought-experiment in detail, in the Hilbert space model, we demonstrate that this is very closely related to the form of conditional iteration used in the theory of reversible computation — the 'particle-style' categorical trace (see Section 8).

2. Hilbert space formalism, and the linear optics toolkit

This section presents the standard Hilbert space formalism for quantum information and computation, together with an exposition of how single-photon experiments using linear optics gates may be described within this formalism.

2.1. Basics of quantum computation and information.
We use the standard Hilbert space formalism to model our thought-experiments. We emphasise that although we take a very categorical approach, the constructions of this paper live within the traditional Hilbert space formalism, rather than the abstract categorical formalism of [**AbCo05**].

DEFINITION 2.1. Given a vector space V over $\mathbb{C}$, an *inner product* is a Hermitian symmetric form (i.e. a map $\langle \_ | \_ \rangle : V \times V \to \mathbb{C}$ that is linear in the first variable and conjugate-linear in the second) that satisfies $\langle x|x \rangle \geq 0$ and $\langle x|x \rangle = 0$ iff $x = \mathbf{0}$. A *complex Hilbert Space* is then a Banach space (i.e. complete normed vector space) over $\mathbb{C}$ whose norm is defined by an inner product, $\|x\| = (\langle x|x \rangle)^{\frac{1}{2}}$.

By the Reisz representation theorem [**Har83**], for every bounded linear map $L : H \to K$ of Hilbert spaces, there exists a unique bounded linear map $L^* : K \to H$ such that, for all $k \in K$ and $h \in H$,

$$\langle k | L(h) \rangle = \langle L^*(k) | h \rangle$$

This is called the **Hermitian adjoint** of L, and is often denoted by either $L^\dagger$ (quantum-mechanical notation) or L^H (functional-analysis notation).

---

[2]It should be emphasised that, in encodings of quantum algorithms, the resources required grow exponentially. Thus such optical circuits are primarily useful either as demonstrations of quantum information, and tests of underlying principles, or in quantum-mechanical communications protocols.

We use Dirac notation for vectors, so $|\psi\rangle \in \mathcal{H}$ is a linear map from $\mathbb{C}$ to $\mathcal{H}$ defined by $|\psi\rangle(z) = z.\psi \in \mathcal{H}$. As is traditional, we abuse notation and refer to $|\psi\rangle$ as a vector of $\mathcal{H}$ (see [**Ab05**] for this concept from a categorical perspective). In quantum computation (especially the *circuit model* [**NC00**]), it is standard to assume that Hilbert spaces are equipped with some fixed orthonormal basis set (the **computational basis**). We use notation derived from a categorical perspective (The l_2 functor of [**Ba92**]), so given some set $X = \{x_i\}_{i \in I}$, the space $l_2(X)$ is the space with a distinguished orthonormal basis $\{|x_i\rangle\}_{i \in I}$. Of particular interest is the **qubit space** $Q = l_2(\{0,1\})$ that plays an analogous role to the set of bits $\{0,1\}$ in classical computation. The assumption that each space is equipped with a fixed orthonormal basis allows us to use matrix representations of linear maps: these are used heavily throughout this paper.

Composite systems are modelled using the tensor product of Hilbert spaces: given spaces $\{\mathcal{H}_k\}_{k=1...n}$ modelling systems $S_1, \ldots S_n$, the composite system is modelled by the space $\otimes_{k=1}^{n} \mathcal{H}_k$. When each space is a copy of the qubit space Q, tensor products of this form are called **quantum registers** of k qubits.

Two key differences between classical and quantum information are the phenomena of superposition and entanglement. We present these mathematically, and refer to any introductory quantum computing text (e.g. [**NC00**]) for physical interpretations.

DEFINITION 2.2. Given a space $\mathcal{H}$ with computational basis $\{|b_0\rangle, \ldots, |b_n\rangle\}$, a state $|\psi\rangle$ is a **superposition** when it is a non-trivial linear combination of computational basis vectors, $|\psi\rangle = \sum_{j=0}^{n} \alpha_j |b_j\rangle$. Note that this concept is *basis-dependent*.

Given another Hilbert space $\mathcal{K}$, with computational basis $\{|c_0\rangle, \ldots |c_m\rangle\}$, the tensor product space $H \otimes K$ has computational basis $\{|b_j c_k\rangle\}_{j=0..n, k=0..m}$. A state $|\phi\rangle = \sum_{j,k} \alpha_{j,k} |b_j c_k\rangle$ is **entangled** when it cannot be written as $|\phi_1\rangle \otimes |\phi_2\rangle$, for any $|\phi_1\rangle \in \mathcal{H}$ and $|\phi_2\rangle \in \mathcal{K}$. Note that, unlike superposition, entanglement is basis-independent.

Physical operations on quantum systems are divided into 2 classes: measurements, and coherent operations. Coherent operations are modelled by *unitary maps*, and measurements are modelled by either *projectors* or *Hermitian operators*. Again, we present these mathematically, and refer to [**NC00**] for physical intuitions.

DEFINITION 2.3. A linear map between Hilbert spaces, $U : \mathcal{H} \to \mathcal{K}$ is **unitary** when it is an inner-product preserving isomorphism[3]. A linear map $P : \mathcal{H} \to \mathcal{K}$ is a **projector** when it is a self-adjoint idempotent. A *measurement* is determined by a self-adjoint operator, or Hermitian matrix. By the spectral decomposition theorem, every (finite) Hermitian matrix has a unique decomposition as the sum of projection operators — in this way a Hermitian matrix describes a set of projections, labelled by eigenvalues, and these are taken to be the experimental outcomes of a measurement.

The class of all Hilbert spaces forms a category, as follows:

---

[3] A useful characterisation of unitary maps is that, given some matrix representation for $U : \mathcal{H} \to \mathcal{H}$, the *conjugate transpose* of this matrix is the matrix representation for U^{-1}. It is a straightforward exercise to show that this characterisation is basis-independent, and equivalent to the definition given above.

DEFINITION 2.4. The category **Hilb** has all (separable) Hilbert spaces as objects, and all continuous (i.e. bounded) linear maps as arrows. An important subcategory is **Hilb$_{\mathbf{FD}}$**, which is simply the above category, restricted to finite-dimensional spaces. The category **Hilb** has two distinct symmetric monoidal tensors: the familiar **tensor product** $\otimes$, and the **direct sum** $\oplus$. In [**Hal58**], the direct sum is defined for arbitrary indexed families of Hilbert spaces, as follows: given an indexed family of spaces $\{H_i\}_{i \in I}$, the direct sum $\oplus_{i \in I} H_i$ has elements given by functions $\alpha : I \to \biguplus_{i \in I} H_i$ such that $\alpha(i) \in H_i$, and $\sum_{i \in I} \|x_i\|_{H_i}^2 < \infty$. exists. The inner product of two elements $\alpha, \beta \in \bigoplus_{i \in I} H_i$ is then given by $\langle \alpha | \beta \rangle = \sum_{i \in I} \langle \alpha(i) | \beta(i) \rangle$.

When the indexing set is finite, it is straightforward that the direct sum is indeed a monoidal tensor. Infinitary direct sums can also be given a categorical interpretation, but this is more subtle.

By contrast, the tensor product is defined for finite families only. Although infinitary analogues have also been considered [**JvN38**], these more naturally live within the theory of C^* and von Neumann algebras, and play no part in this paper.

2.2. The optics toolkit. *This section follows very closely the introduction to linear optics given in* [**GK05**], *and in particular the single-photon case presented in* [**Be05**]. The basic linear optics devices we require are the *Beam Splitter* (**BS**), the *Polarising Beam Splitter* (**PBS**), the *Half Wave Plate* (**HWP**), and the *Phase Plate* (**PP**). These all implement coherent operations, and have standard schematics, as shown in Figure 1.

FIGURE 1. The linear optics toolbox

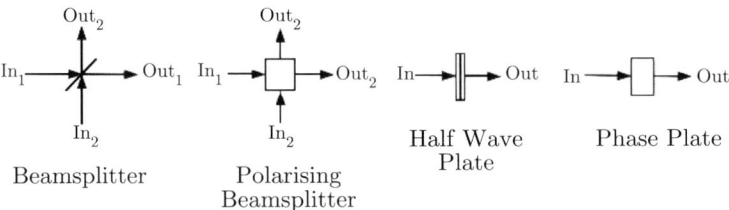

These all have either 1 or 2 input / output channels[4], and their behaviour may be dependent on the photon polarisation – thus the quantum properties we consider are the *polarisation*, and the *'which channel?'* information.

We adopt the convention that horizontal (resp. vertical) polarisation is denoted $|H\rangle$ (resp. $|V\rangle$), and input (resp. output) channel j is denoted $|in_j\rangle$ (resp. $|out_j\rangle$). Thus, a horizontally polarised photon in input channel 1 corresponds to the state vector $|H\rangle |in_1\rangle$, and a vertically polarised photon in an even superposition of both output channels corresponds to the state vector $\frac{1}{\sqrt{2}} |V\rangle (|out_1\rangle + |out_2\rangle)$.

Their action is then described by unitary maps, as follows:

---

[4]In fact, all these devices are completely reversible, and the designation of channels as either *input* or *output* depends on the direction of the incident photon. This is important in the analyses of Section 3.

The beamsplitter The beamsplitter is one of the most standard linear optics devices. An input may be in either channel 1 or channel 2, and may be either horizontally or vertically polarized (or, of course, an arbitrary superposition of any of these properties). Given an input in either of the input channels, the output is a superposition of output channels (however, see Remark 2.5). We emphasise that we are only considering the case where there is a single input photon[5].

The behaviour of the beamsplitter is described by a unitary map, defined by its action on (orthonormal) basis vectors as follows:

$$|H\rangle |in_1\rangle \mapsto \tfrac{1}{\sqrt{2}}|H\rangle (|out_1\rangle + |out_2\rangle) \qquad |V\rangle |in_1\rangle \mapsto \tfrac{1}{\sqrt{2}}|V\rangle (|out_1\rangle + |out_2\rangle)$$

$$|H\rangle |in_2\rangle \mapsto \tfrac{1}{\sqrt{2}}|H\rangle (|out_1\rangle + |out_2\rangle) \qquad |V\rangle |in_2\rangle \mapsto \tfrac{1}{\sqrt{2}}|V\rangle (||out_1\rangle - |out_2\rangle))$$

Note that the action of the beamsplitter is independent of the polarisation of the input photon.

The polarising beamsplitter This is closely related to the above example; however, its behaviour is conditional on the polarisation of the input photon. Intuitively, it transmits photons with horizontal polarisation and reflects (through $\pi/2$) photons with vertical polarisation.

Using the same notation as above, the behaviour of the polarising beamsplitter is given by the unitary map defined by:

$$|H\rangle |in_1\rangle \mapsto |H\rangle |out_1\rangle \qquad |V\rangle |in_1\rangle \mapsto |V\rangle |out_2\rangle$$

$$|H\rangle |in_2\rangle \mapsto |H\rangle |out_2\rangle \qquad |V\rangle |in_2\rangle \mapsto |V\rangle |out_1\rangle\rangle$$

The phase plate This transmits all photons on the input channel, and rotates the phase by an angle of θ. Given an arbitrary incoming photon $|\psi\rangle$, the action of the phase plate is simply

$$|\psi\rangle \mapsto e^{i\theta}|\psi\rangle$$

The half-wave plate This again transmits all photons on the input channel, and adds a $\frac{\pi}{4}$ rotation to the polarisation. The action of the half wave plate is given by a unitary map defined by its action on basis vectors as follows:

$$|H\rangle \mapsto \tfrac{1}{\sqrt{2}}(|H\rangle + |V\rangle)$$

$$|V\rangle \mapsto \tfrac{1}{\sqrt{2}}(-|H\rangle + |V\rangle)$$

REMARK 2.5. **A note on reflection and phases** Readers familiar with the standard optics toolkit will note that the phases in the outputs of the beamsplitters are non-standard — in linear optics experiments, the reflected path (whether from a

---

[5]Precisely, we allow for a photon in Channel 1, or Channel 2, or a photon in a superposition of these locations. We do *not* allow for an input photon in each channel – not only would this require a more sophisticated mathematical treatment, but would also take us away from the underlying motivation of single-particle interference. For readers familiar with the usual Fock space description, our treatment also neglects the vacuum states, for simplicity of notation.

mirror, beamsplitter, polarised beamsplitter, or whatever) actually picks up a phase factor of i. We emphasise that all the devices presented can be made to behave exactly as described, simply by using appropriately placed phase plates [**CAK05**]. However, we do not do this explicitly – partly for simplicity of diagrams, and partly to make the connection with the standard circuit model of quantum computation (given in Appendix A) more immediate.

3. The Sagnac Interferometer

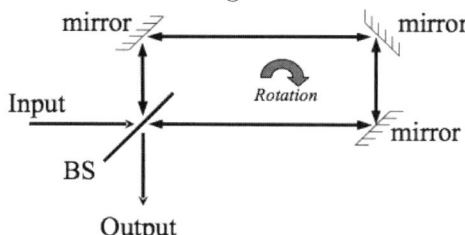

FIGURE 2. The Sagnac interferometer

A schematic of the standard Sagnac interferometer is shown in Figure 2. Intuitively, its action is straightforward: incoming light is split into two counter-rotating paths by the beamsplitter indicated. These travel around the optical loop, and recombine at the beamsplitter. This then produces an interference pattern at the output. When the whole apparatus is rotated, the relative length of the respective paths changes, shifting the interference pattern. Thus, absolute rotation is readily detected by changes in the observed interference pattern.

This apparatus has been analysed in detail, by a number of authors, for at least the past 100 years – we do not attempt to add yet another analysis to the literature. A single-photon analysis is conceptually more interesting, but only in that it requires the strongly quantum-mechanical phenomenon of *single-particle interference*. Mathematically this is trivial; we allow for arbitrary superpositions and (complex) phase differences in the 'which path' information for a single photon, as in Section 2. However, the physical interpretation is more remarkable: as stated in [**PSM96**],

> " In his famous introduction [**FLS65**] to the single particle superposition principle, Feynman stated that, '. . . it has in it the heart of quantum mechanics. In fact, it contains the only mystery.' ".

We leave a single-photon analysis as an interesting exercise, and refer to [**PSM96**] for a demonstration of why the many (entangled) particle case is qualitatively different to the single-photon case.

4. A modified Sagnac Interferometer

We now make the following modifications to the Sagnac interferometer, as shown in Figure 3.

(1) We replace the beamsplitter by a polarising beamsplitter.

FIGURE 3. A modified Sagnac interferometer (MSI)

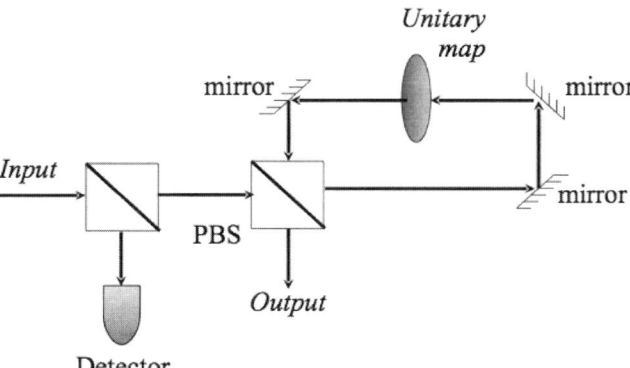

(2) We introduce an arbitrary unitary operation within the optical loop that acts non-trivially on photon polarisation as
$$|H\rangle \mapsto a|H\rangle + c|V\rangle \quad , \quad |V\rangle \mapsto b|H\rangle + d|V\rangle$$
for some $a, b, c, d \in \mathbb{C}$.

(3) We post-select for experiments where the detector shown does *not* record a measurement. This ensures that the input to the second PBS (and hence the optical loop) is horizontally polarised.

Modifications 1. and 3. ensure that the only possible path traversed around the optical loop is the counter-clockwise path. Thus, this device no longer displays the Sagnac interferometer's extraordinary sensitivity to rotation. However, at any time (possibly excluding the start of the experiment), the photon will be in a non-trivial superposition of locations — both within the feedback loop, and on the output channel. In particular, the 'number of times the photon has traversed the feedback loop' is not a well-defined quantity. Thus, although we no longer have two distinct counter-rotating paths in the optical loop, the phenomenon of single-particle interference still has a large part to play in any formal description.

Because of this temporal aspect, it is not immediate how to give a treatment in terms of input and output spaces. Instead, we describe this apparatus in terms of a unitary operator that is repeatedly applied a space describing the entire state of the system — how to translate this into input / output behaviour, and the correct categorical interpretation forms a substantial part of this paper.

In order to analyse the above thought-experiment, we first use the assumption of discrete space and time — a common assumption used in (for example) the 'toy models' of [**Gr02**] or [**Pen04**]. We make the further assumption that, at the very beginning of the experiment, the output stream is empty — the single photon is not in a superposition of input and output modes. With these assumptions, we may draw the individual time-steps as shown in Figure 4.

Note that when we analyse this experiment using these conventions, we do not have a unitary map from a single space to itself – rather, at each step, the unitary evolution F_j is from space $\mathcal{S}_j$ to space $\mathcal{S}_{j+1}$. This is simply a labelling convention – however, it makes the analysis significantly simpler. From Figure 4,

FIGURE 4. Input / output streams in a modified Sagnac interferometer

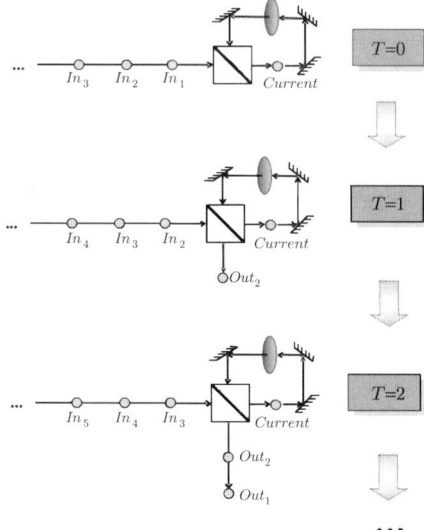

and the fact that input and output modes must be horizontally polarised, we may give orthonormal bases for the spaces $\{\mathcal{S}_j\}_{j=0}^{\infty}$, as follows

- the space $\mathcal{S}_0$ has basis

$$\{|current\rangle|H\rangle, |current\rangle|V\rangle, |in_1\rangle|H\rangle, |in_2\rangle|H\rangle, |in_3\rangle|H\rangle, |in_4\rangle|H\rangle, \ldots\}$$

- the space $\mathcal{S}_1$ has basis

$$\{|out_1\rangle|H\rangle, |current\rangle|H\rangle|current\rangle|V\rangle, |in_2\rangle|H\rangle, |in_3\rangle|H\rangle, |in_4\rangle|H\rangle \ldots\}$$

- the space $\mathcal{S}_2$ has basis

$$\{|out_1\rangle|H\rangle, |out_2\rangle|H\rangle, |current\rangle|H\rangle|current\rangle|V\rangle, |in_3\rangle|H\rangle, |in_4\rangle|H\rangle \ldots\}$$

- ...

Using the description of the actions of the individual components, we may write down the unitary maps $\{F_i : \mathcal{S}_i \to \mathcal{S}_{i+1}\}_{i \in \mathbb{N}}$ as follows :

- $F_0 = \begin{pmatrix} b & a & 0 & 0 & \ldots \\ d & c & 0 & 0 & \ldots \\ 0 & 0 & 1 & 0 & \ldots \\ 0 & 0 & 0 & 1 & \ldots \\ \vdots & \vdots & \vdots & \vdots & \ddots \end{pmatrix}$

- $F_1 = \begin{pmatrix} 1 & 0 & 0 & 0 & \ldots \\ 0 & b & a & 0 & \ldots \\ 0 & d & c & 0 & \ldots \\ 0 & 0 & 0 & 1 & \ldots \\ \vdots & \vdots & \vdots & \vdots & \ddots \end{pmatrix}$

- $F_2 = \begin{pmatrix} 1 & 0 & 0 & 0 & \dots \\ 0 & 1 & 0 & 0 & \dots \\ 0 & 0 & b & a & \dots \\ 0 & 0 & d & c & \dots \\ \vdots & \vdots & \vdots & \vdots & \ddots \end{pmatrix}$

- $F_3 = \dots$

After T timesteps, the overall state of the system is the state in $\mathcal{S}_T$ given by applying the map $F_{T-1} F_{T-2} \dots F_0$ to the initial state in $\mathcal{S}_0$. Rather than analysing this directly, we consider a generalisation, where a, b, c, d in the above series of unitary maps are not simply complex numbers, but are themselves block matrices.

5. The twisted dagger construction

We now generalise the above analysis to the case where the specified unitary operation is a block matrix, rather than simply a 2×2 complex matrix, and consider the limit as the number of timesteps tends to infinity.

LEMMA 5.1. *Let* $L = \begin{pmatrix} A & B \\ C & D \end{pmatrix} : X \oplus U \to Y \oplus U$ *be a unitary map, and let the unitary maps*

$$\{F_i : Y^{\oplus i} \oplus U \oplus X^{\oplus \omega} \to Y^{\oplus (i+1)} \oplus U \oplus X^{\oplus \omega}\}_{i=1}^{\infty}$$

be defined by

$$F_0 = \begin{pmatrix} B & A & 0 & \dots \\ D & C & 0 & \dots \\ 0 & 0 & I & \dots \\ \vdots & \vdots & \vdots & \ddots \end{pmatrix} \quad F_1 = \begin{pmatrix} I & 0 & 0 & 0 & \dots \\ 0 & B & A & 0 & \dots \\ 0 & D & C & 0 & \dots \\ 0 & 0 & 0 & I & \dots \\ \vdots & \vdots & \vdots & \vdots & \ddots \end{pmatrix}$$

$$F_2 = \begin{pmatrix} I & 0 & 0 & 0 & 0 & \dots \\ 0 & I & 0 & 0 & 0 & \dots \\ 0 & 0 & B & A & 0 & \dots \\ 0 & 0 & D & C & 0 & \dots \\ 0 & 0 & 0 & 0 & I & \dots \\ \vdots & \vdots & \vdots & \vdots & \vdots & \ddots \end{pmatrix} \quad \dots$$

Then the composition (product of matrices) $F_n F_{n-1} \dots F_1 F_0$ *has the block matrix*

$$\begin{pmatrix} B & A & 0 & 0 & 0 & \dots & 0 & 0 & 0 & \dots \\ BD & BC & A & 0 & 0 & \dots & 0 & 0 & 0 & \dots \\ BD^2 & BDC & BC & A & 0 & \dots & 0 & 0 & 0 & \dots \\ BD^3 & BD^2C & BDC & BC & A & \dots & 0 & 0 & 0 & \dots \\ \vdots & \vdots & \vdots & \vdots & \vdots & & \vdots & \vdots & \vdots & \\ BD^{n-1} & BD^{n-2}C & BD^{n-3}C & BD^{n-4}C & BD^{n-5}C & \dots & A & 0 & 0 & \dots \\ D^n & D^{n-1}C & D^{n-2}C & D^{n-3}C & D^{n-4}C & \dots & C & 0 & 0 & \dots \\ 0 & 0 & 0 & 0 & 0 & & 0 & I & 0 & \dots \\ 0 & 0 & 0 & 0 & 0 & & 0 & 0 & I & \\ \vdots & \vdots & \vdots & \vdots & \vdots & & \vdots & \vdots & \vdots & \ddots \end{pmatrix}$$

PROOF. We prove this by induction. (Notation: we abbreviate the above infinite matrix, omitting final columns and rows denoted by dots.) As a first step, note that

$$F_1 F_0 = \begin{pmatrix} B & A & 0 & 0 \\ BD & BC & A & 0 \\ D^2 & DC & C & 0 \\ 0 & 0 & 0 & I \end{pmatrix}$$

as required.

Now assume that for some $k \geq 1$,

$$F_k F_{k-1} \ldots F_0 = \begin{pmatrix} B & A & 0 & 0 & \cdots & 0 & 0 \\ BD & BC & A & 0 & \cdots & 0 & 0 \\ BD^2 & BDC & BC & A & \cdots & 0 & 0 \\ BD^3 & BD^2C & BDC & BC & \cdots & 0 & 0 \\ \vdots & \vdots & \vdots & \vdots & & \vdots & \\ BD^{k-1} & BD^{k-2}C & BD^{k-3}C & BD^{k-4}C & \cdots & A & 0 \\ D^k & D^{k-1}C & D^{k-2}C & D^{k-3}C & \cdots & C & 0 \\ 0 & 0 & 0 & 0 & \cdots & 0 & I \end{pmatrix}$$

Then direct calculation gives that $F_{k+1} F_k F_{k-1} \ldots F_0 =$

$$\begin{pmatrix} B & A & 0 & 0 & 0 & \cdots & 0 & 0 \\ BD & BC & A & 0 & 0 & \cdots & 0 & 0 \\ BD^2 & BDC & BC & A & 0 & \cdots & 0 & 0 \\ BD^3 & BD^2C & BDC & BC & A & \cdots & 0 & 0 \\ \vdots & \vdots & \vdots & \vdots & \vdots & & \vdots & \\ BD^k & BD^{k-1}C & BD^{k-2}C & BD^{k-3}C & BD^{k-4}C & \cdots & A & 0 \\ D^{k+1} & D^kC & D^{k-1}C & D^{k-2}C & D^{k-3}C & \cdots & C & 0 \\ 0 & 0 & 0 & 0 & 0 & \cdots & 0 & I \end{pmatrix}$$

Our result thus follows by induction. $\square$

REMARK 5.2. The above matrix calculations are based on a slight generalisation of the thought experiment of Section 4 — this more general case can be thought of as describing a single-particle interferometry experiment where the particle in question carries a number of quantum properties in addition to the polarisation. Let us assume that this particle may, as before, be horizontally or vertically polarised (H or V), and has further independent quantum properties $(k_1, k_2, \ldots, k_n)$. Let the 'polarisation space' P have orthonormal basis $\{|H\rangle, |S_V\rangle\}$ and let the 'additional properties space' K have orthonormal basis $\{|k_1\rangle, |k_2\rangle, \ldots, |k_n\rangle\}$. The state vector for a single particle in such an experiment is then a member of $S = P \otimes K$. We may take a direct sum decomposition of S as

$$S = S_H \oplus S_V \quad \text{where} \quad \begin{cases} S_H & \text{has basis} \quad |H\rangle|k_1\rangle, \, |H\rangle|k_2\rangle, \, \ldots, |H\rangle|k_n\rangle \\ S_V & \text{has basis} \quad |V\rangle|k_1\rangle, \, |V\rangle|k_2\rangle, \, \ldots, |V\rangle|k_n\rangle \end{cases}$$

and using this direct sum decomposition, give a block matrix representation for some unitary $U : S \to S$, as $U = \begin{pmatrix} A & B \\ C & D \end{pmatrix} : S_H \oplus S_V \to S_H \oplus S_V$. Such a matrix may be represented schematically as

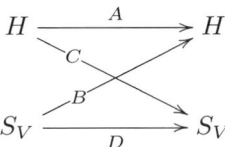

and using similar conventions, the polarising beam splitter is represented as

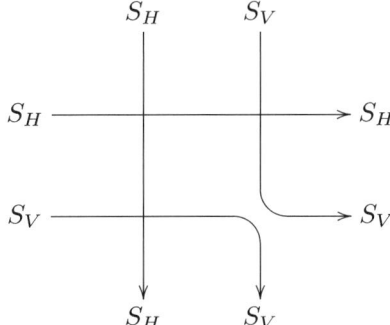

(Note that in the above two schematic diagrams, lines are separated by an implicit direct sum, rather than tensor product. Thus they should be interpreted as categorical string diagrams for the direct sum structure — an interpretation within the quantum circuit paradigm is not appropriate).

We may now compose these two diagrams, with the output fed back into the input as in the thought experiments of Section 4, to produce the schematic diagram shown in figure 5. As well as structure very similar to the usual diagram for a particle-style categorical trace, note the presence of the symmetry map for the direct sum.

DEFINITION 5.3. Let $L : X \oplus U \to Y \oplus U$ be a finite-dimensional linear map, with block matrix $L = \begin{pmatrix} A & B \\ C & D \end{pmatrix}$. We define the **twisted dagger** of L, w.r.t. this decomposition, to be the matrix

$$\dagger^U(L) = \begin{pmatrix} B & A & 0 & 0 & 0 & 0 & 0 & 0 & \ldots \\ BD & BC & A & 0 & 0 & 0 & 0 & 0 & \ldots \\ BD^2 & BDC & BC & A & 0 & 0 & 0 & 0 & \ldots \\ BD^3 & BD^2C & BDC & BC & A & 0 & 0 & 0 & \ldots \\ BD^4 & BD^3C & BD^2C & BDC & BC & A & 0 & 0 & \ldots \\ BD^5 & BD^4C & BD^3C & BD^2C & BDC & BC & A & 0 & \ldots \\ BD^6 & BD^5C & BD^4C & BD^3C & BD^2C & BDC & BC & A & \ldots \\ \vdots & \vdots & \vdots & \vdots & \vdots & \vdots & \vdots & \vdots & \ddots \end{pmatrix}$$

REMARK 5.4. The terminology 'twisted dagger' comes from the similarity of the above construction (especially the finite approximations of Lemma 5.1) with the Elgot dagger — up to an additional twist. Note that we do *not* claim the twisted

FIGURE 5. A schematic diagram for a generalised Sagnac interferometer

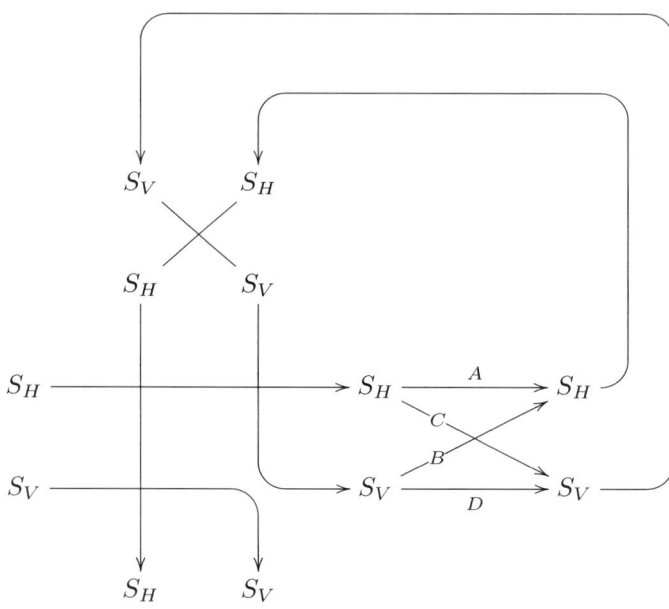

dagger is, in every case, the matrix representation of a continuous (i.e. bounded) linear map. In general, it is simply a formal matrix.

We now investigate the existence and properties of this matrix, with particular emphasis on when it describes either a continuous, or a unitary, linear map. We first require a trivial result:

LEMMA 5.5. *Given* $L = \begin{pmatrix} A & B \\ C & D \end{pmatrix} \in \mathbf{Hilb_{FD}}(X \oplus U, Y \oplus U)$, *and arbitrary* $\phi = \begin{pmatrix} \phi_0 \\ \phi_1 \\ \phi_2 \\ \vdots \end{pmatrix} \in U \oplus X^{\oplus \omega}$, *let us define the formal matrix* $\zeta = \begin{pmatrix} \zeta_0 \\ \zeta_1 \\ \zeta_2 \\ \vdots \end{pmatrix}$ *to be given by the formal matrix product* $\zeta = \dagger^U(L)(\phi)$, *so*

$$\zeta_n = \begin{cases} BD^n(\phi_0) + \sum_{i=1}^n BD^{n-i}C(\phi_i) + A(\phi_{n+1}) & (n > 0) \\ B(\phi_0) + A(\phi_1) & (n = 0) \end{cases}$$

Then ζ_n *exists, for all* $n \in \mathbb{N}$.

PROOF. Observe that ζ_n is given by a finite sum of continuous linear maps applied to a finite vector of elements. □

Note that we do not claim that $\sum_{n=0}^{\infty} \|\zeta_n\|^2$ exists in general, or (equivalently) that ζ is an element of $Y^{\oplus \omega}$ — a sufficient condition for this is given in Theorem 5.6 below.

THEOREM 5.6. *Let $L = \begin{pmatrix} A & B \\ C & D \end{pmatrix} : X \oplus U \to Y \oplus U$ be a linear map between finite-dimensional Hilbert spaces. Then:*

(1) *A sufficient condition for $\dagger^U(L)$ to be the matrix representation of a bounded linear map between Hilbert spaces is that the component D is a strict contraction (i.e. $Sup_{\|\psi\|=1} \|D(\psi)\| < 1$).*
(2) *When L is a unitary map, a sufficient condition for $\dagger^U(L)$ to be unitary is that the component D is a strict contraction.*

PROOF. Let the family of unitary maps $\{F_i\}_{i \in \mathbb{N}}$ be as defined in Lemma 5.1, and define

$$\{G_n : U \oplus X^{\oplus \omega} \to Y^{\oplus(n+1)} \oplus U \oplus X^{\oplus \omega}\}_{n=0}^{\infty}$$

by $G_n = F_n F_{n-1} F_{n-2} \ldots F_0$. (we refer to Lemma 5.1 for explicit fomulæ for G_n). It is immediate that F_i is a well-defined linear map for all $i \geq 0$, and is unitary exactly when L is unitary. Similarly, the maps $\{G_k\}_{k \in \mathbb{N}}$ are bounded linear maps, and unitary exactly when L is unitary.

We now use these preliminaries to prove (1) and (2) above:

(1) Consider arbitrary $\phi \in U \oplus X^{\oplus \omega}$. We now study the sequence

$$\phi = \phi^{(0)} \xrightarrow{F_0} \phi^{(1)} \xrightarrow{F_1} \phi^{(2)} \xrightarrow{F_2} \phi^{(3)} \xrightarrow{F_3} \ldots$$

so $\phi^{(n)} = G_n(\phi)$. We write $\phi^{(n)}$ explicitly as

$$\phi^{(n)} = \begin{pmatrix} \phi_0^{(n)} \\ \phi_1^{(n)} \\ \phi_2^{(n)} \\ \phi_3^{(n)} \\ \vdots \end{pmatrix} \quad \text{where} \quad \begin{cases} \phi_n^{(n+i)} \in Y \\ \phi_n^{(n)} \in U \\ \phi_{n+i}^{(n)} \in X \end{cases} \quad \text{for all } n \in \mathbb{N} \text{ , } i > 0$$

In particular, we make the identification $\phi_i^{(0)} = \phi_i$, for all $i \in \mathbb{N}$.

From the explicit description of $\{F_i\}_{i \in \mathbb{N}}$, we may use standard diagrammatic notation for matrix composition, and draw the calculation of the components of $\phi^{(n)}$ as shown in Figure 6.

From either this diagram, or by direct calculation, we may inductively calculate these components for all $p, q > 0$, as follows:

$$\phi_q^{(p)} = \begin{cases} \phi_q^{(0)} & q > p \\ B\left(\phi_{p-1}^{(p-1)}\right) + A\left(\phi_p^{(0)}\right) & q = p-1 \\ D\left(\phi_{p-1}^{(p-1)}\right) + C\left(\phi_p^{(0)}\right) & p = q \\ \phi_q^{(q+1)} & p > q+1 \end{cases}$$

FIGURE 6. Calculating components of $\phi^{(n)}$

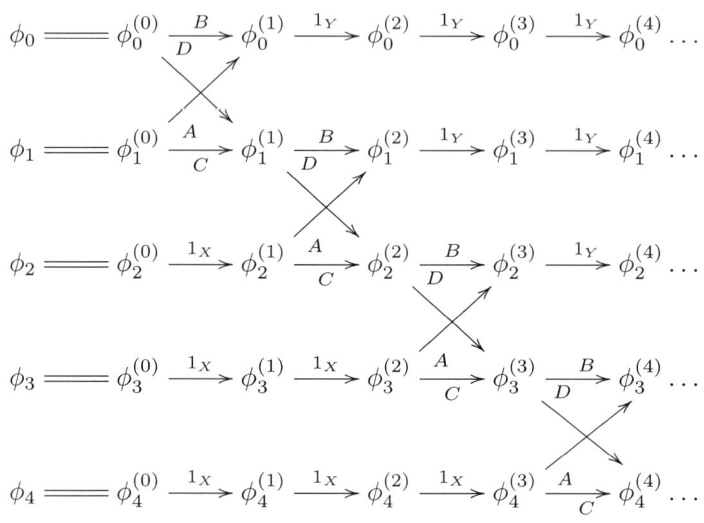

$$\phi_0 = \phi_0^{(0)} \xrightarrow[D]{B} \phi_0^{(1)} \xrightarrow{1_Y} \phi_0^{(2)} \xrightarrow{1_Y} \phi_0^{(3)} \xrightarrow{1_Y} \phi_0^{(4)} \cdots$$

$$\phi_1 = \phi_1^{(0)} \xrightarrow[C]{A} \phi_1^{(1)} \xrightarrow[D]{B} \phi_1^{(2)} \xrightarrow{1_Y} \phi_1^{(3)} \xrightarrow{1_Y} \phi_1^{(4)} \cdots$$

$$\phi_2 = \phi_2^{(0)} \xrightarrow{1_X} \phi_2^{(1)} \xrightarrow[C]{A} \phi_2^{(2)} \xrightarrow[D]{B} \phi_2^{(3)} \xrightarrow{1_Y} \phi_2^{(4)} \cdots$$

$$\phi_3 = \phi_3^{(0)} \xrightarrow{1_X} \phi_3^{(1)} \xrightarrow{1_X} \phi_3^{(2)} \xrightarrow[C]{A} \phi_3^{(3)} \xrightarrow[D]{B} \phi_3^{(4)} \cdots$$

$$\phi_4 = \phi_4^{(0)} \xrightarrow{1_X} \phi_4^{(1)} \xrightarrow{1_X} \phi_4^{(2)} \xrightarrow{1_X} \phi_4^{(3)} \xrightarrow[C]{A} \phi_4^{(4)} \cdots$$

By comparing these elements with the formal matrix $\zeta = \begin{pmatrix} \zeta_0 \\ \zeta_1 \\ \zeta_2 \\ \vdots \end{pmatrix}$ from Lemma 5.5, it is immediate that that $\zeta_i = \phi_i^{(j)}$ for all $i < N$ and $j > i$.

By direct calculation, and the Cauchy-Bunyakovski-Schwarz inequality,

$$\|\phi_k^{(k)}\| \leq \|D^k\|.\|\phi_0^{(0)}\| + \sum_{n=0}^{k-1} \|D^n\|.\|C\|.\|\phi_{k-n}^{(0)}\|$$

However, by assumption $D : U \to U$ is a strict contraction map, so $\|D\| < 1$. Also, $\phi \in X^{\oplus \omega}$ and so $\sum_{i=0}^{\infty} \|\phi_i^{(0)}\|^2 < \infty$. Therefore, we deduce that $\sum_{k=0}^{\infty} \|\phi_k^{(k)}\|^2 < \infty$, and hence the 'diagonal element'

$$\Delta_\phi = \begin{pmatrix} \phi_0^{(0)} \\ \phi_1^{(1)} \\ \phi_2^{(2)} \\ \vdots \end{pmatrix}$$

is a member of $U^{\oplus \omega}$. Finally, observe that

$$\zeta = \begin{pmatrix} B & 0 & 0 & \cdots \\ 0 & B & 0 & \cdots \\ 0 & 0 & B & \cdots \\ \vdots & \vdots & \vdots & \ddots \end{pmatrix} \begin{pmatrix} \phi_0^{(0)} \\ \phi_1^{(1)} \\ \phi_2^{(2)} \\ \vdots \end{pmatrix} + \begin{pmatrix} A & 0 & 0 & \cdots \\ 0 & A & 0 & \cdots \\ 0 & 0 & A & \cdots \\ \vdots & \vdots & \vdots & \ddots \end{pmatrix} \begin{pmatrix} \phi_0^{(0)} \\ \phi_1^{(0)} \\ \phi_2^{(0)} \\ \vdots \end{pmatrix}$$

and hence $\zeta \in Y^{\oplus \omega}$, as required.

To show that the condition $\|D\| < 1$ is not a necessary condition, consider the simplest possible counterexample – the identity matrix $\begin{pmatrix} 1_X & 0 \\ 0 & 1_U \end{pmatrix}$. It is immediate that

$$\dagger^U(L) = \begin{pmatrix} 0 & 1_X & 0 & 0 & 0 & \cdots \\ 0 & 0 & 1_X & 0 & 0 & \cdots \\ 0 & 0 & 0 & 1_X & 0 & \cdots \\ 0 & 0 & 0 & 0 & 1_X & \cdots \\ \vdots & \vdots & \vdots & \vdots & \vdots & \ddots \end{pmatrix}$$

this is clearly not unitary, but is a partial isometry – the *shift map*.

(2) *In this part of the proof, we use the characterisation of unitary maps as, "Partial isometries, with full initial and final subspaces". This is immediate from the definition of partial isometries* [**Hal58**].

We know from 1/ above that $\dagger^U(L)$ exists, for all unitary $L = \begin{pmatrix} A & B \\ C & D \end{pmatrix}$ satisfying $\|D\| < 1$. We now need to show that:

(a) $\dagger^U(L)$ is a partial isometry,
(b) The initial and final subspaces of $\dagger^U(L)$ are the whole of $U \oplus X^{\oplus \omega}$ and $Y^{\oplus \omega}$ respectively.

These results may be seen as follows :

(a) We first define $Term_N : Y^{\oplus(N+1)} \oplus U \oplus X^{\oplus \omega} \to Y^{\oplus \omega}$ for all $N \in \mathbb{N}$, by

$$Term_N \begin{pmatrix} y_0 \\ \vdots \\ y_{N-1} \\ u \\ x_1 \\ \vdots \end{pmatrix} = \begin{pmatrix} y_0 \\ \vdots \\ y_N \\ 0_Y \\ 0_Y \\ \vdots \end{pmatrix}$$

Clearly, $Term_N$ is a linear map, and is a partial isometry, with initial subspace $Y^{\oplus(N+1)} \subseteq Y^{\oplus(N+1)} \oplus U \oplus X^{\oplus \omega}$. Hence, as $G_N : U \oplus X^{\oplus \omega} \to Y^{\oplus(N+1)} \oplus U \oplus X^{\oplus \omega}$ is unitary, the composite $Term_N G_N : U \oplus X^{\oplus \omega} \to Y^{\oplus \omega}$ is a partial isometry.

Now consider arbitrary fixed $\phi \in U \oplus X^{\oplus \omega}$. From part 1/ above, for all $\epsilon > 0$, there exists $M \in \mathbb{N}$ such that

$$\|Term_M(G_M(\phi)) - \dagger^U(L)(\phi)\| < \epsilon$$

By completeness, $\lim_{N \to \infty} Term_N(G_N(\phi)) = \dagger^U(L)(\phi)$, and so in the space $\mathbf{Hilb}(U \oplus X^{\oplus \omega}, Y^{\oplus \omega})$, the series of partial isometries $\{Term_N G_N\}_{N=0}^{\infty}$ converges to $\dagger^U(L)$. By [**AnCo04, AnCo05**], the set of partial isometries between spaces H_1, H_2 forms a smooth closed submanifold of the space $\mathbf{Hilb}(H_1, H_2)$ Therefore, we deduce that the limit $\dagger^U(L)$ is a partial isometry.

(b) *We prove that the inital subspace is full by contradiction.*
Assume there exists some $u \in U$ such that $B(u) = 0$. Then
$$\begin{pmatrix} A & B \\ C & D \end{pmatrix} \begin{pmatrix} 0 \\ u \end{pmatrix} = \begin{pmatrix} 0 \\ D(u) \end{pmatrix}$$
However, $\left\| \begin{pmatrix} 0 \\ u \end{pmatrix} \right\| = \|u\|$, and by the assumption that D is a strong contraction, $\left\| \begin{pmatrix} 0 \\ D(u) \end{pmatrix} \right\| = \|D(u)\| < \|u\|$. This is a contradiction of the unitarity of L, so we deduce that $B(u) \neq 0$, for all $u \in U$.

Now let $\chi \in U \oplus X^{\oplus \omega}$ be in the complement of the initial subspace of $\dagger^U(L)$, so $\dagger^U(L)(\chi) = 0$. As $\lim_{n \to \infty} Term_n(G_n(\phi)) = \dagger^U(L)(\phi)$, we deduce that $\{\chi^{(n)} = Term_n(G_n(\chi))\}_{n=0}^{\infty}$ is a series of elements of $Y^{\oplus \omega}$ that converges to 0. Writing these explicitly as
$$\chi^{(n)} = \begin{pmatrix} \chi_0^{(n)} \\ \chi_1^{(n)} \\ \chi_2^{(n)} \\ \vdots \end{pmatrix}$$
We observe from part 1. that $\chi_{n+k}^{(n)} = \chi_{n+2}^{(n)}$ for all $k \geq 2$. Hence $\chi = 0$ implies that $\chi_{n+2}^{(n)} = 0$, for all $n \in \mathbb{N}$. However, by close inspection of Figure 6, this is only possible when $B(u) = 0$, for some $u \in U$, contradicting the preliminary result above.

We now demonstrate that the final subspace is full
Consider arbitrary $\zeta \in Y^{\oplus \omega}$, written as
$$\zeta = \begin{pmatrix} \zeta_0 \\ \zeta_1 \\ \zeta_2 \\ \vdots \end{pmatrix}$$
As $\zeta \in Y^{\oplus \omega}$, for all $\epsilon > 0$, there exists some $N \in \mathbb{N}$ such that $\sum_{i=N}^{\infty} \|\zeta_i\|^2 < \epsilon$. Using the adjoint of the partial isometry $Term_M$ above, it is immediate that
$$Term_M^*(\zeta) = \begin{pmatrix} \zeta_0 \\ \vdots \\ \zeta_M \\ 0_U \\ 0_X \\ \vdots \end{pmatrix}$$
and, for all $M > N$, $\|\zeta\|^2 - \|Term_M^*(\zeta)\|^2 < \epsilon$. We now define $\lambda^{(M)} \in U \oplus X^{\oplus \omega}$ by $\lambda^{(M)} = G_M^{-1}(Term_M^*(\zeta))$, where the unitary map G_M^{-1} is given by $F_M^{-1} F_{M-1}^{-1} \ldots F_0^{-1}$, for F_i as defined in part 1. Since G_M^{-1} is unitary, $\|\lambda^{(M)}\| = \|Term_M^*(\zeta)\|$. By taking sufficiently large $M > N \in \mathbb{N}$, it follows that $\dagger^U(L)(\lambda^{(M)}) \to \zeta$ as $M \to \infty$, and

as ζ was chosen arbitrarily, the terminal subspace of $\dagger^U(L)$ is exactly $Y^{\oplus\omega}$. This then completes our proof of unitarity. □

6. Setting initial conditions, and compositionality

The intention of this paper is to use the twisted dagger to motivate general categorical constructions — thus, we need an appropriate setting in which such operations give rise to arrows in a category.

As a motivating example, we consider the special case where the apparatus of Figure 4 satisfies the further initial condition, that that the 'internal state' of the optical loop is empty: at the start of the experiment, the probability that an observation of the internal state detects a photon is 0. This will allow us to treat the apparatus of Figure 4 (via the associated 'twisted dagger') as defining a map from the **input space** $l_2(\{in_j\}_{j=0}^\infty)$ to the **output space** $l_2(\{out_j\}_{j=0}^\infty)$. We then generalise this to arbitrary twisted dagger operations.

6.1. Initial conditions, and inclusion maps.
In the apparatus of Figure 4, we wish to impose the initial condition that the 'internal state' of the optical loop is empty. i.e. at the start of the experiment, the probability that an observation on the internal state detects a photon is 0.

Given a state representing an 'input stream' (i.e. a single photon in a superposition of input modes),
$$|in\rangle = \alpha_0 |in_0\rangle + \alpha_1 |in_1\rangle + \alpha_2 |in_2\rangle + \ldots \in l_2(\{in_j\}_{j=0}^\infty)$$
the state in the larger space S_0, describing both the input stream and (empty) 'internal states' is $\iota(|in\rangle) = \alpha_0 |in_0\rangle + \alpha_1 |in_1\rangle + \alpha_2 |in_2\rangle + \ldots \in S_0$, where
$$\iota : l_2(\{in_j\}_{j=0}^\infty) \to S_0 = l_2(\{in_j\}_{j=0}^\infty) \oplus l_2(\{(Current,H),(Current,V)\})$$
is simply the canonical inclusion map associated with the direct sum. This trivially satisfies the measurement condition of Section 6, since
$$\langle Current, H| \cdot \iota \cdot |in\rangle = 0 = \langle Current, V| \cdot \iota \cdot |in\rangle$$

Thus, the input-output map associated with the apparatus of Figure 4, along with this initial condition, is the composite
$$\dagger^U \begin{pmatrix} a & b \\ c & d \end{pmatrix} \circ \iota : l_2(\{in_j\}_{j=0}^\infty) \to l_2(\{out_k\}_{k=0}^\infty)$$

We now generalise this intuition to the general case, for arbitrary finite-dimensional Hilbert spaces, as follows:

DEFINITION 6.1. Let $L = \begin{pmatrix} A & B \\ C & D \end{pmatrix} \in \mathbf{Hilb_{FD}}(X \oplus U, Y \oplus U)$ be a linear map where $\dagger^U(L)$ is defined (i.e. is an arrow of $\mathbf{Hilb}(U \oplus X^{\oplus\omega}, Y^{\oplus\omega})$). We define the **input-output behaviour** of $\dagger^U(L)$ to be the map
$$\Delta^U(L) = \dagger^U(L) \cdot \iota \in \mathbf{Hilb}(X^{\oplus\omega}, Y^{\oplus\omega})$$
where $\iota : X^{\oplus\omega} \to U \oplus X^{\oplus\omega}$ is the canonical inclusion associated with the direct sum.

The following facts about such input-output behaviours are straightforward, but will be essential in the following sections:

LEMMA 6.2. *Let $L = \begin{pmatrix} A & B \\ C & D \end{pmatrix} \in \mathbf{Hilb_{FD}}(X \oplus U, Y \oplus U)$ be a linear map such that $\dagger^U(L) \in \mathbf{Hilb}(U \oplus X^{\oplus \omega}, Y^{\oplus \omega})$, and let $\Delta^U(L) \in \mathbf{Hilb}(X^{\oplus \omega}, Y^{\oplus \omega})$ be as defined above. Then*

(1) *$\Delta^U(L)$ is a continuous (i.e. bounded) linear map.*
(2) *$\Delta^U(L)$ has a matrix representation of the form $\Delta^U(L) = [M_{ij}]$, where*

$$M_{ij} = \begin{cases} 0_{XY} & j - i < 0 \\ p_{j-i} & j - i \geq 0 \end{cases}$$

for some family of linear maps $\{p_k \in \mathbf{Hilb_{FD}}(X, Y)\}_{k=0}^{\infty}$.

PROOF.

(1) This is immediate: $\Delta^U(L) = \dagger^U(L) \cdot \iota$ is the composite of two arrows in the same category, and hence is also an arrow in this category.
(2) From the explicit matrix for the twisted dagger given in Definition 5.3, the composite $\Delta^U(L) \cdot \iota : X^{\oplus \omega} \to Y^{\oplus \omega}$ has the following matrix:

$$\Delta^U(L) = \begin{pmatrix} A & 0 & 0 & 0 & 0 & 0 & 0 & \cdots \\ BC & A & 0 & 0 & 0 & 0 & 0 & \cdots \\ BDC & BC & A & 0 & 0 & 0 & 0 & \cdots \\ BD^2C & BDC & BC & A & 0 & 0 & 0 & \cdots \\ BD^3C & BD^2C & BDC & BC & A & 0 & 0 & \cdots \\ BD^4C & BD^3C & BD^2C & BDC & BC & A & 0 & \cdots \\ BD^5C & BD^4C & BD^3C & BD^2C & BDC & BC & A & \cdots \\ \vdots & \vdots & \vdots & \vdots & \vdots & \vdots & \vdots & \ddots \end{pmatrix}$$

Thus our result follows from this explicit matrix description, by taking

$$p_k = \begin{cases} A & k = 0 \\ BD^{k-1}C & k > 0 \end{cases}$$

□

6.2. Input-output behaviours, and compositionality. We now consider how to compose the input-output behaviours, as defined above. We again motivate this by the single-photon linear optics case, where our interpretation must correspond to treating the input stream of one such experiment as the output stream of another, as shown in Figure 7. (Note that, in this figure, we omit the detectors and post-conditioning, for clarity. Rather, we simply assume that the input photon is guaranteed to be horizontally polarised. Of course, from the experimental setup, the output of the first MSI - and hence the input of the second MSI - is also horizontally polarised).

Trivially, the output stream of the upper MSI becomes the input stream of the lower MSI. Thus, if the input-output behaviour of the upper MSI is $P = [P_{ij}]$, and the input-output behavior of the lower MSI is $Q = [Q_{ij}]$, then the input-output behaviour of the entire apparatus is simply given by the matrix product QP.

6.3. A relevant subcategory of Hilb. We now demonstrate that bounded linear maps with matrices satisfying the special form given in part (2) of Lemma 6.2 are the arrows of a subcategory of **Hilb**.

FIGURE 7. Composing MSIs

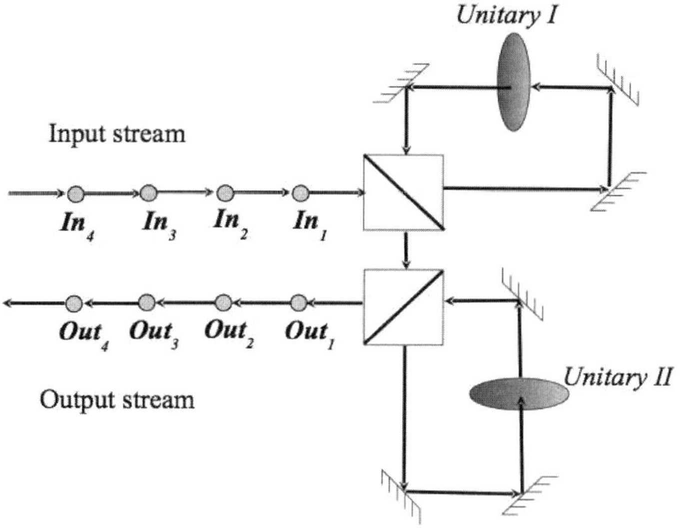

PROPOSITION 6.3. *Arrows in* **Hilb** *with matrix representations of the form*
$$M_{ij} = \begin{cases} 0_{XY} & j-i < 0 \\ p_{j-i} & j-i \geq 0 \end{cases} \quad \text{for some } \{p_k \in \mathbf{Hilb_{FD}}(X,Y)\}_{k=0}^{\infty}$$
are closed under composition, and include identity maps, and hence define a subcategory of **Hilb**.

PROOF. Consider arrows $L \in \mathbf{Hilb}(X^{\oplus\omega}, Y^{\oplus\omega})$ and $M \in \mathbf{Hilb}(Y^{\oplus\omega}, Z^{\oplus\omega})$ where
$$M_{ij} = \begin{cases} 0_{XY} & j-i < 0 \\ q_{j-i} & j-i \geq 0 \end{cases} \quad \text{for some } \{q_k \in \mathbf{Hilb_{FD}}(X,Y)\}_{k=0}^{\infty}$$
$$L_{ij} = \begin{cases} 0_{XY} & j-i < 0 \\ p_{j-i} & j-i \geq 0 \end{cases} \quad \text{for some } \{p_k \in \mathbf{Hilb_{FD}}(X,Y)\}_{k=0}^{\infty}$$
Then from the standard formula for matrix multiplication,
$$[ML]_{i,k} = \sum_{j=0}^{\infty} [M]_{i,j}[L]_{j,k}$$
their composite $ML \in \mathbf{Hilb}(X^{\oplus\omega}, Z^{\oplus\omega})$ has matrix representation
$$[ML]_{i,k} = \begin{cases} 0_{XY} & k-i < 0 \\ r_{k-i} & k-i \geq 0 \end{cases} \quad \text{where } r_c = \sum_{c=b+a} q_b p_a$$
and hence is of the required form. Finally, observe that the identity matrix is trivially of this form. Thus, bounded linear maps of this form specify a subcategory of **Hilb**. □

We observe the similarity between the composition in Proposition 6.3 above, and the *Cauchy product* of power series. This motivates the general categorical setting below.

7. A categorical setting for twisted daggers

We now introduce a categorical setting for twisted daggers (or rather, their input-output behaviours), based on the above observation that composition may be expressed as a Cauchy product (i.e. the usual composition of single-variable power series).

7.1. A category of formal power series.

DEFINITION 7.1. We define the category $\mathbf{Hilb_{FD}[z]}$ of **formal power series** over $\mathbf{Hilb}$ as follows:
- Objects of $\mathbf{Hilb_{FD}[z]}$ are finite-dimensional complex Hilbert spaces.
- The hom-set $\mathbf{Hilb_{FD}[z]}(H, K)$ is the set of all formal power series in z over $\mathbf{Hilb}(H, K)$. Thus $p \in \mathbf{Hilb_{FD}[z]}(H, K)$ may be written as

$$p = \sum_{n=0}^{\infty} p_n . z^n \quad \text{where} \quad p_n \in \mathbf{Hilb}(H, K) \;\; \forall n \in \mathbb{N}$$

We will equivalently and interchangeably refer to the formal power series

$$p = p_0 + p_1.z + p_2.z^2 + \ldots \in \mathbf{Hilb_{FD}[z]}(\mathbf{H}, \mathbf{K})$$

and the function $p : \mathbb{N} \to \mathbf{Hilb_{FD}}(H, K)$ where $p(k) \in \mathbf{Hilb_{FD}}(H, K)$ is the coefficient of z^k in $p_0 + p_1.z + p_2.z^2 + \ldots$.
- Composition is the usual *Cauchy product* [**Ti83**]. Given $p \in \mathbf{Hilb_{FD}[z]}(H, K)$ and $q \in \mathbf{Hilb_{FD}[z]}(K, L)$, then their composite $r = qp \in \mathbf{Hilb_{FD}[z]}(H, L)$ is the formal power series

$$r = r_0 + r_1.z + r_2.z^2 + r_3.z^3 + \ldots$$

where $r_c = \sum_{c=b+a} q_b p_a \in \mathbf{Hilb}(H, L)$, for all $c \in \mathbb{N}$.

We emphasize that these power series are *formal*, rather than convergent. Indeed, we may identify them with an infinite sequence $(p_n)_{n \in \mathbb{N}}$ of bounded linear operators. We do not assume that the sum $p_0 + p_1\zeta + p_2\zeta^2 + \ldots$ converges for any non-zero $\zeta \in \mathbb{C}$ or even that $\sum_{k=0}^{\infty} \|p_k\|_\nu$ converges, for any particular operator norm $\| \|_\nu$ — although it may be observed that imposing such requirements defines various subcategories of $\mathbf{Hilb_{FD}[z]}$.

PROPOSITION 7.2. $\mathbf{Hilb_{FD}[z]}$, *as defined above, is a category.*

PROOF. First note that, for all $p \in \mathbf{Hilb_{FD}[z]}(H, K)$ and $q \in \mathbf{Hilb_{FD}[z]}(K, L)$, their composite

$$r = r_0 + r_1.z + r_2.z^2 + r_3.z^3 + \ldots \quad \text{where} \quad r_c = \sum_{c=b+a} q_b p_a \in \mathbf{Hilb}(H, L)$$

is defined, since for all $k \in \mathbb{N}$

$$r_k = q_k p_0 + q_{k-1} p_1 + \ldots + q_1 p_{k-1} + q_0 p_k$$

is a finite sum of linear maps. It remains to show that composition is associative, and has identities at each object. However, associativity of composition of formal power series is long-established ([**Ti83**]), and the identity map I_H at an object H is simply the formal power series $1_H + 0_H.z + 0_H.z^2 + 0_H.z^3 + \ldots$. Thus, $\mathbf{Hilb_{FD}[z]}$ is a category. □

LEMMA 7.3. *There exists a canonical inclusion* $\iota : \mathbf{Hilb_{FD}} \to \mathbf{Hilb_{FD}[z]}$.

PROOF. The canonical inclusion is simply given by
- **Objects** $\iota(A) = A \in Ob(\mathbf{Hilb_{FD}[z]})$, for all $A \in Ob(\mathbf{Hilb_{FD}})$.
- **Arrows** Given $f \in \mathbf{Hilb_{FD}}(X,Y)$, then $\iota(f) \in \mathbf{Hilb_{FD}[z]}(X,Y)$ is defined by

$$\iota(f)(n) = \begin{cases} f \in Hilb_{FD}(X,Y) & n = 0 \\ 0_{X,Y} & \text{otherwise.} \end{cases}$$

It is immediate from the definition of composition in $\mathbf{Hilb_{FD}[z]}$ that this is an injective functor. □

We are now able to identify a suitable category in which the input-output behaviour of twisted daggers (as in Section 6.3) lives, and by extension a category suitable for reasoning about the thought-experiment of Section 4.

DEFINITION 7.4. Let us denote the category of Proposition 6.3 by **THilb**, so
- $H^{\oplus \omega} \in Ob(\mathbf{THilb})$, for all $H \in Ob(\mathbf{Hilb_{FD}})$.
- $M \in \mathbf{Hilb}(H^{\oplus \omega}, K^{\oplus \omega})$ is an arrow of **THilb** iff M has a block matrix representation of the form

$$M_{ij} = \begin{cases} 0_{XY} & j - i < 0 \\ p_{j-i} & j - i \geq 0 \end{cases} \quad \text{for some } \{p_k \in \mathbf{Hilb_{FD}}(H,K)\}_{k=0}^{\infty}$$

THEOREM 7.5. *There exists an embedding of* **THilb** *into* $\mathbf{Hilb_{FD}[z]}$.

PROOF. This is immediate by mapping matrices of the form

$$M_{ij} = \begin{cases} 0_{HK} & j - i < 0 \\ p_{j-i} & j - i \geq 0 \end{cases} \quad \text{for some } \{p_k \in \mathbf{Hilb_{FD}}(H,K)\}_{k=0}^{\infty}$$

to power series defined by

$$M = p_0 + p_1.z + p_2.z^2 + \ldots \in \mathbf{Hilb_{FD}[z]}(H,K)$$

□

REMARK 7.6. *The intuition behind categories of power series* Abstractly, an arrow in a category $f \in \mathcal{C}(X,Y)$ may be thought of as a process that transforms data of type X into data of type Y. Categories of formal power series extend this intuition by considering the time associated with such transformations, and associating with each input-output pairing a discrete number of time-steps. Thus, the arrow $f(n) \in \mathcal{C}(X,Y)$ describes the input-output mappings of f that take exactly n timesteps. Given a suitable notion of summability of arrows (as in $\mathbf{Hilb_{FD}}$), the Cauchy product of power series then has a natural interpretation: the input-output mappings of gf that take c timesteps arise as the sum of all processes that take b timesteps in g, and a timesteps in f, where $c = b + a$.

7.2. Additional structure on the category of power series.
Our category of formal power series has a natural notion of summation on its hom-sets, derived from the familiar summation of arrows of $\mathbf{Hilb_{FD}}$, as we now demonstrate:

DEFINITION 7.7. Let $\{f_j \in \mathbf{Hilb_{FD}[z]}(H,K)\}_{j \in J}$ be a countably indexed family of arrows. We say that this family is **summable** when $\sum_{j \in J}(f_j(n))$ exists (in the sense of absolute convergence of countable families of linear maps), for all $n \in \mathbb{N}$.

When the family $\{f_j\}_{j\in J}$ above is summable, its **formal sum**

$$F = \sum_{j\in J} f_j \in \mathbf{Hilb_{FD}[z]}(H,K)$$

is the function $F : \mathbb{N} \to \mathbf{Hilb_{FD}}(H,K)$ given by

$$F(n) = \sum_{j\in J} f_j(n) \in \mathbf{Hilb}_{FD}(H,K)$$

Note that, as this notion of summation is based on absolute convergence at each power of z, we have distributivity of composition over arbitrary sums. Also, it is straightforward from its definition in terms of absolute convergence that this notion of summation satisfies a one-sided version of the *partition-associativity* axiom of [**Ha00, HS04**], which itself arose from the theory of partially-additive semantics in [**MA86**].

The weak partition-associativity property: Let $\{f_i \in \mathbf{Hilb_{FD}[z]}(H,K)\}_{i\in I}$ be a countably indexed summable family, and let $\{I_j\}_{j\in J}$ be a countable partition[6] of I. Then $\{f_i\}_{i\in I_j}$ is summable for every $j \in J$, as is $\{\sum_{i\in I_j} f_i\}_{j\in J}$, and

$$\sum_{i\in I} f_i = \sum_{j\in J}\left(\sum_{i\in I_j} f_i\right)$$

Informally, this may be phrased as: 'sub-families of summable families are themselves summable, and replacing any sub-sum by its sum neither affects summability nor changes the result'. Finally, note that, by contrast with the axioms of [**Ha00, HS04**], the implication in the above property is strictly one-way.

Given this notion of summation, we may give matrix representations of such power series, as follows:

DEFINITION 7.8. Consider some formal power series $p \in \mathbf{Hilb_{FD}[z]}(A,B)$. Further assume that A and B are given as (finite) direct sum decompositions, so

$$A = \bigoplus_{i=1}^{m} A_i \quad \text{and} \quad B = \bigoplus_{j=1}^{n} B_j$$

Since the direct sum is a biproduct on $\mathbf{Hilb_{FD}}$, $p(t)$ may be written as an $(m\times n)$ matrix, where $[p(t)]_{x,y} : A_y \to B_x$, for all $t \in \mathbb{N}$. Using this matrix decomposition for each $p(t) \in \mathbf{Hilb_{FD}}(A,B)$, we define the **matrix** of $p \in \mathbf{Hilb_{FD}[z]}(A,B)$ to be the $n \times m$ matrix of arrows of $\mathbf{Hilb_{FD}[z]}$ defined by

$$[p]_{x,y}(t) \stackrel{def.}{=} [p(t)]_{x,y} \in \mathbf{Hilb_{FD}}(A_y, B_x)$$

Composition of such matrices of power series is defined in the natural way, with composition defined by the Cauchy product of power series (i.e. composition in $\mathbf{Hilb_{FD}[z]}$), and summation is as in Definition 7.7 above.

As an illustrative example, this is simply the familiar interchangeability of formal power series whose coefficients are matrices of linear maps with matrices

---

[6]Following [**MA86**], we also allow countably many I_j to be empty.

whose coefficients are formal power series of linear maps, e.g. the equivalence of the power series of matrices

$$\begin{pmatrix} i & 0 \\ 0 & 1 \end{pmatrix} + \begin{pmatrix} 0 & 1 \\ 1 & 0 \end{pmatrix}.z + \frac{1}{\sqrt{2}}\begin{pmatrix} 1 & 1 \\ 1 & -1 \end{pmatrix}.z^2$$

with the matrix of power series

$$\begin{pmatrix} i + \frac{1}{\sqrt{2}}z^2 & z + \frac{1}{\sqrt{2}}z^2 \\ z + \frac{1}{\sqrt{2}}z^2 & 1 - \frac{1}{\sqrt{2}}z^2 \end{pmatrix}$$

PROPOSITION 7.9. *The interpretation of formal power series as matrices above is compatible with composition: the matrix of the composite is the product of the matrices. Precisely, given*

$$q \in \mathbf{Hilb_{FD}[z]}\left(\bigoplus_{j=1}^{m} B_j, \bigoplus_{k=1}^{n} C_k\right) \quad \text{and} \quad p \in \mathbf{Hilb_{FD}[z]}\left(\bigoplus_{i=1}^{l} A_i, \bigoplus_{j=1}^{m} B_j\right)$$

then $[q][p] = [qp]$.

PROOF. This is a simple-index-chasing argument, that follows from comparing the definition of matrix multiplication of Definition 7.8 with the definition of composition in $\mathbf{Hilb_{FD}[z]}$.

□

7.3. A monoidal tensor on $\mathbf{Hilb_{FD}[z]}$. We use the above notion of matrices in $\mathbf{Hilb_{FD}[z]}$ to provide a monoidal tensor:

DEFINITION 7.10. We define $\oplus : \mathbf{Hilb_{FD}[z]} \times \mathbf{Hilb_{FD}[z]} \to \mathbf{Hilb_{FD}[z]}$ as follows:

- **Objects** Given $A, B \in Ob(\mathbf{Hilb_{FD}[z]})$, then $A \oplus B \in Ob(\mathbf{Hilb_{FD}[z]})$ is simply the direct sum of A and B, as in Definition 2.4.
- **Arrows** Given $p \in \mathbf{Hilb_{FD}[z]}(H, J)$ and $q \in \mathbf{Hilb_{FD}[z]}(L, M)$, then

$$(p \oplus q) : \mathbb{N} \to \mathbf{Hilb}(H \oplus L, J \oplus M)$$

is simply defined by

$$(p \oplus q)(k) = p(k) \oplus q(k) \in \mathbf{Hilb}(H \oplus L, J \oplus M)$$

THEOREM 7.11. *The map* $\oplus : \mathbf{Hilb_{FD}[z]} \times \mathbf{Hilb_{FD}[z]} \to \mathbf{Hilb_{FD}[z]}$ *defined above is a symmetric monoidal tensor.*

PROOF. *For clarity, this proof uses matrix notation for both linear maps and formal power series. This is justified by Proposition 7.8 above.*

- **compositionality** Consider arrows

$$p = \sum_{i=0}^{\infty} p_i z^i \in \mathbf{Hilb_{FD}[z]}(H, J) \quad \text{and} \quad q = \sum_{j=0}^{\infty} q_j z^j \in \mathbf{Hilb_{FD}[z]}(L, M)$$

and similarly

$$r = \sum_{i=0}^{\infty} r_i z^i \in \mathbf{Hilb_{FD}[z]}(J, K) \quad \text{and} \quad s = \sum_{j=0}^{\infty} s_j z^j \in \mathbf{Hilb_{FD}[z]}(M, N)$$

Then
$$(p \oplus q) = \sum_{i=0}^{\infty} \begin{pmatrix} p_i & 0 \\ 0 & q_i \end{pmatrix} z_i \text{ and } (r \oplus s) = \sum_{j=0}^{\infty} \begin{pmatrix} r_j & 0 \\ 0 & s_j \end{pmatrix} z_j$$

By definition of composition in $\mathbf{Hilb_{FD}}[\mathbf{z}]$,
$$((r \oplus s)(p \oplus q))(c) = \sum_{c=b+a} \begin{pmatrix} r(b) & 0 \\ 0 & s(b) \end{pmatrix} \begin{pmatrix} p(a) & 0 \\ 0 & q(a) \end{pmatrix}$$
$$= \sum_{c=b+a} \begin{pmatrix} r(b)p(a) & 0 \\ 0 & s(b)q(a) \end{pmatrix} = \begin{pmatrix} \sum_{c=b+a} r(b)p(a) & 0 \\ 0 & \sum_{c=b+a} s(b)q(a) \end{pmatrix}$$

However, this is simply $(rp \oplus sq)(c)$, as required.

- **Identities** The identity $1_A \in \mathbf{Hilb_{FD}}[\mathbf{z}](A, A)$ is given by
$$1_A(n) = \begin{cases} 1_A \in \mathbf{Hilb}(A, A) & n = 0 \\ 0_{A,A} & \text{otherwise.} \end{cases}$$

It is immediate from the definition that $1_A \oplus 1_B \in \mathbf{Hilb_{FD}}[\mathbf{z}](A \oplus B, A \oplus B)$ satisfies
$$(1_A \oplus 1_B)(n) = \begin{cases} 1_{A \oplus B} \in \mathbf{Hilb}(A \oplus B, A \oplus B) & n = 0 \\ 0_{A \oplus B, A \oplus B} & \text{otherwise.} \end{cases}$$

- **Associativity** Let us denote the canonical associativity isomorphisms for $(\mathbf{Hilb_{FD}}, \oplus)$ by $t_{A,B,C}$. The corresponding associativity isomorphisms for $\mathbf{Hilb_{FD}}[\mathbf{z}]$ are given by $\iota(t_{A,B,C})$, where $\iota : \mathbf{Hilb_{FD}} \to \mathbf{Hilb_{FD}}[\mathbf{z}]$ is given in Lemma 7.3. It is straightforward to verify that $\iota : \mathbf{Hilb_{FD}} \to \mathbf{Hilb_{FD}}[\mathbf{z}]$ is a monoidal functor, and MacLane's pentagon condition then follows by functoriality.

- **Symmetry** Similarly to the associativity isomorphisms, the symmetry isomorphisms for $\mathbf{Hilb_{FD}}[\mathbf{z}]$ are given by $\iota(s_{A,B})$, where $s_{A,B}$ is the family of symmetry isomorphisms for $\mathbf{Hilb_{FD}}$. The commutativity hexagon again follows by functoriality.

- **Units objects and arrows** Finally, it is immediate that the unit object for $(\mathbf{Hilb_{FD}}, \oplus)$ is also the unit object for $(\mathbf{Hilb_{FD}}[\mathbf{z}], \oplus)$.

$\square$

8. Partial traces in symmetric monoidal categories

The notion of categorical trace was introduced by Joyal, Street and Verity in an influential paper [**JSV96**]. The motivation for their work arose in algebraic topology and knot theory, although the authors were aware of applications in Computer Science, where they include such notions as feedback, fixedpoints, etc. The starting point for our investigation of the categorical structures associated with the Twisted Dagger operation of Section 5 was the observation in Remark 5.2 that the columns of this formal matrix bear a close similarity with the summands of the particle-style categorical trace; for some of the history of these ideas, see [**Ab96, AHS02, HS10**]. There are also deep connections of traced monoidal categories with the proof theory of linear logic, and Girard's Geometry of Interaction program, as described in [**Hi98, AHS02, HS04, HS10a**], which motivated the introduction of the axioms of partial traces below.

In categories associated with linear maps on Hilbert spaces it is more natural to consider *partial traces* defined in terms of infinite sums (or, in the case of [**HS10**], the invertibility of certain operators). We follow the definitions of [**HS10**], where a partial particle-style trace on Hilbert spaces with direct sums is exhibited. We also contrast this with an alternative definition of partial trace on Hilbert spaces with tensor products given in [**ABP98**].

Recall, following Joyal, Street, and Verity [**JSV96**], a (parametric) trace in a symmetric monoidal category $(C, \otimes, I, s)$ is a family of maps

$$Tr^U_{X,Y} : C(X \otimes U, Y \otimes U) \to C(X, Y),$$

satisfying various naturality equations. A *partial* (parametric) trace requires instead that each $Tr^U_{X,Y}$ be a partial map (with domain denoted $\mathbb{T}^U_{X,Y}$) satisfying various closure conditions.

The following definitions are taken from [**HS10, HS10a**].

DEFINITION 8.1. Let $(C, \otimes, I, s)$ be a symmetric monoidal category. A *(parametric) trace class* in C is a choice of a family of subsets, for each object U of C, of the form

$$\mathbb{T}^U_{X,Y} \subseteq C(X \otimes U, Y \otimes U) \text{ for all objects } X, Y \text{ of } C$$

together with a family of functions, called a *(parametric) partial trace*, of the form

$$Tr^U_{X,Y} : \mathbb{T}^U_{X,Y} \to C(X, Y)$$

subject to the following axioms. Here the parameters are X and Y and a morphism $f \in \mathbb{T}^U_{X,Y}$, by abuse of terminology, is said to be *trace class*.

(1) **Naturality** in X and Y: For any $f \in \mathbb{T}^U_{X,Y}$ and $g : X' \to X$ and $h : Y \to Y'$,
$$(h \otimes 1_U)f(g \otimes 1_U) \in \mathbb{T}^U_{X',Y'},$$
and $\quad Tr^U_{X',Y'}((h \otimes 1_U)f(g \otimes 1_U)) = h \, Tr^U_{X,Y}(f) \, g.$

(2) **Dinaturality** in U: For any $f : X \otimes U \to Y \otimes U', g : U' \to U$,
$$(1_Y \otimes g)f \in \mathbb{T}^U_{X,Y} \text{ iff } f(1_X \otimes g) \in \mathbb{T}^{U'}_{X,Y},$$
and $\quad Tr^U_{X,Y}((1_Y \otimes g)f) = Tr^{U'}_{X,Y}(f(1_X \otimes g)).$

(3) **Vanishing I**: $\mathbb{T}^I_{X,Y} = C(X \otimes I, Y \otimes I)$, and for $f \in \mathbb{T}^I_{X,Y}$
$$Tr^I_{X,Y}(f) = \rho_Y f \rho_X^{-1}.$$

Here $\rho_A : A \otimes I \to A$ is the right unit isomorphism of the monoidal category.

(4) **Vanishing II**: For any $g : X \otimes U \otimes V \to Y \otimes U \otimes V$, if $g \in \mathbb{T}^V_{X \otimes U, Y \otimes U}$, then
$$g \in \mathbb{T}^{U \otimes V}_{X,Y} \text{ iff } Tr^V_{X \otimes U, Y \otimes U}(g) \in \mathbb{T}^U_{X,Y},$$
and $\quad Tr^{U \otimes V}_{X,Y}(g) = Tr^U_{X,Y}(Tr^V_{X \otimes U, Y \otimes U}(g)).$

(5) **Superposing**: For any $f \in \mathbb{T}^U_{X,Y}$ and $g : W \to Z$,
$$g \otimes f \in \mathbb{T}^U_{W \otimes X, Z \otimes Y},$$
and $\quad Tr^U_{W \otimes X, Z \otimes Y}(g \otimes f) = g \otimes Tr^U_{X,Y}(f).$

(6) **Yanking**: $s_{U,U} \in \mathbb{T}^U_{U,U}$, and $Tr^U_{U,U}(s_{U,U}) = 1_U$.

A symmetric monoidal category $(C, \otimes, I, s)$ with such a trace class is called a *partially traced category*, or a *category with a trace class*. If we let X and Y be I (the unit of the tensor), we get a family of operations $Tr_{I,I}^U : \mathbb{T}_{I,I}^U \to C(I,I)$ defining what we call a *non-parametric* (or *scalar-valued*) trace.

In the case when the domain of Tr is the entire hom-set $C(X \otimes U, Y \otimes U)$, the axioms above reduce to the original notion of traced monoidal category in [**JSV96**]. It is also important to notice the extremely subtle "conditional" nature of Vanishing II in the partial case.

In [**HS10**], there are a number of examples of such partial traces, of both "particle" as well as "wave" style (using the terminology of [**Ab96**, **AHS02**]). One relevant example here, with many variations is the following:

DEFINITION 8.2. consider the symmetric monoidal category $(\mathbf{Hilb_{FD}}, \oplus)$ of finite dimensional complex Hilbert spaces.

We shall say an $f : X \oplus U \to Y \oplus U$ is *trace class* iff $(I - f_{22})$ is invertible, where I is the identity matrix, and I and f_{22} have size $dim(U)$. In that case, we can define
$$Tr_{X,Y}^U(f) = f_{11} + f_{12}(I - f_{22})^{-1} f_{21}$$

9. Weak partial traces

We have observed the subtle 'conditional' nature of Vanishing II in the above axioms for a partial trace. However, it is not uncommon to find examples that satisfy all the axioms except the *existence* conditions of Vanishing II. We first present an axiomatisation of this situation, and then a naturally occurring example closely related to the twisted dagger construction.

DEFINITION 9.1. Let $(C, \otimes, I, s)$ be a symmetric monoidal category. We define a *weak parametric trace class* to be a parametrised family of subsets
$$\mathbb{W}_{X,Y}^U \subseteq C(X \otimes U, Y \otimes U) \text{ for all } U, X, Y \in Ob(\mathcal{C})$$
together with a family of functions, called a *weak (parametric) partial trace*, of the form
$$wTr_{X,Y}^U : \mathbb{W}_{X,Y}^U \to C(X,Y)$$
. These are required to satisfy axioms (1)-(3) and (5)-(6) of Definition 8.1 above, and the following weaker version of Vanishing II:

(3') *Weak vanishing II* Let $g : X \otimes U \otimes V \to Y \otimes U \otimes V$ be an arrow in the intersection of $\mathbb{W}_{X \otimes U, Y \otimes U}^V$ and $\mathbb{W}_{X,Y}^{U \otimes V}$ satisfying
$$Tr_{X \otimes U, Y \otimes U}^V(g) \in \mathbb{W}_{X,Y}^U$$
then
$$wTr_{X,Y}^{U \otimes V}(g) = wTr_{X,Y}^U(wTr_{X \otimes U, Y \otimes U}^V(g))$$

Note that any partial trace is trivially a weak partial trace. Also, when the domain of $wTR_{X,Y}^U$ is the entire homset $\mathcal{C}(X \otimes U, Y \otimes U)$, for all X, Y, U, then wTr is exactly a categorical trace in the sense of [**JSV96**].

We may find weak partial trace classes as subsets of partial trace classes, as the following example in the category of finite-dimensional Hilbert spaces demonstrates:

DEFINITION 9.2. An arrow
$$F = \begin{pmatrix} a & b \\ c & d \end{pmatrix} \in \mathbf{Hilb_{FD}}(X \oplus U, Y \oplus U)$$
is *(strictly) lower-right contractive (LRC)* when $|d| < 1$ (using the operator norm). Let us use the notation:
$$F \in LRC_{X,Y}^U \subseteq \mathbf{Hilb_{FD}}(X \oplus U, Y \oplus U)$$

REMARK 9.3. Note that the LRC condition is exactly that required by Theorem 5.6, to ensure that the twisted dagger of a unitary map is itself unitary. For a physical interpretation, recall the generalised interferometry experiment of Remark 5.2. in this setting, the LRC condition may be interpreted as stating that, when prepared with a particle in state $|\psi\rangle$ that is with probability 1 within the feedback loop of Figure 5, the probability $\|d^T |\psi\rangle\|^2$ of observing the particle within this loop at some later time T tends to zero, as T increases.

We now demonstrate that the LRC arrows define a weak partial trace class.

PROPOSITION 9.4. *In* $\mathbf{Hilb_{FD}}$, *the parametric family of arrows*
$$LRC_{X,Y}^U \subseteq \mathbf{Hilb_{FD}}(X \oplus U, Y \oplus U)$$
defined above, together with the parametric function $wTr_{X,Y}^U : LRC(X \oplus U, Y \oplus U) \to \mathbf{Hilb_{FD}}(X,Y)$ *given by*
$$wTr_{X,Y}^U \begin{pmatrix} a & b \\ c & d \end{pmatrix} = a + \sum_{j=0}^{\infty} b d^j c$$
specifies a weak partial trace.

PROOF. First recall that, in any Banach algebra B, for any element $x \in B$ satisfying $\lim_{n \to \infty} x^n = 0$, the element $(1 - x)$ is invertible, with inverse given by $(1-x)^{-1} = \sum_{j=0}^{\infty} x^j$. Thus, in *finite-dimensional* Hilbert spaces, for any strictly contractive map $|d| < 1$, the element $I - d$ is invertible, and $(I - d)^{-1} = \sum_{j=0}^{\infty} d^j$.

From this, we deduce that $LRC_{X,Y}^U \subset \mathbb{T}_{X,Y}^U$, for all $X, Y, U \in Ob(\mathbf{Hilb_{FD}})$ (this inclusion is strict, since $(I_U - \alpha I_U)$ is invertible, for arbitrary $\alpha > 1 \in \mathbb{C}$), and for all $f \in LRC_{X,Y}^U$,
$$Tr_{X,Y}^U(f) = wTr_{X,Y}^U(f) = a + b(I_U - d)^{-1}c = a + b\left(\sum_{j=0}^{\infty} d^j\right)c = a + \sum_{j=0}^{\infty} bd^j c$$

It is then almost immediate that axioms (1)-(3) and (5)-(6) for a partial categorical trace are satisfied, and the fact that wTR and Tr coincide when both are defined is enough to establish the equality required for the weak Vanishing II axiom of Definition 9.1.

To see that LRC, wTr does *not* define a partial trace in the sense of Definition 8.1, consider the 3×3 complex matrix
$$M = \begin{pmatrix} -1 & 1 & 0 \\ 1 & -2 & 1 \\ 0 & 1 & \frac{1}{2} \end{pmatrix} : X \oplus U \oplus V \to Y \oplus U \oplus V$$
where $X \cong Y \cong U \cong V \cong \mathbb{C}$.

Then $M \in LRC^V_{X \oplus U, Y \oplus U}$, and

$$wTr^V_{X \oplus U, Y \oplus U}(M) = \begin{pmatrix} -1 & 1 \\ 1 & 0 \end{pmatrix} \in LRC^U_{X,Y}$$

Thus the hypothesis for the ($\Leftarrow$) implication of Vanishing II is satisfied. However,

$$\begin{pmatrix} -2 & 1 \\ 1 & \frac{1}{2} \end{pmatrix} : U \oplus V \to U \oplus V$$

is clearly not strictly contractive, and thus M is *not* a member of $LRC^{U \oplus V}_{X,Y}$.
(The above counterexample was motivated by a similar calculation taken from [**MSS11**]). $\square$

9.1. A weak partial trace ($\mathbf{Hilb_{FD}[z]}, \oplus$). We now use the result of Proposition 9.4 to give a weak partial trace on $\mathbf{Hilb_{FD}[z]}$ satisfying the axioms of Definition 9.1 above.

DEFINITION 9.5. For each $X, Y, U \in Ob(\mathbf{Hilb_{FD}[z]})$, we define

$$\mathbb{W}^U_{X,Y} \subseteq \mathbf{Hilb_{FD}[z]}(X \oplus U, Y \oplus U)$$

as follows: An arrow $f = \begin{pmatrix} a & b \\ c & d \end{pmatrix} \in \mathbf{Hilb_{FD}[z]}(X \oplus U, Y \oplus U)$ is in $\mathbb{W}^U_{X,Y}$ when the following condition is satisfied:

- For all norm-1 vectors $|\psi\rangle \in U$,

$$\sum_{j=0}^{\infty} \|d^j |\psi\rangle\| < 1$$

Note that the above condition implies both that d_j is strictly contractive, for all $j \in \mathbb{N}$, and that $\sum_{j=0}^{\infty} d^j$ exists.

Given $f = \begin{pmatrix} a & b \\ c & d \end{pmatrix} \in \mathbb{W}^U_{X,Y}$, we define

$$wTr^U_{X,Y}(f) = a + \sum_{j=0}^{\infty} bd^j c$$

(The existence of this sum is a straightforward corollary of Proposition 9.4 above).

REMARK 9.6. Note that the condition of Definition 9.5 above implies, but it is not implied by, the condition we might assume from a physical motivation – that $\sum_{j=0}^{\infty} \|d^j |\psi\rangle\|^2 \leq 1$ for all norm-1 vectors $|\psi\rangle$. Thus, in using this as the definition of our weak trace class, we are not simply requiring that arrows be amenable to a physical interpretation.

THEOREM 9.7. *Definition 9.5 above specifies a weak partial trace on* $\mathbf{Hilb_{FD}[z]}$, *as defined in Definition 9.1.*

PROOF. By direct calculation, it is relatively straightforward to verify axioms (1)-(3) and (5)-(6) of Definition 8.1, with reference to Propositions 7.9 and 9.4. As before, the only non-trivial point is the weak Vanishing II axiom of Definition 9.1.

Consider some $P \in \mathbf{Hilb}[\mathbf{z}](X \oplus U \oplus V, Y \oplus U \oplus V)$, given explicitly in matrix form as

$$P = \begin{pmatrix} a & b & c \\ d & e & f \\ g & h & k \end{pmatrix} : X \oplus U \oplus V \to Y \oplus U \oplus V$$

and assume further that P is a member of both $\mathbb{W}^V_{X \oplus U, Y \oplus U}$ and $\mathbb{W}^{U \oplus V}_{X,Y}$. Thus, both $\sum_{j=0}^{\infty} k^j$ and $\sum_{j=0}^{\infty} \begin{pmatrix} e & f \\ h & k \end{pmatrix}$ exist.

We may give an explicit formula for $wTr^V(P)$; however, it is more instructive to give P itself as the following digraph:

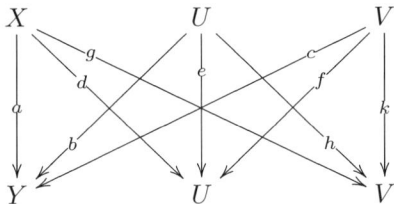

and observe that the matrix of $wTr(P)$ may be found by summing over all paths in the following diagram.

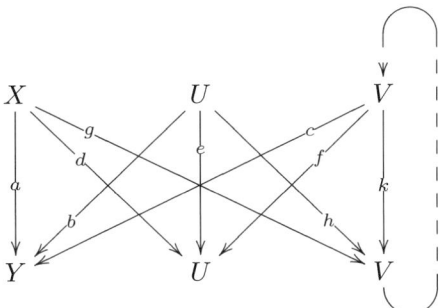

giving $wTr^V_{X \oplus U, Y \oplus U}(P) = \begin{pmatrix} a + \sum_{j=0}^{\infty} ck^j g & b + \sum_{j=0}^{\infty} ck^j h \\ p + \sum_{j=0}^{\infty} fk^j g & e + \sum_{j=0}^{\infty} fk^j h \end{pmatrix}$. Using similar graphical notation, we draw this as

$\begin{array}{c} X \quad\quad U \\ \downarrow \,\searrow r \, \nearrow \downarrow \\ {}_p \; \nearrow \;\; q \; \searrow \;{}_s \\ Y \quad\quad U \end{array}$ where $\begin{cases} p = a + \sum_{j=0}^{\infty} ck^j g & q = b + \sum_{j=0}^{\infty} hk^j c \\ r = d + \sum_{j=0}^{\infty} gk^j f & s = c + \sum_{j=0}^{\infty} hk^j f \end{cases}$

We will now make the assumption (for the *weak* version of vanishing II) that $wTr^V_{X \oplus U, Y \oplus U}(P) \in \mathbb{W}^U_{X,Y}$, and give the following diagrammatic representation for

$wTr^U_{X,Y}\left(wTr^V_{X\oplus U, Y\oplus U}(P)\right)$

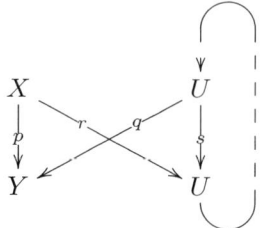 where p, q, r, s are as above.

It is straightforward, although unenlightening, to derive an explicit formula for $wTr^U_{X,Y}\left(wTr^V_{X\oplus U, Y\oplus U}(P)\right)$ from this diagram.

We now use the same formalism to calculate $Tr^{U\oplus V}_{X,Y}(P)$. Let us draw $P: X\oplus (U\oplus V) \to Y\oplus (U\oplus V)$ (together with the appropriate feedback loop) as

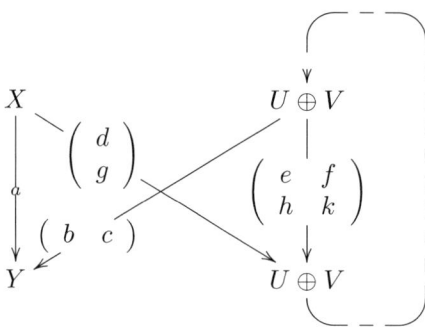

Let first us replace all matrix-labelled arrows by the appropriate digraphs, giving

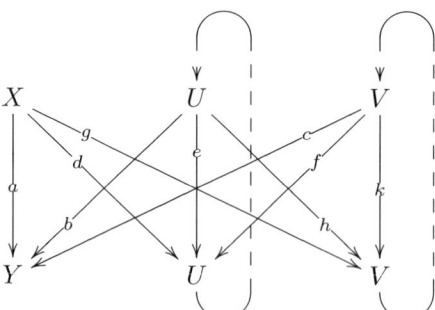

We may then sum over all paths from X to Y, giving an explicit formula for $Tr^{U\oplus V}_{X,Y}(P)$. It may be verified – either by diagrammatic manipulations, or converting these into explicit calculations, that under the *existence* conditions imposed by the weak version of Vanishing II, that $Tr^U_{X,Y}(Tr^V_{X\oplus U, Y\oplus U}(P)) = Tr^{U\oplus V}_{X,Y}(P)$. □

Note that this weak trace of formal power series has a useful interpretation as the weak trace of Proposition 9.4, where components are 'split up' according to the number of iterative cycles they require.

9.2. Relating the trace and the thought-experiment.

We are now able to relate the weak categorical trace on $\mathbf{Hilb_{FD}}[\mathbf{z}]$ to the thought-experiment of Section 4 and its generalisation given in Section 5. Consider a unitary map

$$L = \begin{pmatrix} a & b \\ c & d \end{pmatrix} : X \oplus U \to Y \oplus U$$

where d is strictly contractive. Using the interpretation of categories of formal power series given in Remark 7.6, we assume that a single application of this unitary map takes a single timestep. Thus, we consider the formal power series

$$p = 0 + L.z + 0.z^2 + 0.z^3 + \ldots$$

We observe that $p \in \mathbf{Hilb_{FD}}[\mathbf{z}](X \oplus U, Y \oplus U) \in \mathbb{W}_{X,Y}^U$, and by definition

$$wTr_{X,Y}^U(p) = 0.z^0 + a.z^1 + bc.z^2 + bdc.z^3 + bd^2c.z^4 + \ldots$$

Now compare this to the formal matrices of Lemma 6.2, and the embedding of Theorem 7.5. We observe that the input-output behaviour of the general form of our thought-experiment is given by the weak trace on $\mathbf{Hilb_{FD}}[\mathbf{z}]$. However, the trace has an additional leading zero (the component of z^0). This fits in well with the intuition of formal power series as 'timing' iteration. A non-zero coefficient of z^0 interprets as a computation (or physical process) that takes no time at all. Due to the labelling conventions of Section 4, this is not immediately apparent from a straightforward analysis, but drops naturally out of the categorical description.

10. Conclusions and applications

It is perhaps unsurprising that the theory of conditional iteration, whether in the quantum or the classical setting, should be related to the theory of categorical traces. What is more unexpected is that not only is this apparent from minor modifications to a 100 year old experiment, but that this gives a previously unobserved decomposition of the usual particle-style categorical trace. Moreover, we still find it slightly mysterious that there should be an apparent connection with the Elgot dagger.

In terms of applications, quantum circuits that implement the operations of Lemma 5.1, and thus provide finitary approximations to the twisted dagger in the quantum circuit model, are used heavily in [**Hi09**]. Given the original motivation for the twisted dagger in terms of a simple linear optics experiment, and the encoding of the standard circuit model into this optical framework presented in Appendix A, it is not unreasonable to suppose that the (rather complicated) circuits of [**Hi09**] have a simple, almost trivial, realisation in terms of similar experiments.

From a more mathematical point of view, a great deal of theory remains to be developed. The construction of a category of formal power series from $\mathbf{Hilb_{FD}}$ is clearly a special case of a general categorical construction. This is developed further, in a much more general framework, in [**Hi10**]. However, although the analogous construction in [**Hi10**] is shown to be functorial, its interaction with monoidal tensors is not considered. Many other questions remain open. The general theory of partially traced categories considered above (from [**HS10**]) is developed in detail in the thesis of Octavio Malherbe [**Mal10**] with an eye towards models of quantum programming languages. Such partial traces are completely characterised in the paper [**MSS11**].

Finally, this paper has been phrased very concretely in terms of physical experiments. Similarly, the applications proposed above are very concrete constructions involving quantum circuits. Despite this, we should not forget the motivation from the Geometry of Interaction program in linear logic [**HS10, HS10a**], which attempts to model the invariants for the dynamics of normalization (rewriting) of formal proofs.

References

[Ab96] S. Abramsky 1996, Retracing some paths in Process algebra, In *CONCUR 96, Lecture Notes in Computer Science*, pp.1-17, Springer-Verlag.

[ABP98] S. Abramsky, R. Blute, P. Panangaden 1998, Nuclear and Trace Ideals in Tensor *-Categories, *Journal of Pure and Applied Algebra*, Special edition in honour of M. Barr.

[AHS02] S. Abramsky, E. Haghverdi, P. Scott 2002 Geometry of interaction and linear combinatory algebras, *Math. Struct. Comput. Sci. 12 (5)* 625-665

[Ab05] S. Abramsky 2005, Abstract Scalars, Loops, and Free Traced and Strongly Compact Closed Categories, in *Proceedings of CALCO 2005, Springer LNCS 3629*, 1–31

[AbCo05] S. Abramsky, B. Coecke 2005, A categorical semantics of quantum protocols. In *Proc. 19th Annual IEEE Symp. on Logic in Computer Science (LICS 2004)*, IEEE Computer Soc. Press, 415-425.

[AnCo04] E. Andruchow, G. Corach 2004, Differential geometry of partial isometries and partial unitaries, *Illinois J. Math. 48(1)*, 97-120

[AnCo05] E. Andruchow, G. Corach 2005, Metrics in the set of partial isometries with finite rank, *Rend. Mat. Acc Lincei s 9, v.16* (31-44)

[Ba92] M. Barr 1992, Algebraically Compact Functors *Journal of Pure and Applied Algebra 82* Elsevier (1992) 211-231

[Be05] A. Beige 2005 Quantum Optics, CONQUEST research network Lectures, available as http://www.quniverse.sk/conquest/lectures.php

[Br02] K. Brown, Reflections on Relativity, *Mathpages.com online texts* (2002). Available as: http://www.mathpages.com/rr/refrel2.htm

[CAK05] N. Cerf, C. Adami, P. Kwiat 1998, Optical Simulation of quantum logic, *Physical Review A 57(3)*

[FLS65] R. Feynman, R. Leighton, M. Sands 1965, *The Feynman lectures on Physics, Vol. 3*, Addison-Wesley, Reading, Mass.

[GK05] C. Gerry, P. Knight 2005 *Introductory quantum optics*, Cambridge University Press.

[Gr02] R. Griffiths 2002 *Consistent Quantum Theory*, Cambridge University Press

[Ha00] E. Haghverdi 2000, A categorical approach to linear logic, geometry of proofs and full completeness, *PhD Thesis, Univ. Ottawa*

[HS04] Haghverdi, E. and Scott, P.J. A categorical model for the Geometry of Interaction, *Theoretical Computer Science* Volume 350, Issues 2-3 , Feb 2006, pp. 252-274.

[HS10] E. Haghverdi, P. Scott 2010 Towards a Typed Geometry of Interaction, *Mathematical Structures in Comp. Science*, Vol. 20, Camb. U. Press, pp. 473-521. (preliminary version in: CSL2005 (Computer Science Logic), Luke Ong, Ed. SLNCS 3634, pp. 216-231.)

[HS10a] Haghverdi, E. and Scott, P.J. 2010. Geometry of Interaction and the Dynamics of Proof Reduction: a tutorial, in: *New Structures for Physics*, B. Coecke (ed), Lecture Notes in Physics, Vol. 813, Springer-Verlag, pp. 339–397.

[Hal58] P.R. Halmos 1958, *Introduction to Hilbert Space and the Theory of Spectral Multiplicity*, 2nd Ed. Chelsea Publishing, New York

[Har83] D. G. Hartig 1983, The Reisz Representation Theorem Revisited, *American Mathematical Monthly (90)4* 277-280

[Hi98] P. Hines 1998, The algebra of self-similarity and its applications, *PhD Thesis , University of Wales Bangor*

[Hi09] P. Hines 2010, Quantum circuit oracles for abstract machine computations *Theoretical Computer Science 411* (2010) 1501-1520

[Hi10] P. Hines, A categorical analogue of monoid semirings *(under revision)* Available as http://www.peterhines.net/downloads/papers/Cauchy.pdf

[Je97] A. Jeffrey (1997), Premonoidal categories and a graphical view of programs. *Electr. Notes Theor. Comput. Sci. 10*

[JSV96] A. Joyal, R. Street, D. Verity 1996 Traced Monoidal categories, *Math. Proc. Camb. Phil. Soc.* 425-446

[McL98] S. MacLane 1998, *Categories for the working mathematician*, Graduate texts in mathematics 5, 2nd ed. Springer-Verlag, New York

[Mal10] O. Malherbe 2010, *Phd thesis*, Dept. of Mathematics, U. Ottawa. http://www.fing.edu.uy/~malherbe/octavio/personal.html

[MSS11] O. Malherbe, P. Scott, P. Selinger 2011, Partially Traced Categories, to appear, *J. Pure and Applied Algebra.*

[MA86] E. Manes, M. Arbib 1986, *Algebraic Approaches to Program Sematics* Springer-Verlag

[NC00] M. Nielsen, I. Chuang 2000 *Quantum Computation and Quantum Information*, Cambridge University Press

[Pen04] R. Penrose 2004 *The Road To Reality: A complete guide to the laws of the universe* Random House Ltd., London.

[PSM96] T. B. Pittman, D. V. Strekalov, A. Migdall, M. H. Rubin, A. V. Sergienko, Y. H. Shih 1996 Can Two-Photon Interference be Considered the Interference of Two Photons?, *Physical Review Letters 77 (10)*, 1917-1920

[Sa13i] G. Sagnac, *L'éther lumineux dèmontré par éffet du vent relatif d'éther dans un interféromètre en rotation uniforme*, Comptes Rendus 157 (1913) 708-710

[Sa13ii] G. Sagnac *Sur la preuve de la réalité de l'éther lumineux paer l'experience de l'interférographe tournant* Comptes Rendus 157 (1913), S. 1410-1413

[St97] G.E. Stedman *Ring-laser tests of fundamental physics and geophysics* Rep. Prog. Phys. 60 (1997) 615688

[Ti83] E. C.Titchmarsh (1983) *The theory of Functions*, 2nd Edition, Oxford University Press

[Lo93] O. Lodge *Abberation Problems* Philosophical Transactions of the Royal Society (London) 184A 727 (1893)

[JvN38] J. von Neumann *On infinite direct products* Compos. Math. 6: 177 (1938)

Appendix A: linear optics devices as quantum logic gates

This appendix is expository, and details how the optics toolkit presented in Section 2 may be used to implement a universal set of quantum logic gates[7]. For a fuller treatment, and applications (including linear optics circuits for teleportation), we refer to [**CAK05**]).

Recall that — for single-photon experiments — the basic optics devices in Section 2 are modelled by unitary operations on a 4-dimensional Hilbert space. By considering this 4-dimensional space as the tensor product of two 2-dimensional spaces, we may give a treatment in terms of qubits, and the standard quantum computational logic gates.

The quantum information is encoded on:

(1) The choice of channel.
(2) The photon polarisation.

Thus, we consider the first qubit to be encoded on the choice of channel, with channel 1. (resp. channel 2.) corresponding to $|0\rangle$ (resp. $|1\rangle$). Similarly, the second qubit is encoded on the photon polarisation, with horizontal (resp. vertical) polarisation corresponding to $|0\rangle$ (resp. $|1\rangle$).

This then gives a straightforward encoding of 2-qubit states, e.g.

- The pure state $|0\rangle |1\rangle$ corresponds to a vertically polarised photon in channel 1.

---

[7]We emphasise, as noted in Section 1.2, this encoding is *not* efficiently scalable.

- The superposition $\frac{1}{\sqrt{2}}(|0\rangle + |1\rangle)|0\rangle$ corresponds to a horizontally polarised photon in a superposition of channels 1. and 2.
- The entangled state $\frac{1}{\sqrt{2}}(|00\rangle + |11\rangle)$ corresponds to a photon that is in a superposition of horizontal polarisation in channel 1. and vertical polarisation in channel 2. This is the highly important *Bell state*, used in both quantum teleportation and cryptography [**NC00**].

This encoding also gives a neat realisation of the standard quantum logic gates in terms of linear optics devices, as in the following examples:

- **The Hadamard gate**
  The action of the beamsplitter is simply to apply a Hadamard gate $H = \frac{1}{\sqrt{2}}\begin{pmatrix} 1 & 1 \\ 1 & -1 \end{pmatrix}$ to the first qubit (encoded on the choice of channel) and leave the second one (encoded on the polarisation) alone. In the standard quantum circuit formalism, this is drawn as:

  'channel' qubit   —[H]—   'channel' qubit
  'polarisation' qubit————————'polarisation' qubit

- **The controlled-not gate**
  The polarised beamsplitter applies a NOT gate $X = \begin{pmatrix} 0 & 1 \\ 1 & 0 \end{pmatrix}$ to the channel qubit when the polarisation qubit is $|1\rangle$, and leaves the channel qubit unchanged otherwise. This is the controlled-not, or CNOT gate, drawn in the standard circuit formalism as

  'channel' qubit   —⊕—   'channel' qubit
  'polarisation' qubit——•——'polarisation' qubit

- **The phase shift gate**
  By placing a phase plate in the second channel only of such an experiment, a phase shift gate $R_k = \begin{pmatrix} 1 & 0 \\ 0 & e^{2\pi i/2^k} \end{pmatrix}$ may be implemented. Note also that this has no effect on the polarisation. In the standard circuit formalism, this is drawn as:

  'channel' qubit   —[R_k]—   'channel' qubit
  'polarisation' qubit————————'polarisation' qubit

Using this encoding, it is easy to verify that the Bell state $\frac{1}{\sqrt{2}}(|00\rangle + |11\rangle)$ may be produced by introducing a horizontally polarised photon into channel 1. of the apparatus shown in Figure 8.

FIGURE 8. Linear-optics apparatus to produce the Bell state

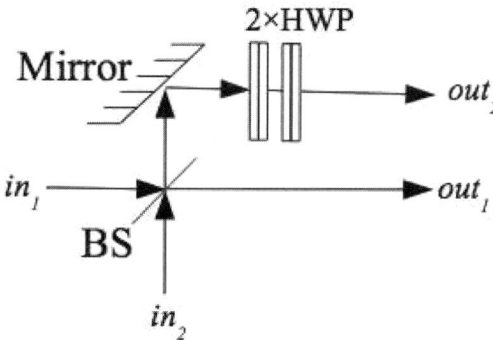

Once the Bell state has been produced, it is a short step to an implementation of quantum teleportation. The theoretical details of how to implement teleportation using linear optics are given in [**CAK05**].

COMPUTER SCIENCE, UNIVERSITY OF YORK
Current address: Department of Computer Science, University of York, York, U.K.
E-mail address: `Peter.Hines@cs.york.ac.uk`

DEPARTMENT OF MATHEMATICS & STATISTICS, UNIVERSITY OF OTTAWA, CANADA
E-mail address: `phil@site.uottawa.ca`

Compact Affine Monoids, Harmonic Analysis and Information Theory

Karl H. Hofmann and Michael Mislove

ABSTRACT. This paper revisits the theory of compact affine semigroups and monoids, an area of intense research activity in earlier phases of compact semigroup theory. We outline the basic structure theory, highlighting results that we believe will be of interest to researchers in quantum and classical information theory. We include an updated complete presentation of one of the early accomplishments of this theory, a proof of the existence of Haar measure on a compact group using semigroup-theoretical arguments, first derived by WENDEL. Finally, a new approach to analyzing the capacity of a classical channel is presented, based on compact affine monoids and domain theory.

Introduction

Recently, interest has been revived in the theory of compact affine monoids because of the role they play as models of channels in classical and quantum communication. Indeed, Martin, *et al.* [26, 25, 24] have illustrated that compact monoids are central to understanding both classical and quantum channels. But compact affine monoids have a rich history dating back to the decades from 1950 to the 1970s, much of which has been overlooked or forgotten. Our purpose here is to recount some of the salient points of that theory, highlighting those results that we believe are most relevant for researchers now investigating classical and quantum information. We also update some of the results, placing them in a categorical framework, and we add some new results that are inspired by problems from classical and quantum information.

We begin with an overview of the structure theory of compact affine monoids, pointing out some of the basic structural results about compact semigroups and monoids in general, and focusing on how those results can be refined in the case of affine semigroups.

One of the most fundamental results in the theory of compact affine monoids is the degeneracy of compact affine groups. In this paper, we present a new proof of this result that eliminates the hypothesis that the group is embedded in a locally convex space. To put this in perspective, we also include two other proofs, one

---

2000 *Mathematics Subject Classification.* Primary 22A15, 22A20, 60B15, 94A40.

The second author gratefully acknowledges the support of the US Office of Naval Research.

relies on a deep result of Borel's on contractible groups, and the other that uses the traditional approach that relies on the Krein-Milman Theorem.

Any compact semigroup, affine or not, has a smallest two-sided ideal, called the *minimal ideal*, which is closed and which is a union of groups. In the case of an affine semigroup, however, each of these groups is also affine, and since they are compact, each must be degenerate. This implies, for example, that an abelian compact affine semigroup has a zero element as its minimal ideal. More generally, we show that the monoid of classical channels of any (finite) dimension is a compact affine monoid whose minimal ideal consists of right zeroes.

We also present two results about the role of probability measures in the theory of compact affine monoids. The first result shows that the monoid of probability measures on a compact group is the free compact locally convex affine monoid over this group. The second result generalizes this to show that the monoid of probability measures over a compact monoid is the free, locally convex, compact affine monoid over the original monoid. These results are folklore, certainly among researchers in compact monoids, but we believe this is the first place where the proofs are provided in detail. In particular, we believe this is the first instance where the crucial role of local convexity *internal to the monoid* is made explicit.

Our interest in classical and quantum channels motivates the focus on probability measures on a compact monoid. Indeed, in Shannon's classical approach, (noisy) channels are conduits taking input data to output data, while subjecting the data to random noise. In the case of a finite set of input "ports" or "values", and an equal number of output "ports" or "values", the random noise generated by the channel can be represented as a square matrix whose entries are the conditional probabilities $p(y|x)$, which should be read as the "probability that output y occurs, given that the input was x". Thus, each row of the channel matrix is a subprobability distribution on the outputs for the input x that heads the row. Assuming the channel is lossless – i.e., each input must generate some output – implies each row is a probability distribution. In short, a classical noisy, lossless channel on finite inputs is a stochastic matrix. It is well-known that the family of stochastic $n \times n$-matrices is a compact convex subset of $[0,1]^{n^2}$ that forms a compact monoid under matrix multiplication. Thus, the $n \times n$-stochastic matrices are a compact affine monoid, for each n. Analyzing the structure of such monoids is another goal of this paper.

One of the most fundamental results in the structure and representation theory of compact groups is the existence and uniqueness of a left invariant normalized measure, the so-called *Haar measure*. One of the early accomplishments of the theory of compact monoids was to provide a proof of the existence of Haar measure using mainly semigroup-theoretical arguments. In the second part of this paper we present a self-contained proof of this fact. The essence of the proof is due to Wendel [**40**], and it relies on an exploration of the compact affine monoid of probability measures over the group and includes an application of the degeneracy of compact affine groups.

The paper is structured as follows. In Part 1, we survey results about compact semigroups and monoids, highlighting how they reveal the structure theory of compact affine semigroups and monoids. This part culminates with a proof that a compact affine group is a point. Actually, we present several proofs of this result. The main one emphasizes the minimal hypotheses that are required, justifying, among other things our choice of definition of a compact affine semigroup, one that is more economical than the existing definitions in the literature. Only when we introduce *convexity* do we refer to an embedding real vector space. Even then, we do not postulate a topology on the ambient vector space, let alone that it be locally convex. Instead we define *intrinsically to the given monoid* which elements of structure we require.

Part 2 of the paper is devoted to an application of compact monoid theory: a self-contained proof of the existence of Haar measure on compact groups in the spirit of Wendel. The treatment is from first principles, touching on most of the background that is needed to develop the proof not just of Haar measure, but indeed, of probability measures on compact semigroups and groups. Our intention is to provide the reader with the details needed to prove the main result, by explicitly exhibiting the ingredients needed to construct the proof and other results closely related to it. This part culminates with results about the structure of one of the principal examples of compact affine monoids: the family of stochastic matrices of fixed size.

Part 3 of the paper is devoted to applying the tools of compact monoid theory to a study of classical channels. We include several examples. We also indicate the role of *domain theory* [1] in this area, following an idea originally pioneered by Martin et al. [26]. Our hope is that the results we obtain—and the methods we use to prove them—will prove useful in understanding classical and quantum channels. The area of quantum channels is very much in flux; while, after Shannon's seminal work, we understand the theory of classical channels very well, our understanding of the area of quantum channels is only now emerging. For example, the question of constitutes a quantum channel is still unsettled, as is the question of what the "right notion" of capacity of such a channel should be.

The style of presentation is briefly characterized as follows: In Part 1, we include full details, in an effort to introduce the reader to the methods used in the compact monoid theory. Our aim is to allow readers to apply both these techniques as well as the enumerated results. In Part 2, we cover a broad range of material, from probability measures on a compact space all the way to convolutions of measures on a compact semigroup. Here we include numerous exercises, accompanying them with detailed outlines of their proof. Finally, in Part 3 we return to presenting all proofs, however citing established results where appropriate and where easily accessible references are available.

Part 1. Compact Monoids and Semigroups

1. Compact Semigroups

To begin, an *algebraic semigroup* (*semigroup* for short) is a non-empty set S together with an associative binary operation

$$(x, y) \mapsto xy : S \times S \to S.$$

The authoritative reference for the structure theory of algebraic semigroups is [6]. If S is a semigroup, then a non-empty subset $I \subseteq S$ is an *ideal* if $IS \cup SI \subseteq I$. An *idempotent* is an element $e \in S$ satisfying $e^2 = e$, and the set of idempotents is denoted $E(S)$. To each $e \in E(S)$, there is associated a largest subgroup of S for which e is the identity element. This subgroup is denoted $H(e)$, and semigroup theoreticians call $H(e)$ *the unique largest subgroup associated with* e.

We begin with a basic result about when a semigroup is actually a group.

LEMMA 1.1. *If S is a semigroup, then the following are equivalent:*
(1) *S is a group.*
(2) *$Ss = S = sS$ for all $s \in S$.*

In particular, a commutative semigroup S is a group if and only if it satisfies $Ss = S$ for all $s \subset S$.

PROOF. That 1) implies 2) is clear. Conversely, suppose 2) holds. If $s \in S$, then $sS = S$ implies $se = s$ for some $e \in S$. If $s' \in S$, then $s' \in S = Ss$ implies $s' = ys$ for some $y \in S$, and so $s'e = yse = ys = s'$. It follows that $s'e = s'$ for all $s' \in S$; in particular, $ee = e$ so e is an idempotent. By 2) we know $eS = S = Se$, which implies e is an identity element for S, and for any $s \in S$, $sS = S = Ss$ implies there are $s', s'' \in S$ with $ss' = e = s''s$. But then, $s' = es' = (s''s)s' = s''(ss') = s''e = s''$, which shows s has an inverse, and so S is a group. □

We now turn to an overview of the "basics" of compact monoid theory. A *compact semigroup* is a compact Hausdorff space S that admits a continuous associative binary operation

$$(x, y) \mapsto xy : S \times S \to S.$$

In passing we note that there exist well-established lines of research that make different assumptions about semigroups endowed with a topology. In these more general approaches, the multiplication is sometimes assumed to be *separately continuous* (i.e., it is continuous as a function of each input separately), while other approaches assume multiplication is continuous in only one variable (e.g., the multiplication is left continuous or right continuous). Most of our discussion will focus on semigroups whose multiplication is jointly continuous.

A good reference for the theory of compact semigroups is [3] (1983, 1986); the earliest monographs are [31] (1964) and [17] (1966), the former being a dissertation that is still good introductory reading, while the latter also includes a number of more advanced topics such as transformation group-theoretical aspects. A good source for compact semigroups with separately continuous multiplication is [34] (1984). Finally in this regard, compact affine semigroups will be a focus of much our discussion (cf. Sections 1.1, 1.3, 7 and 8). These semigroups were first studied in the seminal work of Cohen and Collins [8]; see also [9].

In the theories of compact semigroups and groups, workers often use the language of nets, that is, families $\{x_i\}_{i \in I}$ whose index set I is a directed set more general than the ordered set of natural numbers. A reference that is still in use is [20] dating from 1956; however, for the convenience of the reader, in Subsection 1.4 we provide a brief rundown on the essential features of nets.

Our next result shows that every compact semigroup S contains an idempotent $e \in E(S)$, a fact that holds even in compact semigroups whose multiplication is a continuous function of only one of its inputs. This was published by Ellis in 1958 (cf. [12], Lemma 1, p. 402). While he writes that this fact "is probably well-known", we know of no earlier record. Ellis' own reference is the survey article on compact semigroups by A. D. Wallace [37] cited frequently at the time. But Wallace never considered anything but jointly continuous semigroups. Nevertheless, Ellis' result was republished, for instance thirty years later in [13] and it was reproduced in monographs (e.g., [34]). We record the succinct, elementary but still nontrivial approach of 1958:

LEMMA 1.2 (Ellis' Lemma). *If S is a compact Hausdorff space and a semigroup such that the functions $x \mapsto sx : S \to S$ are continuous for all $s \in S$, then S contains an idempotent.*

PROOF. Since the intersection of a filter basis of compact non-empty subsemigroups is again compact, non-empty and a subsemigroup, Zorn's Lemma yields a minimal compact subsemigroup $K \subseteq S$. Let $e \in K$. Then eK is a subsemigroup contained in K, and eK is compact because $x \mapsto ex$ is continuous; thus $eK = K$ by minimality of K. Then $F \stackrel{\text{def}}{=} \{f \in K : ef = e\}$ is (i) non-empty since $eK = K$, (ii) compact by the continuity of $x \mapsto ex$ again, (iii) a subsemigroup since $ef_1 f_2 = ef_2 = e$ for $f_1, f_2 \in F$, and (iv) contained in K by definition of F. The minimality of K implies $F = K$ and so $eK = \{e\}$, yielding $ee = e$. □

The essay [16] is an informative source for comparing the effects of different degrees of continuity of the multiplication in semigroups endowed with a *Conv* Hausdorff topology. We now turn to semigroups in which the multiplication is jointly continuous, and stay with them for the remainder of this text.

LEMMA 1.3. *A compact semigroup S has a unique smallest semigroup ideal, $\mathcal{M}(S)$, which is a closed, hence compact subset of S. Moreover, if $x \in \mathcal{M}(S)$, then $SxS = \mathcal{M}(S)x\mathcal{M}(S) = \mathcal{M}(S)$.*

PROOF. Let $\mathcal{I} = \{I \subseteq S \mid \emptyset \neq I = \overline{I} \supseteq IS \cup SI\}$ be the family of closed ideals of S. If $I, J \in \mathcal{I}$, then $IJ = \{xy \mid x \in I, y \in J\}$ is again as ideal, and it is closed because S is compact. Hence $\mathcal{I}$ is filtered, since $IJ \subseteq I, J$. Since S is compact, $\mathcal{M}(S) \equiv \cap \mathcal{I} \neq \emptyset$ is closed, and it is an ideal, being an intersection of such. $\mathcal{M}(S)$ clearly is the smallest closed ideal of S.

Next, if $x \in \mathcal{M}(S)$, then $\mathcal{M}(S)x\mathcal{M}(S) \subseteq SxS \subseteq \mathcal{M}(S)$, and $\mathcal{M}(S)x\mathcal{M}(S)$ is a closed ideal, since it is the continuous image of the compact set $\mathcal{M}(S) \times \{x\} \times \mathcal{M}(S)$, so all three are equal.

Finally, returning to $\mathcal{I}$, if J is any ideal and $x \in J$, then $\mathcal{M}(S)x\mathcal{M}(S) \in \mathcal{I}$ and $\mathcal{M}(S)x\mathcal{M}(S) \subseteq J$, so $\mathcal{M}(S) \subseteq J$, which shows $\mathcal{M}(S)$ is the smallest ideal of S. □

PROPOSITION 1.4. *Let S be a compact semigroup, and let $x \in S$. Then $\Gamma(x) \stackrel{\text{def}}{=} \overline{\{x^n \mid n \geq 1\}}$, the smallest compact subsemigroup of S containing x has a minimal ideal $\mathcal{M}(x)$ which is a group whose identity e satisfies $ex = xe = exe \in \mathcal{M}(x) \subseteq H(e)$.*

PROOF. For any $x \in S$, $\Gamma(x)$ is a *Conv* commutative semigroup. So, $\Gamma(x)$ has a minimal ideal $\mathcal{M}(x)$. If $y \in \mathcal{M}(x)$, then $y\mathcal{M}(x) \subseteq \mathcal{M}(x)$ is also an ideal:

$$(y\mathcal{M}(x))\Gamma(x) = y(\mathcal{M}(x)\Gamma(x)) \subseteq y\mathcal{M}(x),$$

and similarly,

$$\Gamma(x)(y\mathcal{M}(x)) = (y\mathcal{M}(x))\Gamma(x) \subseteq y\mathcal{M}(x),$$

the first equality following from commutativity. Since $\mathcal{M}(x)$ is minimal, $y\mathcal{M}(x) = \mathcal{M}(x)$, and commutativity implies $\mathcal{M}(x)y = \mathcal{M}(x)$ as well. The fact that $y \in \mathcal{M}(x)$ is arbitrary implies $\mathcal{M}(x)$ is a group by Lemma 1.1. Now if $e = e^2 \in \mathcal{M}(x)$ is the identity of $\mathcal{M}(x)$, then $\mathcal{M}(x) \subseteq H(e)$ since $\mathcal{M}(x)$ is a group, and $xe \in \mathcal{M}(x)$, so $xe = ex = exe$, by commutativity. □

NOTATION. If S is a compact monoid and $x \in S$, then we write $e(x)$ for the identity of the minimal ideal of $\Gamma(x)$.

THEOREM 1.5. *Let S be a compact semigroup. Then*
  (1) *Each maximal subgroup $H(e)$ of S is closed, and hence compact. The restriction of the semigroup operation of S to $H(e)$ makes $H(e)$ into a compact group (in particular, inversion also is continuous).*
  (2) *The set of idempotents, $E(S)$, is closed, and hence compact.*
  (3) *If $e \in \mathcal{M}(S) \cap E(S)$, then $H(e) = eSe$.*

PROOF. For (1), we claim that $H(e) = \{x \in eSe \mid (\exists y \in eSe)\ xy = yx = e\}$. Let's call the right side K, and we first show K is a compact subgroup of S. If $x, x' \in K$, with associated $y, y' \in eSe$, then $(xx')(y'y) = e = (y'y)(xx')$, so K is a subsemigroup of S. Clearly, $e \in K$, and if $x \in K$ with associated $y \in eSe$, then clearly $y \in K$, and y is the inverse of x in K. Hence K is a subgroup of S, and since e is the identity of K, we have $K \subseteq H(e)$. But equally, $H(e) \subseteq K$, by the definition of K. Thus $K = H(e)$ is the largest subgroup of S having e as its identity.

Next, we show $H(e)$ is closed. if $\{x_\alpha\}_{\alpha \in A} \subseteq H(e)$ is a net, then since S is compact, $\{x_\alpha\}_{\alpha \in A} \subseteq H(e) \subseteq eSe$ has a convergent subnet, as does the net of inverses $\{y_\alpha\}_{\alpha \in A}$, and by passing to subnets if necessary, we assume $\lim x_\alpha = x, \lim y_\alpha = y$ for some $x, y \in S$. Now eSe is compact, being the continuous image of S via the map $s \mapsto ese \colon S \to S$, and so $x, y \in eSe$. Finally, $xy = (\lim_\alpha x_\alpha)(\lim_\alpha y_\alpha) = \lim_\alpha (x_\alpha y_\alpha) = e$, and similarly, $yx = e$, so, $x, y \in H(e)$. Thus $H(e)$ is closed, hence compact.

Multiplication restricted to $H(e)$ is obviously continuous. Let $x = \lim x_\alpha$ in $H(e)$. If x'_α is the inverse of x_α in $H(e)$, then we just saw that any cluster point of $\{x'_\alpha\}_{\alpha \in A}$ is an inverse of x, and since inverses are unique in a group, $\lim x_\alpha$ exists and equals x', the inverse of x in $H(e)$. Hence inversion also is continuous, so $H(e)$ is a compact group.

The assertion that $E(s)$ is closed follows from the fact that multiplication is continuous, since $E(S) = \{x \in S \mid x = x^2\}$ is the set on which two continuous functions—the identity map and the squaring map—agree.

For (3), let $e \in \mathcal{M}(S)$ be an idempotent. Then $H(e) \subseteq eSe \subseteq \mathcal{M}(S)$, since $x \in eSe$ implies $x = ex \in \mathcal{M}(S)$. If $I \subseteq eSe$ is an ideal of eSe, then $SIS = \mathcal{M}(S)$ since SIS is an ideal of S. So, $e \in SIS$. But then $e \in e(SIS)e = (eSe)I(eSe) \subseteq I$, since I is an ideal of eSe. It follows that $I = eSe$ since e is the identity of eSe. But this implies that eSe has no proper ideals and so it must be a group; i.e., $eSe = H(e)$. □

If S is a monoid, the maximal subgroup, $H(1_S)$ of the identity 1_S of S is called *the group of units of S*.

EXAMPLE 1.6. We illustrate Theorem 1.5 with some examples:
  (1) If A is any set, then A^*, the family of finite words over A, together with the empty word, is a monoid under concatenation of words. It has a non-compact topology (as long as $A \neq \emptyset$) generated by the so-called *Nivat metric:* $d(s,t) = 2^{-n}$, where $n = \min\{j \mid s_j \neq t_j\}$. In this topology, A^* can be completed into the compact monoid $A^\omega \stackrel{\text{def}}{=} A^* \cup A^\infty$, by adding the infinite words. The infinite words form the minimal ideal $\mathcal{M}(A^\omega)$, which consists of left zeros. All groups in A^ω are degenerate.

(2) The next example concerns the unit interval, $\mathbb{I} = [0,1]$. This enjoys several monoid structures relative to which it is a compact monoid:
- Usual multiplication: $(x,y) \mapsto xy$. Under this operation, the only idempotents are 1, which is the identity, and 0, which comprises the minimal ideal. Both idempotents have maximal subgroups that are degenerate.
- If $0 < r < 1$, then the subset $\mathbb{I}_r \stackrel{\text{def}}{=} [r,1]$ of $\mathbb{I}$ is a monoid under the multiplication $(x,y) \mapsto \max\{xy,r\}$. The mapping $x \mapsto \max\{x,r\}\colon \mathbb{I} \to \mathbb{I}_r$ is a continuous surmorphism (surjective homomorphism) of compact monoids whose kernel congruence is $\Delta(\mathbb{I}) \cup ([0,r] \times [0,r])$, where $\Delta(\mathbb{I})$ is the diagonal in $\mathbb{I} \times \mathbb{I}$. All non-identity elements of the monoid $\mathbb{I}_r$ have finite order. All monoids of the form $\mathbb{I}_r$ for $0 < r < 1$ are isomorphic.
- We can also endow $[0,1]$ with the multiplication $(x,y) \mapsto \min\{x,y\}$, under which $\mathbb{I}$ is an inf-semilattice. Then every element is idempotent, and again, all groups are degenerate. Dually, we can define $(x,y) \mapsto \max\{x,y\}$, which gives $\mathbb{I}$ a sup-semilattice structure.

(3) Of course, any compact group is a compact monoid with no proper ideals. In particular, $\mathbb{R}/\mathbb{Z}$, the circle group is a compact monoid.

Since compact monoids are closed under products, we can form the cylinder $\mathbb{I} \times \mathbb{R}/\mathbb{Z}$ and endow it with the product multiplication: $(r,g) \cdot (r',g') = (rr',gg')$. Under this operation, $\mathbb{I} \times \mathbb{R}/\mathbb{Z}$ has a group of units $\{1\} \times \mathbb{R}/\mathbb{Z} \simeq \mathbb{R}/\mathbb{Z}$. The only other idempotent is $(0,1)$, whose maximal subgroup is again $\{0\} \times \mathbb{R}/\mathbb{Z} \simeq \mathbb{R}/\mathbb{Z}$. Of course, any compact group can be used in place of $\mathbb{R}/\mathbb{Z}$.

(4) The monoid $S \stackrel{\text{def}}{=} \mathbb{I} \times \mathbb{R}/\mathbb{Z}$ has some interesting submonoids. Let $\mathbb{R}_+$ denote the additive topological monoid of nonnegative real numbers. Then the function $\phi\colon \mathbb{R}_+ \to S$ defined by $\phi(t) = (e^{-t}, t + \mathbb{Z})$ is a homomorphism of topological monoids. Its image is a spiral "winding down" from the identity $(1,\mathbb{Z}) \in S$ onto the minimal ideal $\mathcal{M}(S) = \{0\} \times \mathbb{R}/\mathbb{Z}$ and containing $\mathcal{M}(S)$ in its closure. The projection $p\colon S \to [0,1] \times [-1,1]$ by $p(s, t + \mathbb{Z}) = (s, \sin 2\pi t)$ maps the image of ϕ onto a graph similar to the $\sin \frac{1}{x}$-graph calculus students know, with $p(\mathcal{M}(S))$ representing its set of limit points.

If $\mathbb{I}_r = [r,1]$ denotes the nil thread with $0 < r < 1$ defined in (2) above, then $\phi_r\colon \mathbb{R}_+ \to S_r \stackrel{\text{def}}{=} \mathbb{I}_r \times \mathbb{R}/\mathbb{Z}$ by $\phi_r(t) = (\max\{r, e^{-t}\}, t + \mathbb{Z})$ is a morphism of compact monoids. Its image is a "lasso": a curve homeomorphic to $[r,1]$ merging at $\phi(-\log r)$ into the circle $\{r\} \times \mathbb{R}/\mathbb{Z} = \mathcal{M}(S_r)$.

The monograph [17] describes a large catalogue of examples of connected compact monoids in Chapter D, pp. 236–278.

(5) Section 8 is devoted to the study of the compact affine monoid of stochastic matrices, which brings in a whole class of examples.

COROLLARY 1.7.

(1) Swelling Lemma: *If S is a compact semigroup and $X \subseteq S$ is closed, then for any $x \in S$, if $X \subseteq Xx$, then $X = Xx$.*

(2) If S is a compact monoid and if $x, y \in S$ satisfy $xy \in H(1_S)$, then $x, y \in H(1_S)$.

PROOF. For the first part, assume $X \subseteq S$ is closed and non-empty and $X \subseteq Xx$ for some $x \in S$. Then $X \subseteq Xx$ implies $Xx \subseteq (Xx)x = Xx^2$, and so $\{Xx^n \mid n \in \mathbb{N}\}$ is an increasing chain of compact subsets of S. In particular, note that $X \subseteq Xx \subseteq Xx^n$ for every $n \geq 1$. Now since S is compact, there is an idempotent $e \in \mathcal{M}(x)$, and $e = \lim_\alpha x^{n_\alpha}$ for some net $\{n_\alpha \mid n_\alpha \in \mathbb{N}\}$. Since multiplication is continuous and the sets Xx^{n_α} are compact, it follows that $\lim_\alpha Xx^{n_\alpha} = Xe$ (where the limit is taken in the Vietoris topology on the hyperspace of compact subsets of S), so $X \subseteq Xx \subseteq Xe$ because $X \subseteq Xx \subseteq Xx^{n_\alpha}$ for each α. But e is idempotent, so $X \subseteq Xe$ implies $y = ye$ for all $y \in X$, so $X = Xe$. It follows that $X = Xx$.

Now, suppose $x, y \in S$ with $xy \in H(1_S)$. Then there is some $h \in H(1_S)$ with $(xy)h = 1_S$. Next, Sx is compact and $Sx \subseteq S = S(xy)h = (Sx)(yh)$, so $Sx = S$ by the first part. Thus there is some $z \in S$ with $zx = 1_S$, and then $z = z1_S = z(x(yh)) = (zx)(yh) = 1_S(yh) = yh$, which means $z = yh = x^{-1}$ and so $x \in H(1_S)$ and $yh = x^{-1}$ is also in $H(1_S)$. But $h \in H(1_S)$ then implies $y = (yh)h^{-1} \in H(1_S)$ as well. $\square$

Notice that (2) implies that $S \setminus H(1)$ is the largest proper ideal if $S \neq H(1)$. Thus if a compact monoid has no proper ideals, it is a group.

REMARK 1.8. This corollary illustrates the difference between the theory of compact monoids and that of, say, monoids of operators on a Banach space. Indeed, in the latter setting, the right-shift operator $\langle x_0, x_1, \ldots \rangle \mapsto \langle 0, x_0, x_1 \ldots \rangle$ is in the monoid of operators on $\ell^1(\mathbb{R})$, and it has a left inverse, the left-shift operator, but it is not invertible. The Corollary shows such an example cannot arise in the theory of compact monoids.

We now delve a little deeper into the fine structure of $\mathcal{M}(S)$ for a compact semigroup S.

THEOREM 1.9. *Let S be a compact semigroup. Then*
1. *S contains minimal left ideals, each of which is a compact subset of $\mathcal{M}(S)$, and distinct left ideals are disjoint.*
2. *Likewise, S contains minimal right ideals, each of which is a compact subset of $\mathcal{M}(S)$, and distinct right ideals are disjoint.*
3. *If L is a minimal left ideal and R a minimal right ideal, then $LR = \mathcal{M}(S)$, while $L \cap R$ is a maximal subgroup of S.*
4. *If S is commutative, then $\mathcal{M}(S)$ is a commutative abelian group.*
5. *If $\mathcal{M}(S)$ has a single idempotent, then $\mathcal{M}(S)$ is a group.*

PROOF. For (1), note that S has closed left ideals, since S itself is one. If we choose a maximal chain (under reverse inclusion) of closed left ideals, then the intersection is non-empty and compact, since S is compact and Hausdorff. If L is the intersection of such a maximal chain, then L is a left ideal, being an intersection of such. Also, if $x \in L$, then $Sx \subseteq L$ is a left ideal: $S(Sx) = S^2x \subseteq Sx$. Hence $Sx = L$ as L is the intersection of a maximal chain of such ideals. It follows that L is a minimal left ideal. Moreover, $\mathcal{M}(S)L = L$ since L is a minimal left ideal, and so $L = \mathcal{M}(S)L \subseteq \mathcal{M}(S)$.

If L' is any left ideal and $L \cap L' \neq \emptyset$, then for any $x \in L \cap L'$, $L = Sx \subseteq L \cap L'$, so $L \subseteq L'$. It follows that distinct minimal left ideals are disjoint.

Part (2) follows by an analogous argument.

For (3), we have $LR \subseteq \mathcal{M}(S)$ is an ideal of S, and since $\mathcal{M}(S)$ is minimal, it follows that $LR = \mathcal{M}(S)$.

On the other hand, $RL \subseteq L \cap R$, so $L \cap R \neq \emptyset$. If $x \in L \cap R$, then $\Gamma(x) \subseteq L \cap R$, since each of L and R is a compact subsemigroup of S. Now $\mathcal{M}(x)$ is a group by Proposition 1.4, and $\mathcal{M}(x) \subseteq \Gamma(x) \subseteq L \cap R$. If $e \in \mathcal{M}(x)$ is the identity, then $Se \subseteq L$ implies $Se = L$. Likewise, $eS = R$, so e is an identity for $L \cap R$, i.e., $L \cap R \subseteq eSe$. But $eSe \subseteq Se = L$ and $eSe \subseteq eS = R$ imply $eSe \subseteq L \cap R$, and so $L \cap R = eSe$. Since $eSe = H(e)$ by part (3) of Theorem 1.5, it follows that $L \cap R = H(e)$ is a maximal subgroup of S.

For (4), we know from the above that $\mathcal{M}(S)$ contains a minimal left ideal L and a minimal right ideal R, but then these are both ideals since S is commutative. Hence $\mathcal{M}(S) = L = R$, and so $L \cap R = \mathcal{M}(S)$ as well. But $L \cap R = H(e)$ for some idempotent e.

Finally, for (5), let e be the unique idempotent in $\mathcal{M}(S)$. Then since minimal left ideals are pairwise disjoint, $\mathcal{M}(S)$ can have only one minimal left ideal L, and similarly, only one right ideal R. Moreover, if e is the unique idempotent in $\mathcal{M}(S)$, then $e \in L \cap R$, which implies $L \cap R = H(e)$. Now, $L = Le$ since L is a minimal left ideal, and if $x \in L$, then $x = xe$, and $e \in \mathcal{M}(x)$ since e is the unique idempotent in $\mathcal{M}(S)$. Hence $x = xe \in \mathcal{M}(x) \subseteq H(e)$. It follows that $L = H(e)$, and similarly for R. But then $\mathcal{M}(S) = LR = H(e)$ as well. $\square$

REMARK 1.10. Our proof of part (2) of the Theorem appeals to the analogy between left and right ideals in a semigroup. To carry out the proof (which is left to the reader), one must go through the arguments given in the proof of (1) and validate that they can be transferred from reasoning about left ideals to reasoning about right ideals. Another, perhaps more elegant approach is to reason by *duality*: each semigroup $(S, \cdot)$ defines a dual semigroup $(S, *)$, where the multiplication is given by $x * y = y \cdot x$ for all $x, y \in S$. The results in $(S, \cdot)$ about left (respectively, right) ideals then transfer *verbatim* to ones about right (respectively, left) ideals in $(S, *)$, so if a result is established about right ideals in general, then this duality shows it also holds for left ideals. This line of reasoning can be applied in other results we present in this paper, notably Proposition 1.12, Corollary 1.21, Proposition 1.34 and Proposition 1.34.

LEMMA 1.11. *Let L be a minimal left ideal of a compact semigroup. Then*
(1) *$L = Le = Se$ for every idempotent e contained in L;*
(2) *if L has only one idempotent e, then L is a group, that is $L = H(e)$.*

PROOF. (1) Note that $S(Le) \subseteq Le \subseteq Se \subseteq L$, so Le is a left ideal contained in L; the minimality of L implies $Le = L$, and so $Le = Se = L$. Next, e is a right identity for L, that is $xe = x$ for all $x \in L$.

(2) Now assume that e is the only idempotent in L and let $x \in L$. Then $e = e(x) \in \Gamma(x)$ since L has only one idempotent. Then $xe(x) = xe = x$ and so $\Gamma(x) = \mathcal{M}(\Gamma(x))$, which is a group by Proposition 1.4. Therefore x has an inverse with respect to e and so L is a group with identity e. Thus $L \subseteq H(e)$, and since $e \in L$ and L is a left ideal, $H(e) \subseteq L$. Hence $L = H(e)$. $\square$

PROPOSITION 1.12. *For a compact semigroup, the following statements are equivalent:*

(1) The minimal ideal $\mathcal{M}(S)$ is a group.
(2) Every minimal left ideal contains only one idempotent, and every minimal right ideal contains only one idempotent.
(3) Every minimal left ideal is a group, and every minimal right ideal is a group.

If these conditions are satisfied and e is the unique idempotent of $\mathcal{M}(S)$, then the function $s \mapsto se \colon S \to H(e)$ is a retractive homomorphism; i.e., $s \mapsto se$ is a retraction of S onto $H(e)$.

PROOF. (1) $\Rightarrow$ (2). By Theorem 1.9(1), any minimal left ideal L is contained in $\mathcal{M}(S)$. Hence by (1) it has at most one idempotent. As a compact semigroup, it has at least one idempotent.

(2) $\Rightarrow$ (3). This follows from Lemma 1.11.

(3) $\Rightarrow$ (1). By Theorem 1.9, parts (1) and (2), $\mathcal{M}(S)$ has a minimal left ideal L and a minimal right ideal R. Then $L \cap R$ (containing RL) is a compact semigroup and thus contains an idempotent e. By (2) and Theorem 1.9(3) $L = H(e) = R$. Since $H(e) = H(e)H(e) = LR = \mathcal{M}(S)$ by Theorem 1.9(3). This is (1).

Now assume $\mathcal{M}(S) = H(e)$. Let $f \colon S \to H(e)$ be defined by $f(s) = se$. Then f is a well-defined continuous function such that $f(f(s)) = se^2 = se = f(s)$. Moreover, $f(st) = ste$. Since $te \in H(e)$ we have $ete = te$ and so $ste = sete = f(s)f(t)$ and thus f is a homomorphism. $\square$

If $\mathcal{M}(S)$ is a group with identity e, then the homomorphism $s \mapsto se \colon S \to H(e)$ has been called the *Clifford-Miller endomorphism* (cf. [5]).

PROPOSITION 1.13. *Let G be a topological group and S a compact subset. Then the following two statements are equivalent:*
(1) *S is a subgroup.*
(2) *$SS \subseteq S$.*

PROOF. Trivially (1) $\Rightarrow$ (2). For a proof of (2) $\Rightarrow$ (1) assume that S is a compact subsemigroup of a topological group G. Then $\mathcal{M}(S)$ is a compact subsemigroup of G as well, and since G has only one idempotent—the identity—the same holds for $\mathcal{M}(S)$. Then Proposition 1.12(1) implies $\mathcal{M}(S)$ is a group. Then, the identity $e \in \mathcal{M}(S)$ is the identity of G, so for any $s \in S$, $s = se \in \mathcal{M}(S)$, so $S = \mathcal{M}(S)$. Thus S is a subgroup of G. $\square$

In particular,

A closed subsemigroup in a compact group is a subgroup.

An element z of a semigroup S is said to be a *zero* if $(\forall s \in S)\, sz = z = zs$. If z and z' are zeroes, then $z = zz' = z'$.

DEFINITION 1.14. On the set of idempotents $E(S)$ of a semigroup we can define a partial order by $e \leq f$ if and only if $ef = fe = e$. (Figuratively, "e is a zero for f" or "f is an identity for e".)

A self-map $s \mapsto s^* \colon S \to S$ is called an *involution* if it satisfies $s^{**} = s$ and $(st)^* = t^*s^*$ for all $s, t \in S$. A semigroup with an involution is said to be *involutive*. For example, the inversion $g \mapsto g^{-1}$ in a group is an involution.

PROPOSITION 1.15.

(1) *Let S be a compact semigroup and assume that for $e_1, e_2 \in E(S)$ there is an $e_3 \in E(S)$ such that $e_3 \leq e_1$ and $e_3 \leq e_2$, that is, assume that $E(S)$ is filtered. Then $E(S)$ has a unique minimal idempotent which is contained in $\mathcal{M}(S)$, so $\mathcal{M}(S)$ is a group.*
(2) *Let S be an involutive semigroup in which all $e \in E(S)$ satisfy $e^* = e$. Then $E(S)$ is filtered. In particular, $\mathcal{M}(S)$ is a group.*

PROOF. (1) Let L be a minimal left ideal and e_1 and e_2 idempotents contained in the compact semigroup L. Then by hypothesis there is an $e \in E(S)$ such that $e \leq e_1$ and $e \leq e_2$. Since $e = ee_1$ we have $e \in L$. Then $L = Le$ by Lemma 1.11(1) and e is a right identity for L. So $e = e_1 e = e_1$. Likewise $e = e_2$. Thus $e_1 = e = e_2$. Thus every minimal left and right ideal is a group by Lemma 1.11(2), and then Proposition 1.12 implies $\mathcal{M}(S) = H(e)$.

(2) Let $e_1, e_2 \in E(S)$. Set $s = e_1 e_2 = e_1 e_1 e_2 \in e_1 S e_2$. Then $\Gamma(s) \subseteq e_1 S e_2$ and so $e_1 e(s) = e(s) = e(s) e_2$. Now, by hypothesis, $e(s) = e(s)^* = (e_1 e(s))^* = e(s)^* e_1^* = e(s) e_1$. Likewise $e_2 e(s) = e(s)$. Therefore $e(s) \leq e_j$, $j = 1, 2$. The remainder now follows from (1). □

1.1. Compact affine semigroups. Compact affine semigroups arise most frequently as compact convex subspaces of topological vector spaces, such as $\mathbb{R}^n$ or $\mathbb{C}^n$, for some n (cf. [8, 9]). They are, however, compact semigroups in their own right with an additional affine structure implied by convexity.

We with a purely topological definition: An *affine space* X is a convex subset of a real (but not necessarily topological!) vector space together with a topology on X such that the function

$$(r, s, t) \mapsto r \cdot s + (1-r) \cdot t \colon [0,1] \times X \times X \to X,$$

called the *affine structure of X*, is continuous. (We use "·" to denote scalar multiplication of the surrounding vector space.)

A continuous map $f \colon X \to Y$ between affine spaces is *affine* if it preserves the affine structure:

$$(\forall x, x' \in X, \forall r \in [0,1]) \quad f(r \cdot x + (1-r) \cdot x') = r \cdot f(x) + (1-r) \cdot f(x'),$$

DEFINITION 1.16. A compact semigroup S is *affine* if:
- S is an affine space,
- the self maps $s \mapsto su$ and $s \mapsto us$ of S are affine for all $u \in S$

A *morphism* $f \colon S \to T$ *of affine semigroups* is a continuous mapping between affine semigroups that preserves both the semigroup operation and the affine structure, that is, $\forall s, s' \in S$ and $\forall r \in [0,1]$,

$$f(ss') = f(s)f(s') \quad \text{and} \quad f(r \cdot s + (1-r) \cdot s') = r \cdot f(s) + (1-r) \cdot f(s').$$

The explicit expressions for the affinity of the left and right multiplications of S read
- $(r \cdot s + (1-r) \cdot t)u = r \cdot su + (1-r) \cdot tu$ for all $r \in [0,1]$ and $s, t, u \in S$;
- $u(r \cdot s + (1-r) \cdot t) = r \cdot us + (1-r) \cdot ut$ for all $r \in [0,1]$ and $s, t, u \in S$.

EXAMPLE 1.17. Here are some examples of affine semigroups:

(1) The unit interval, $\mathbb{I} \stackrel{\text{def}}{=} [0,1]$ in the usual topology and with the usual product and affine structure forms the canonical example of a compact affine monoid.
(2) More generally, any product of compact affine semigroups is a compact affine semigroup, and so any power $\mathbb{I}^X$ is a compact affine monoid in the product topology, for any set X.
(3) The open half-line $((0,\infty),\cdot)$ under multiplication and the affine structure induced by $\mathbb{R}$ is a locally compact affine topological *group*.
(4) If $n > 0$, let $\mathsf{ST}(n)$ denote the set of real $n \times n$-matrices $M = (m_{ij})_{i,j=1,\ldots,n}$ such that $0 \le m_{ij}$ and $\sum_{j=1}^n m_{ij} = 1$ for all $i,j \le n$. Endowed with the topology induced from the ring of all $n \times n$ real matrices, $\mathsf{ST}(n)$ is a compact affine monoid under (matrix) multiplication. The members of $\mathsf{ST}(n)$ are called *stochastic matrices* of degree n.

Later, after we have developed a sufficient body of measure theory we shall see that the probability measures on a compact semigroup form a compact affine semigroup which is universal in a sense to be specified.

In a compact affine semigroup we have a significant ergodic theorem which is an affine parallel to Proposition 1.4:

PROPOSITION 1.18 ((Chow's Lemma) [4]). *For each element x of a compact affine semigroup S, the closed convex hull $C(x) \stackrel{\text{def}}{=} \overline{\text{conv}}(\Gamma(x))$ of $\Gamma(x)$ is a compact commutative affine subsemigroup with a zero z, where*

$$z = \lim_{n \to \infty} \frac{1}{n} \cdot (x + x^2 + \cdots + x^n).$$

PROOF. It's straightforward to show that the convex hull of a commutative subsemigroup in an affine topological semigroup is a commutative subsemigroup, and since the closure of a commutative, affine subsemigroup is another such, it follows that $C(x)$ is a compact commutative affine subsemigroup of S for each $x \in S$.

Fix an $x \in S$, and define $x_n \stackrel{\text{def}}{=} \frac{1}{n} \cdot (x + x^2 + \cdots + x^n)$ for $n = 1, 2, \ldots$. In the compact space $\Gamma(x) \times C(x)$, the sequence $(x^n, x_n)_{n \in \mathbb{N}}$, has a cluster point (h, z). Thus there is a directed set J and a cofinal function $j \mapsto n(j): J \to \mathbb{N}$ such that such that $\lim_{j \in J}(x^{n(j)}, x_{n(j)}) = (h, z)$. Now

(i) $$x_{n+1} = \frac{1}{n+1} \cdot (n \cdot x_n + x^{n+1}) = \frac{n}{n+1} \cdot x_n + \frac{1}{n+1} \cdot x^{n+1}.$$

Also,

(ii) $$x_{n+1} = \frac{1}{n+1} \cdot (x + n \cdot x x_n) = \frac{1}{n+1} \cdot x + \frac{n}{n+1} \cdot x x_n, \quad x x_n = x_n x.$$

In (i) and (ii) we replace n by $n(j)$ and pass to the limit as j ranges through J. Then, since G is a compact affine semigroup, from (i) we obtain

$$\lim_{j \in J} x_{n(j)+1} = z$$

and from (ii) we get

$$\lim_{j \in J} x_{n(j)+1} = xz = zx.$$

Therefore, $x = xz = zx$. Since the set of finite convex combinations of the powers of x is dense in $C(x)$, the element z is a zero of $C(x)$.

In the second part of the proof we argue that z is actually the limit of the sequence $(x_n)_{n\in\mathbb{N}}$ by claiming that z is the only cluster point of $(x_n)_{n\in\mathbb{N}}$. So let z' be an arbitrary cluster point of the sequence $(x_n)_{n\in\mathbb{N}}$. Then there is a subnet $(x_{m(i)})_{i\in I}$ converging to z'. Now by the compactness of $\Gamma(x)$, there is a cofinal function α from a directed set K to I such that $(x^{m(\alpha(k))}, x_{m(\alpha(k))})_{k\in K}$ converges to (h', z'). Then the argument in the first part of the proof shows that z' is a zero of $C(x)$. But zeros are unique; hence $z' = z$ and this proves the claim. □

REMARK 1.19. There is more to this lemma than meets the eye. In compact Hausdorff spaces the convergence of nets is an occurrence that is frequent enough to determine the topology; in metric spaces, convergent *sequences* determine the topology, but in compact spaces that are not necessarily metric, the convergence of *sequences* may be too rare to matter. For instance, in a space like the Čech hull $\beta(\mathbb{N})$ of the discrete set of natural numbers, a *sequence* converges if and only if it is finally constant while, due to compactness, every sequence has to have a convergent sub*net*. The spaces $\Gamma(x)$ may indeed be nonmetric (see for instance [**18**], 12.13ff., notably Example 12.18), and thus, while $C(x)$ is separable, that is, it has a countable, dense subset, it may well fail to have a countable basis for its topology. Thus the fact that in any affine compact semigroup S (see 1.16), and for any element $x \in S$, the *sequence* $\frac{1}{n} \cdot (x + x^2 + \cdots + x^n)$, $n = 1, 2, \ldots$ converges is something special indeed.

An obvious question is what a compact affine *group* looks like. In fact, as we now show, such a group is necessarily singleton—in contrast with the locally compact situation where we observed the set of positive reals to be an affine topological group.

THEOREM 1.20. *A compact affine group G is a point.*

PROOF. Let $g \in G$. By Chow's Lemma 1.18, the compact affine subsemigroup $C(g)$ has a zero z; but the identity e of G is the only idempotent of G whence $z = e$ and thus $g = ge = e$. So $G = \{e\}$. □

COROLLARY 1.21. *Let S be a compact affine semigroup. Then*
 (1) *each maximal subgroup in the minimal ideal $\mathcal{M}(S)$ is a point,*
 (2) *each minimal left ideal is a closed convex left zero semigroup,*
 (3) *each minimal right ideal is a closed convex right zero semigroup, and*
 (4) *if $\mathcal{M}(S)$ has a single idempotent, then S has a zero.*

PROOF. (1) If $e = e^2 \in \mathcal{M}(S)$ is idempotent, then $H(e) = eSe$ by Theorem 1.5(3), and eSe is convex, since multiplication on S is an affine map. But then $H(e)$ is a point by Theorem 1.20.

(2) Let L be a minimal left ideal and $e = e^2 \in L$. Then $L = Se$ and thus $x \mapsto xe \colon S \to L$ is an affine retraction, so L is convex. So each subgroup of L is a point by (1). Now, if $x \in L$, then xS is a right ideal of S, so there is a minimal right ideal M with $x \in M \subseteq Sx$. But then $L \cap M$ is a group by Theorem 1.9(3), which by (1) is a point; i.e., $x \in E(S)$. If follows that $L \subseteq E(S)$. Then Lemma 1.11(1) implies every element of L is a right identity for L, and thus every element of L is a left zero for L.

(3) is an analog of (2).

(4) Finally, if $\mathcal{M}(S)$ has only one idempotent, then $\mathcal{M}(S)$ is a group, by Theorem 1.9(5), and so $\mathcal{M}(S)$ is a point; i.e., S has a zero. □

DEFINITION 1.22. If S is an involutive semigroup, an element $s \in S$ is called *involutive* if $s^* = s$.

COROLLARY 1.23. *A compact affine involutive semigroup in which all idempotents are involutive has a zero.*

PROOF. This follows from Proposition 1.15 and Corollary 1.21. □

1.2. Alternate Proofs to Theorem 1.20. We return to the proof of Theorem 1.20. While the degeneracy of compact affine groups was realized early on and is, on a philosophical level, readily accepted by everyone familiar with compact affine semigroups, a closer look reveals that there are some issues in this context that merit further discussion. In this subsection, we present two alternative approaches to proving the result, approaches which rely on very different arguments from the one we used above.

DEFINITION 1.24. A set S is said to be *star shaped* with respect to a point $s_0 \in S$ if there is an action
$$(r, s) \mapsto r \cdot s : \mathbb{I} \times S \to S$$
such that $0 \cdot s = s_0$ and $1 \cdot s = s$ for all $s \in S$. We say that S is a *star shaped space* if there is a topology on S relative to which the action is continuous.

Any star shaped space is contractible (topologists also say: "homotopy equivalent") to a point.

A convex set is star shaped with respect to any of its points, and a convex space is a star shaped space with respect to any of its points. Any star shaped space is contractible, and so any convex space is contractible.

Let G be a Hausdorff topological group acting on a Hausdorff space X, and for $x \in X$, let $G_x = \{g \in G : g \cdot x = x\}$ denote the isotropy group at x. We have a commutative diagram

$$\begin{array}{ccc} G & \xrightarrow{g \mapsto g \cdot x} & X \\ q \downarrow & & \uparrow \text{inc} \\ G/G_x & \xrightarrow[gG_x \mapsto g \cdot x]{} & G \cdot x \end{array}$$

where $q: G \to G/G_x$ is the quotient map $g \mapsto gG_x$ and $\text{inc}: G \cdot x \to X$ is the inclusion map. The map $f: G/G_x \to G.x$ defined unambiguously by $f(gG_x) = g \cdot x$ is a continuous equivariant bijection (for the obvious actions of G on the left on G/G_x and on $G \cdot x$). If G is compact, then f is a homeomorphism.

The following Theorem is due to Armand Borel. The only published version of Borel's original proof (which is highly non-trivial) can be found in [**17**], p. 310; an apparently very short alternative proof due to Michael can be found in [**28**]:

THEOREM 1.25. *Let G be a compact group and K a closed subgroup such that G/K is connected and satisfies*
$$(\forall n = 1, 2, \dots) \ H^n(G/K, \mathbb{Z}) = 0 \text{ and } H^n(G/K, \mathbb{Z}/2\mathbb{Z}) = 0,$$
for the respective integral and mod 2 Čech cohomology groups. Then $K = G$.

The "acyclicity conditions" for Čech cohomology are certainly satisfied if G/K is contractible. In view of what we saw before, we have

COROLLARY 1.26. *Assume that the compact group G acts on a space X. Then for an orbit $S \stackrel{\text{def}}{=} G\cdot x$ the following conditions are equivalent:*

(1) *S is a convex space.*
(2) *S is a star shaped space.*
(3) *S is contractible.*
(4) *S is singleton.*

As far as we can determine, the original definition of a locally convex topological vector space is that each point has a basis of convex, open sets (cf. [**30, 39**]), and then a locally convex space is one that arises as a convex subspace of such a topological vector space. Since we do not insist that a compact affine space or a compact affine semigroup should originate as a closed convex subset of a locally convex topological vector space (see Definition 1.16 and the preceding paragraph), we provide a concept of local convexity that is intrinsic to the compact affine semigroup itself. Our definition is one of a number of equivalent definitions in the literature on compact convex spaces (cf. [**22**] for related results).

If X is a compact affine space, we let $\mathsf{Aff}(X,\mathbb{R}) \subseteq C(X,\mathbb{R})$ denote the set of continuous affine maps $f\colon X \to \mathbb{R}$. Note that $\mathsf{Aff}(X,\mathbb{R})$ is a closed vector subspace of the Banach space $C(X,\mathbb{R})$.

DEFINITION 1.27. A compact affine topological space X is called *locally convex* if the set $\mathsf{Aff}(X,\mathbb{R})$ separates the points of X. This means that

$$(\forall x, x' \in X)\ x \neq x' \Rightarrow (\exists f\colon X \to \mathbb{R}\text{ affine})\ f(x) \neq f(x').$$

Further, X is *sharply locally convex* if $\mathsf{Aff}(X,\mathbb{R})$ separates points and compact convex subsets. (Recall that we do *not* assume a relationship between the topology on X and that of the ambient vector space in which X resides.)

Note that for an affine semigroup S to be locally convex, the affine mappings $f\colon S \to \mathbb{R}$ do *not* have to preserve the semigroup multiplication. Also observe that a compact convex subset of a locally convex topological vector space is locally convex in the sense of Definition 1.27.

The final proof of Theorem 1.20 we present relies on the Krein-Milman Theorem: a non-empty compact convex subset of a locally convex vector space has an *extreme point:* a point which cannot be expressed as the convex combination of other points. The traditional proof of this result relies on the geometry of the ambient space. In the interests of completeness, we present a proof based directly on our definition of local convexity. We also present an extension that shows that every point of a sharply locally convex compact affine space is in the closed convex hull of the extreme points; this result follows from the weaker one, if one assumes the compact space in question inherits its topology from the ambient vector space, but we have chosen *not* to make this assumption.

To begin, recall that a convex subset Y of a convex set X (in a real vector space) is called a *face* of X whenever it is non-empty and if the midpoint of a line segment is in Y, then the whole segment is in Y: i. e.,

$$(\forall x, y \in X)\ \tfrac{1}{2} \cdot (x+y) \in Y \Rightarrow x, y \in Y.$$

If $\{x\}$ is a face of X, then x is an *extreme point* of X (it is straightforward to show this is equivalent to the point not being expressible as a convex combination of other points). The entire convex set X is a face of itself (unless X is empty). We now have:

LEMMA 1.28 (Krein-Milman Theorem).
(1) *A locally convex compact affine space has an extreme point.*
(2) *Moreover, a sharply locally convex compact affine space is the closed convex hull of its extreme points.*

PROOF. For (1), let X be a locally convex compact affine space. The intersection of a tower of closed faces of X is a closed face. Invoking Zorn's Lemma, we find a minimal compact face F. We claim that F is singleton. A proof of this claim will finish the proof of part (1) the Lemma. Now F itself is a compact affine space on which the restrictions $\{f|_F \mid f \in \mathsf{Aff}(X, \mathbb{R})\}$ separate the points. Hence we are done if we assume that X is a minimal closed face of itself and prove that X is singleton. Now suppose that we find $x \neq y$ in X. Consider an $f \in \mathsf{Aff}(X, \mathbb{R})$ such that $f(x) \neq f(y)$ in $\mathbb{R}$. Then $f(X)$ is nondegenerate compact interval (since f is continuous and affine); hence it is of the form $[a, b]$ with $a < b$. Now $F \stackrel{\mathrm{def}}{=} f^{-1}(b)$ is readily seen to be a closed proper face of X, again since f is continuous and affine and since $[a, b] \neq \{b\}$. This contradicts the minimality of X among the closed faces of X, and this contradiction establishes the claim and proves part (1).

For (2), let X be sharply locally convex, and let $E \subseteq X$ be the set of extreme points of X. Then $\overline{\mathrm{conv}}(E)$ is a compact convex subset of X. If $\overline{\mathrm{conv}}(E)$ is a proper subset of X, then given $x \in X \setminus \overline{\mathrm{conv}}(E)$, there is some affine map $f \colon X \to \mathbb{R}$ with $f(x) \notin f(\overline{\mathrm{conv}}(E))$. We assume without loss of generality that $f(x) < f(y)$ for every $y \in \overline{\mathrm{conv}}(E)$. Then $a \stackrel{\mathrm{def}}{=} \min f(X) < f(y)$ also holds for each $y \in \overline{\mathrm{conv}}(E)$. Now $f^{-1}(a)$ is a closed face in X, and the first part implies $f^{-1}(a)$ contains an extreme point of X, which contradicts $E \subseteq \overline{\mathrm{conv}}(E)$. This contradiction implies $\overline{\mathrm{conv}}(E) = X$. □

We are now ready to provide the alternate proofs of Theorem 1.20, which has been the main point of the present discussion.

THEOREM 1.18 *A compact affine group G is a point.*

PROOF. *First Proof.* We have already proved this in our first discussion of Theorem 1.20 above. QED
Second Proof: As a convex space, G is contractible. Since G acts on itself by left translations, the assertion follows from Corollary 1.26. QED
Third Proof (under an additional hypothesis (LC)): Indeed assume that

(LC) *G is a locally convex compact affine group.*

Then by Lemma 1.28, G has an extreme point. Since the group G acts transitively and affinely on the left, every point is an extreme point. But if $e \neq g$, then $\frac{1}{2} \cdot (e + g)$ fails to be an extreme point. Hence $G = \{e\}$. QED □

REMARK 1.29. We note the following about the proofs of Theorem 1.20:
(1) The first proof of 1.20 using Chow's Lemma is by far the most immediate, robust and direct. It has two additional advantages: First, it works for our fairly general definition of affine semigroup which does not require

an embedding into a *topological vector space*, as we confirm by inspecting Definition 1.16. (Chow's original argument relied on such an embedding, but our reformulation of his proof avoids this assumption.) Second, Chow's Lemma (Proposition 1.18) generalizes to separately continuous affine semigroups; however we do not pursue this here.

(2) The second proof yields Theorem 1.20 as a corollary of the more general result Corollary 1.26. It uses the rather sophisticated argument of Borel's Theorem but applies to *all* contractible compact groups.

(3) The third proof is likely more pleasing to functional analysts for whom compact affine semigroups frequently arise as closed convex subspaces of locally convex topological vector spaces, and who likely find the extreme point argument geometrically appealing.

1.3. The affine structure of compact affine monoids. We recall that Theorem 1.20 implies that a minimal left ideal of a compact affine semigroup is a compact affine left zero semigroup; conversely, every compact affine space can be endowed with a left zero multiplication. We shall need the following pieces of information in Section 8.

LEMMA 1.30. *Let S be an affine semigroup and suppose that $e, f \in E(S)$ satisfy $e \leq f$ (see Definition 1.14). Then the line segment $r \cdot f + (1-r) \cdot e$ is a one-parameter semigroup from f to e, that is, the mapping*

$$r \mapsto (1-r) \cdot f + r \cdot e : ([0,1], \cdot) \to S$$

is a monoid homomorphism, where we endow $[0,1]$ with the multiplication $(r,s) \mapsto r + s - rs$.

PROOF. Note that, for $r, s \in [0,1]$,

$$\begin{aligned}(r \cdot f + (1-r) \cdot e)(s \cdot f + (1-s) \cdot e) &= rs \cdot f + (1-r)s \cdot ef + r(1-s) \cdot fe \\ &\quad + (1-r)(1-s) \cdot e \\ &= rs \cdot f + (1-rs) \cdot e,\end{aligned}$$

the second equality following from the fact that $ef = fe = e$. □

The following is an elementary fact of convex geometry.

LEMMA 1.31. *Let C and C' be compact convex subsets of a real topological vector space V. Then the closed, convex hull $\overline{\text{conv}}(C \cup C')$ is the union of the line segments from points of C to points in C'.*

PROOF. Let C, C' be compact and convex. If $x = (1-r)c_1 + rc_1', y = (1-s)c_2 + sc_2'$ satisfy $c_1, c_2 \in C, c_1', c_2' \in C'$ with $0 < r, s < 1$, then, given $u \in (0,1)$, we have

$$\begin{aligned}ux + (1-u)y &= u(1-r)c_1 + (1-u)(1-s)c_2 + urc_1' + (1-u)sc_2' \\ &= (u(1-r) + (1-u)(1-s))\left(\frac{u(1-r)}{(u(1-r) + (1-u)(1-s))}c_1 \right. \\ &\qquad\qquad\qquad\qquad\qquad\quad \left. + \frac{(1-u)(1-s)}{(u(1-r) + (1-u)(1-s))}c_2\right) \\ &\quad + (ur + (1-u)s)\left(\frac{ur}{ur + (1-u)s}c_1' + \frac{(1-u)s}{ur + (1-u)s}c_2'\right)\end{aligned}$$

and $1 = (u(1-r) + (1-u)(1-s)) + ur + (1-u)s$. The cases that one or more of r, s, u are in $\{0, 1\}$ follow by simpler calculations. It follows that the convex hull of $C \cup C'$ is the set of line segments from points of C to points of C'.

Now, since C and C' are compact and the mapping $(r, c, c') \mapsto (1-r) \cdot c + r \cdot c' \colon [0, 1] \times C \times C' \to V$ is continuous, its image is compact, and hence closed. Thus the convex hull of $C \cup C'$ is closed. $\square$

The proof of the following remark is straightforward; in fact we used a special case in the proof of Chow's Lemma (Proposition 1.18).

LEMMA 1.32. *Let S be a compact affine semigroup and T a subsemigroup. Then the closed convex hull $\overline{\mathrm{conv}}(T)$ is a compact affine subsemigroup of S.* $\square$

Recall that in a monoid S the largest subgroup containing the identity is denoted $H(1)$ and is called the group of units. If S is a compact affine monoid, then $\overline{\mathrm{conv}}(H(1))$ is the unique smallest compact affine submonoid of S containing all units. Cohen and Collins [8] have examples of compact monoids in which the minimal ideal is not connected. The following result shows that we can still deduce the structure of the convex hull $H(1)$ and $\mathcal{M}(S)$ generate, using the previous two lemmas.

COROLLARY 1.33. *Let S be a compact affine monoid. Then $\overline{\mathrm{conv}}(\mathcal{M}(S) \cup H(1))$ is a closed affine submonoid of S which is the union of all line segments joining points of $\overline{\mathrm{conv}}(H(1))$ to points of $\overline{\mathrm{conv}}(\mathcal{M}(S))$. In particular, if $\mathcal{M}(S)$ is convex, then $\overline{\mathrm{conv}}(\mathcal{M}(S) \cup H(1))$ is the union of line segments from points of $\overline{\mathrm{conv}}(H(1))$ to points of $\mathcal{M}(S)$.*

PROOF. Note that $\mathcal{M}(S) \cup H(1)$ is the smallest submonoid of S containing all units and the minimal ideal, so Lemma 1.32 implies that $\overline{\mathrm{conv}}(\mathcal{M}(S) \cup H(1)) = \overline{\mathrm{conv}}(\overline{\mathrm{conv}}(\mathcal{M}(S)) \cup \overline{\mathrm{conv}}(H(1)))$ is the smallest compact affine submonoid containing $\mathcal{M}(S)$ and all units of S. Then apply Lemma 1.31 with $C = \overline{\mathrm{conv}}(H(1))$ and $C' = \overline{\mathrm{conv}}(\mathcal{M}(S))$. The last comment is obvious. $\square$

We can also observe the following:

PROPOSITION 1.34. *Let S be a compact affine monoid, L a minimal left ideal of S and $A \subseteq L$ an arbitrary subset. Then the following sets agree and form a compact affine submonoid of S:*

(1) *The closed convex hull of $A \cup \{1\}$,*
(2) *the closed convex hull of the union of all line segments from 1 to points of A,*
(3) *the union of all line segments from 1 to a point of the closed convex hull $\overline{\mathrm{conv}}(A)$ of A,*

and each of these line segments is a one-parameter subsemigroup.

PROOF. By Corollary 1.21(2), L is a compact affine left zero subsemigroup. Then the subset $\overline{\mathrm{conv}}(A)$ is a compact affine left zero subsemigroup. Now $T \stackrel{\text{def}}{=} \overline{\mathrm{conv}}(A \cup \{1\}) = \overline{\mathrm{conv}}(\overline{\mathrm{conv}}(A) \cup \{1\})$. Since $\overline{\mathrm{conv}}(A) \cup \{1\}$ is a submonoid, by Lemma 1.32, T is a compact affine submonoid. By Lemma 1.31, T is the union of all line segments from 1 to the points of $\overline{\mathrm{conv}}(A)$. Each of these line segments is a one parameter subsemigroup by Lemma 1.30. $\square$

A completely analogous proposition applies to the minimal right ideals of a compact affine monoid.

We note further that the previous proposition applies to the case that S and hence also L are sharply locally convex and A is the set of extreme points, in which case $\overline{\operatorname{conv}}(A) = L$ by the Krein-Milman Theorem 1.28(2).

Theorem 1.20 tells us that compact affine groups are singleton, a result with some noteworthy consequences. However, it does not tell us much about subgroups of compact affine monoids in general, such as the group $H(1)$ of units (i. e., invertible elements) in such a monoid. We include here an informative geometric theorem due to Cohen and Collins [8] in a more special form.

THEOREM 1.35. *Let S be a sharply locally convex compact affine monoid. Then every unit is an extreme point.*

PROOF. The group $H(1)$ acts as a group of isomorphisms of the compact affine structure of S under left multiplications $x \mapsto gx$, $g \in H(1)$, permuting the units transitively. Hence it suffices to show that the identity 1 is an extreme point.

Now suppose that $1 = \frac{1}{2}\cdot a + \frac{1}{2}\cdot b$; we have to show $a = b = 1$. By the Krein-Milman Theorem 1.28(1) there is an extreme point p in S. Then $p = 1 \cdot p = (\frac{1}{2}\cdot a + \frac{1}{2}\cdot b)p = \frac{1}{2}\cdot ap + \frac{1}{2}\cdot bp$, and since p is extreme, we conclude $ap = bp = p$. For any finite convex combination $q \stackrel{\text{def}}{=} \sum_{n=1}^{N} r_n \cdot p_n$ of extreme points p_n, $0 \leq r_n$, $n = 1, \ldots, N$, $\sum_{n=1}^{N} r_n = 1$, we deduce $aq = bq = q$. Hence if E is the set of extreme points of S and $s \in \overline{\operatorname{conv}}(E)$ we have $as = bs = s$.

Now, since S is sharply locally convex, the sharp Krein-Milman Theorem 1.28(2) implies $\overline{\operatorname{conv}}(E) = S \ni 1$ and thus $a = a1 = 1 = b1 = b$ as needed. □

1.4. An Appendix: Moore Smith Convergence. In this section, we review the basics of *Moore Smith convergence*, an approach to defining limits in general topological spaces without the restriction that they be sequences. A standard reference for this material is Chapter 5 of [20]. This approach begins with the following:

DEFINITION 1.36. A *directed set* is a non-empty set I, together with a partial order $\leq \,\subseteq I \times I$ satisfying $\alpha, \beta \in I \Rightarrow (\exists \gamma \in I)\, \alpha, \beta \leq \gamma$.

For $J \subseteq I$ we say:
- J is *final* in I if there is some $\alpha_0 \in I$ satisfying $\{\beta \in I \mid \alpha_0 \leq \beta\} \subseteq J$.
- J is *cofinal* in I if $(\forall \alpha \in I.\exists \beta \in J)\, \alpha \leq \beta$.
- a function $f \colon I \to J$ between direct sets is called *cofinal* if the set $f(I)$ is cofinal in I. (Note: monotonicity is not required.)

DEFINITION 1.37. Let X be a topological space. A *net* in X is a mapping $f \colon I \to X$, where I is a directed set. We usually denote a net $f \colon I \to X$ by $\{x_\alpha\}_{\alpha \in I}$, where $x_\alpha = f(\alpha)$ for each $\alpha \in I$.

Let $\{x_\alpha\}_{\alpha \in I}$ be a net in the topological space X. Then
- $\{x_\alpha\}_{\alpha \in I}$ *converges to* $x \in X$ if, for each open set $U \subseteq X$, if $x \in U$, then $\{\alpha \in I \mid x_\alpha \in U\}$ is final in I. We write this as $x_\alpha \to x$, and as $x = \lim_\alpha x_\alpha$.
- $\{x_\alpha\}_{\alpha \in I}$ *clusters to* $x \in X$ if, for each open set $U \subseteq X$, if $x \in U$, then $\{\alpha \in I \mid x_\alpha \in U\}$ is cofinal in I.

EXAMPLE 1.38. Let X be a topological space, and let $A \subseteq X$. Then $\overline{A}$ can be characterized as the set of points $x \in X$ such that there exists a net $\{x_\alpha\}_{\alpha \in I} \subseteq A$ which converges to x. This can be shown as follows. If $I = \{U \mid x \in U \text{ open}\}$, then I is directed by reverse set containment: $U \leq V$ iff $V \subseteq U$. For each $U \in I$, choose $x_U \in U \cap A$. Then $\{x_U\}_{U \in I}$ converges to x.

The concept of a subnet is psychologically convenient because it is reminiscent of the concept of a subsequence of a sequence, but tricky and does require careful handling.

DEFINITION 1.39. Let $\{x_\alpha\}_{\alpha \in I}$ be a net in a space X. A *subnet* of $\{x_\alpha\}_{\alpha \in I}$ is a net $\{y_\beta\}_{\beta \in J}$ in X and a cofinal function $\phi \colon J \to I$ satisfying $(\forall \beta \in J)\ y_\beta = x_{\phi(\beta)}$.

All the basic results of topology can be stated using nets. In particular, the following hold:
- A net $\{x_\alpha\}_{\alpha \in I}$ in X converges to $x \in X$ iff every subnet of $\{x_\alpha\}_{\alpha \in I}$ converges to x.
- If $\{x_\alpha\}_{\alpha \in I}$ is a net in X and $\{x_\alpha\}_{\alpha \in I}$ clusters to x, then $\{x_\alpha\}_{\alpha \in I}$ has a subnet converging to x.
- $f \colon X \to Y$ is continuous iff, for any net $\{x_\alpha\}_{\alpha \in I}$ in X, if $x_\alpha \to x \in X$, then $f(x_\alpha) \to f(x)$.

The language of nets is particularly compatible with the apparatus of compact Hausdorff spaces:
- A space X is compact iff every net in X has a cluster point iff every net in X has a subnet that converges in X.
- A net $(x_\alpha)_{\alpha \in I}$ in compact Hausdorff space converges to x iff the limit of every convergent subnet agrees with x.

The last fact was used in the second part of the proof of Chow's Lemma (Proposition 1.18).

Part 2. Compact Affine Monoids: Probability Measures, Wendel's Theorem and the Monoid of Stochastic Matrices

2. The Definition of Haar Measure

For the moment, let G denote a compact Hausdorff space and let $\mathbb{K} = \mathbb{R}$ or $\mathbb{K} = \mathbb{C}$, and let $C(G, \mathbb{K})$ denote the Banach space of continuous $\mathbb{K}$-valued functions on G. An element μ of the topological dual—that is the vector space of all continuous linear functionals—of the Banach space $C(G, \mathbb{K})$ is a ($\mathbb{K}$-valued) *integral* or *measure*. (It is not uncommon in our context to use the words "integral" and "measure" synonymously; the eventual justification is, as is usual in the case of such an equivocation, a theorem; here it is the Riesz Representation Theorem of measure theory.) The number $\mu(f)$ is also written $\langle \mu, f \rangle$ or indeed $\int f\, d\mu = \int_G f(g)\, d\mu(g)$. It is not our task here to develop or review measure theory in full. What we need is the uniqueness and existence of one and only one particular measure on a compact group G which is familiar from the elementary theory of Fourier series as Lebesgue measure on the circle group $\mathbb{T} = \mathbb{R}/\mathbb{Z}$. The formulation of the existence (and uniqueness theorem) is easily understood. We shall give a proof through a sequence of lemmas.

For a semigroup G, an element $g \in G$, and a function $f \colon G \to \mathbb{K}$ we define $_g f(x) = f(xg)$. If μ is a measure, we define μ_g by $\mu_g(f) = \mu(_g f)$.

DEFINITION 2.1. Let G denote a compact group. A measure μ is called *invariant* if $\mu_g = \mu$, that is, $\mu(_gf) = \mu(f)$ for all $g \in G$ and $f \in E = C(G, \mathbb{K})$. It is called *positive* if it satisfies $\mu(f) \geq 0$ for all $f \geq 0$. A measure is called a *Haar measure* if it is invariant and positive. The measure μ is called *normalized* if $\mu(1) = 1$ where 1 also indicates the constant function with value 1. A normalized positive measure is also called a *probability measure*.

EXAMPLE 2.2.
(1) If $g \in G$, then the function $f \mapsto f(g) : C(G, \mathbb{K}) \to \mathbb{K}$ is a continuous linear functional, written δ_g and called *Dirac measure* in g, or *point measure*, or *point mass* at g.
(2) Let G be a finite group with $|G|$ elements. Then $C(G, \mathbb{K}) = \mathbb{K}^G$. Define a measure γ by $\langle \gamma, f \rangle = \frac{1}{|G|} \cdot \sum_{g \in G} f(g)$. Then γ is a unique normalized Haar measure on G. We note that $\gamma = \frac{1}{|G|} \cdot \sum_{g \in G} \delta_g$, and that therefore γ is a barycenter of the point masses δ_g.
(3) If $p \colon \mathbb{R} \to \mathbb{T}$ denotes the morphism given by $p(t) = t + \mathbb{Z}$ and $C_1(\mathbb{R}, \mathbb{K})$ denotes the Banach space of all continuous functions $f \colon \mathbb{R} \to \mathbb{K}$ with period 1, then $f \mapsto f \circ p \colon C(\mathbb{T}, \mathbb{K}) \to C_1(\mathbb{R}, \mathbb{K})$ is an isomorphism of Banach spaces. The measure γ on $\mathbb{T}$ defined by $\gamma(f) = \int_0^1 (f \circ p)(x)\, dx$ with the ordinary Riemann integral on $[0, 1]$ is a normalized Haar measure on $\mathbb{T}$.

EXERCISE 2.3. Verify the assertions of Example 2.2. Give a normalized Haar measure on $\mathbb{S}^1 = (\{z \in \mathbb{C} \mid |z| = 1\}, \times)$. For $n \in \mathbb{Z}$ define $e_n \colon \mathbb{T} \to \mathbb{C}$ by $e_n(t + \mathbb{Z}) = e^{2\pi i n t}$. Compute $\gamma(e_j \bar{e}_k)$ for $j, k \in \mathbb{Z}$.

We now state the Existence and Uniqueness Theorem on Haar Measure. We shall provide one of its numerous proofs in the following; this one uses compact semigroups and thus is also of independent interest.

THEOREM 2.4 (The Existence and Uniqueness of Haar Measure). *For each compact group G there is one and only one normalized Haar measure.*

EXERCISE 2.5. Use the preceding theorem to show that any Haar measure γ also satisfies the following conditions:
(1) $\int_G f(gt)\, d\gamma(t) = \gamma(f)$ for all $g \in G$ and $f \in C(G, \mathbb{K})$.
(2) $\int_G f(t^{-1})\, d\gamma(t) = \gamma(f)$ for all $f \in C(G, \mathbb{K})$.

DEFINITION 2.6. We shall use the notation $\gamma \in C(G, \mathbb{K})'$ for the unique normalized Haar measure, and we shall also write $\gamma(f) = \int_G f(g)\, dg$.

We now outline how to prove the Existence Theorem of Haar measure on compact groups.

3. The Required Background of Radon Measure Theory

First we observe (or recall), in a self-contained fashion, some basic features of the measure theory we use. Among other sources, the monographs [19], [29] provide reading material on the matters at hand. By definition, for a compact Hausdorff space X, the space $M(X, \mathbb{K})$ of $\mathbb{K}$-valued *measures* is the topological dual $M(X, \mathbb{K}) = C(X, \mathbb{K})'$ of the Banach space of continuous $\mathbb{K}$-valued functions on X. If $\phi\colon Y \to X$ is a continuous map between compact spaces, then it induces a contractive linear map $f \mapsto f \circ \phi\colon C(X, \mathbb{K}) \to C(Y, \mathbb{K})$ (contravariant!) which is also called $C(\phi)$ and induces in turn a contractive linear adjoint operator $M(\phi)\colon M(Y, \mathbb{K}) \to M(X, \mathbb{K})$ (covariant!) via $M(\phi)(\mu)(f) = \mu(f \circ \phi)$.[1] Note that $M(\phi)(\mu)$ is the "push forward" of the measure μ which in the case of simple measures is given by $M(\phi)(\sum_i r_i \delta_{x_i}) = \sum_i r_i \delta_{\phi(x_i)}$. If ϕ is the inclusion map of a closed subset Y into X then $M(\phi)(\mu)$ is the extension of μ to a measure μ_X, that is, $\mu_X(f) = \mu(f|Y)$ for $f \in C(X, \mathbb{K})$.

3.1. Product measures. We need an understanding how measures on products of spaces are to be treated. Indeed for two measures μ_1 on a compact space X_1 and μ_2 on a compact space X_2, define the *product measure* $\mu_1 \otimes \mu_2$ on $X_1 \times X_2$ as follows.

If $f \in C(X_1 \times X_2, \mathbb{K})$ we note that f is uniformly continuous on the compact space $X_1 \times X_2$ and so $x_2 \mapsto f(-, x_2)\colon X_2 \to C(X_1, \mathbb{K})$ is continuous.

EXERCISE 3.1. Prove that for $f \in C(X_1 \times X_2, \mathbb{K})$ for compact spaces X_1 and X_2 the function $x_2 \mapsto f(-, x_2)\colon X_2 \to C(X_1, \mathbb{K})$ is continuous.

From the continuity of μ_2, it follows that

$$x_2 \mapsto \langle \mu_1, f(-, x_2) \rangle = \int_{X_1} f(x_1, x_2)\, d\mu_1(x_1)$$

is a member of $C(X_2, \mathbb{K})$. Therefore

$$f \mapsto \int_{X_2} \left(\int_{X_1} f(x_1, x_2)\, d\mu_1(x_1) \right) d\mu_2(x_2)$$

is a member of $M(X_1 \times X_2, \mathbb{K})$. We provisionally denote it by $\mu_1 \otimes_1 \mu_2$, that is,

$$\langle \mu_1 \otimes_1 \mu_2, f \rangle = \int_{X_1 \times X_2} f\, d(\mu_1 \otimes_1 \mu_2) = \int_{X_2} \left(\int_{X_1} f(x_1, x_2)\, d\mu_1(x_1) \right) d\mu_2(x_2).$$

Quite analogously we define

$$\langle \mu_1 \otimes_2 \mu_2, f \rangle = \int_{X_1 \times X_2} f\, d(\mu_1 \otimes_2 \mu_2) = \int_{X_1} \left(\int_{X_2} f(x_1, x_2)\, d\mu_2(x_2) \right) d\mu_1(x_1).$$

Now consider two functions $f_j, \in C(X_j, \mathbb{K})$, $j = 1, 2$ and define $f_1 \otimes f_2\colon X_1 \times X_2 \to \mathbb{K}$ by $(f_1 \otimes f_2)(x_1, x_2) = f_1(x_1) f_2(x_2)$. The finite linear combinations of these functions form a dense subalgebra

$$C(X_1, \mathbb{K}) \otimes C(X_2, \mathbb{K}) \quad \text{of} \quad C(X_1 \times X_2, \mathbb{K}).$$

Therefore a continuous linear functional from $M(X_1 \times X_2, \mathbb{K})$ on $C(X_1 \times X_2, \mathbb{K})$ is uniquely determined by its values on $C(X_1, \mathbb{K}) \otimes C(X_2, \mathbb{K})$.

---

[1]Sometimes $M(\phi)$ is denoted ϕ^*, but we reserve this notation for the adjoint of an element in an involutive semigroup (cf. remarks following Definition 1.14).

Now we compute

$$\begin{aligned}
\langle \mu_1 \otimes_1 \mu_2, f_1 \otimes f_2 \rangle &= \int_{X_2} \left(\int_{X_1} (f_1 \otimes f_2)(x_1, x_2) \, d\mu_1(x_1) \right) d\mu_2(x_2) \\
&= \int_{X_2} \left(\int_{X_1} f_1(x_1) \cdot f_2(x_2) \, d\mu_1(x_1) \right) d\mu_2(x_2) \\
&= \int_{X_2} \left(\int_{X_1} f_1(x_1) \, d\mu_1(x_1) \cdot f_2(x_2) \right) d\mu_2(x_2) \\
&= \mu_2(\mu_1(f_1) \cdot f_2) \\
&= \mu_1(f_1) \mu_2(f_2).
\end{aligned}$$

In the same spirit we calculate

$$\langle \mu_1 \otimes_2 \mu_2, f_1 \otimes f_2 \rangle = \mu_1(f_1) \mu_2(f_2).$$

We conclude that $\otimes_1 = \otimes_2$ and that

$$\langle \mu_1 \otimes \mu_2, f_1 \otimes f_2 \rangle = \mu_1(f_1) \mu_2(f_2).$$

Moreover, we have, for $f \in C(X_1 \times X_2, \mathbb{K})$, the (small) *Fubini Theorem*

$$\begin{aligned}
\int_{X_1 \times X_2} f(x_1, x_2) \, d(\mu_1 \otimes \mu_2)(x_1, x_2) &= \int_{X_2} \left(\int_{X_1} f(x_1, x_2) \, d\mu_1(x_1) \right) d\mu_2(x_2) \\
&= \int_{X_1} \left(\int_{X_2} f(x_1, x_2) \, d\mu_2(x_2) \right) d\mu_1(x_2).
\end{aligned}$$

So much, for the time being, for product measures and Fubini.

3.2. The Support of a Measure. Let μ be a positive measure on a compact space G. An open subset U of G is a μ-null set if for every positive $f \in C(G, \mathbb{K})$ such that $\{x \in G \mid f(x) > 0\} \subseteq U$ we have $\mu(f) = 0$. The support $\operatorname{supp}(\mu)$ of a positive measure is the complement of the largest open μ-null set.

As a prerequisite of a proof of the following Proposition 3.2 we assume the standard result from topology that for a point x in a compact space X and a closed subset $A \subseteq X$ not containing x there is a continuous function $\phi \colon X \to [0, 1]$ such that $\phi(x) = 1$ and $\phi(A) = \{0\}$. (That is, every compact space is completely regular.) Other than that we only use the definition of an open μ-null set and the support of μ.

PROPOSITION 3.2. *Let X and Y be compact, Hausdorff spaces.*
(1) *If μ is a positive measure on X, then $x \notin \operatorname{supp}(\mu)$ iff there is a nonnegative continuous function f such that*
   (a) $\langle \mu, f \rangle = 0$, *and*
   (b) $f(x) > 0$.
(2) *Suppose $\phi \colon X \to Y$ is a continuous map and that $\mu \in M(X, \mathbb{K})$ is a positive measure on X. Then an open set V in Y is a $M(\phi)(\mu)$-null set iff $\phi^{-1}(V)$ is a μ-null set in X.*
(3) *In the circumstances of (2) we have $\operatorname{supp}(M(\phi)(\mu)) = \phi\bigl(\operatorname{supp}(\mu)\bigr)$.*

PROOF. For (1), if $x \notin \operatorname{supp}\mu$ use complete regularity of X to find a continuous $f \colon X \to [0, 1]$ such that $f(x) = 1$ and $f\bigl(\operatorname{supp}(\mu)\bigr) \subseteq \{0\}$; then the definition of $\operatorname{supp}(\mu)$ implies $\langle \mu, f \rangle = 0$. Conversely, assume that there is an f satisfying (a) and (b). By (b) find a compact neighborhood K of x such that $f(K) \subseteq [f(x)/2, \max f(X)]$. Then $\phi \stackrel{\text{def}}{=} \min\{f, f(x)/2\}$ is nonnegative and satisfies $\phi(\text{supp}(\mu)) \subseteq \{0\}$ and $\phi(K) = \{f(x)/2\}$. Now from (a) we have $0 \leq \langle \mu, \phi \rangle \leq \langle \mu, f \rangle = 0$. Now let V be the interior of K and take any nonnegative continuous function F such that $F(X \setminus V) \subseteq \{0\}$. Since F is bounded there is a $k \geq 0$ such that $F \leq k \cdot \phi$. Hence $\langle \mu, F \rangle \leq k \cdot \langle \mu, \phi \rangle = 0$. Thus V is an open μ-null set containing x and thus $x \notin \text{supp}(\mu)$ by the definition of $\text{supp}(\mu)$.

For (2), let f be a positive member of $C(Y, \mathbb{K})$ such that $f(g) > 0$ implies $g \in V$. Then $\langle M(\phi)(\mu), f \rangle = \langle \mu, f \circ \phi \rangle$ and so an open set V in Y is a $M(\phi)(\mu)$-null set iff $\phi^{-1}(V)$ is a μ-null set; note that (1) is helpful for the harder implication.

For (3), we have

$$\phi^{-1}(Y \setminus \text{supp}(M(\phi)(\mu))) \subseteq X \setminus \text{supp}(\mu), \text{ i.e.,}$$

$$\text{supp}(\mu) \subseteq \phi^{-1}\big(\text{supp}(M(\phi)(\mu))\big) \text{ by (2), and so}$$

$$(*) \qquad \phi\big(\text{supp}(\mu)\big) \subseteq \text{supp}\big(M(\phi)(\mu)\big);$$

but $V \stackrel{\text{def}}{=} Y \setminus \phi\big(\text{supp}(\mu)\big)$ is an open set since $\phi\big(\text{supp}(\mu)\big)$ is compact as a continuous image of a compact set, and $\phi^{-1}(V) \cap \text{supp}(\mu) = \emptyset$, whence V is a $M(\phi)(\mu)$-null set by (2) and thus is contained in $X \setminus \text{supp}\big(M(\phi)(\mu)\big)$ by the definition of the support. Thus equality holds in $(*)$. $\square$

PROPOSITION 3.3. *Let X be a compact space, μ a positive measure, and $f \in C(X, \mathbb{K})$. Assume that $f(x) \geq 0$ for all $x \in \text{supp}(\mu)$. Then*

(1) $\langle \mu, f \rangle \geq 0$, *and*
(2) *if there is an $s \in \text{supp}(\mu)$ such that $f(s) > 0$, then $\langle \mu, f \rangle > 0$.*
(3) *If $g \in C(X, \mathbb{K})$ agrees with f on $\text{supp}(\mu)$, then $\langle \mu, g \rangle = \langle \mu, f \rangle$.*

PROOF. As a first step of the proof define $F = \max\{f, 0\}$. Then F is a continuous nonnegative function such that, by hypothesis on f, we have $F(x) = f(x)$ for $x \in \text{supp}(\mu)$ and $F(x) \geq f(x)$ for $x \in X \setminus \text{supp}(\mu)$. Then

$$(F - f)(x) \begin{cases} = 0 & \text{for } x \in \text{supp}(\mu) \\ \geq 0 & \text{for } x \in X \setminus \text{supp}(\mu) \end{cases} \geq 0.$$

Then, by the definition of the support of μ, we have $\langle F - f, \mu \rangle = 0$. Thus $\langle F, \mu \rangle = \langle f, \mu \rangle$ and the positivity of μ implies $\langle f, \mu \rangle \geq 0$ and that is the proof of (1).

For a proof of (2), assume $f(s) > 0$ for an $s \in \text{supp}(\mu)$. We must argue that $\langle \mu, f \rangle > 0$. We proceed by contradiction and assume that $\langle \mu, f \rangle = 0$.

Let U be an open neighborhood of s such that $f(u) \geq f(s)/2 > 0$ for all $u \in U$. Now let ϕ be any nonnegative continuous function such that $\phi(X \setminus U) = \{0\}$. As a continuous function, ϕ is bounded and after multiplying ϕ with a positive number, we assume without loss of generality that $\phi(X) \subseteq [0, f(s)/2]$. Then

$$0 \leq \phi(x) \begin{cases} \leq f(s)/2 \leq f(x) & \text{for } x \in U \\ = 0 & \text{for } x \in X \setminus U \end{cases} \leq f(x),$$

since $f \geq 0$. Thus $0 \leq \langle \mu, \phi \rangle \leq \langle \mu, f \rangle = 0$ by assumption. But then, by the definition of open μ-null set U is an open μ-null set. By the definition of the support, this implies $U \cap \text{supp}(\mu) = \emptyset$. Yet this contradicts $s \in U \cap \text{supp}(\mu)$. This contradiction completes the proof of (2), and (3) is an immediate consequence of (2) applied to $\max\{f - g, 0\}$ and $\max\{g - f, 0\}$. $\square$

PROPOSITION 3.4. *If $\mu_1 \in M(X_1, \mathbb{K})$ and $\mu_2 \in M(X_2, \mathbb{K})$ are positive measures, then*

(**) $$\operatorname{supp}(\mu_1 \otimes \mu_2) = \operatorname{supp}(\mu_1) \times \operatorname{supp}(\mu_2).$$

PROOF. If U is a μ_1-null set in X_1, then we claim that the product $U \times X_2$ is an open $(\mu_1 \otimes \mu_2)$-null set. For let $F \colon X_1 \times X_2 \to \mathbb{K}$ be a positive function vanishing outside $U \times X_2$, then by Fubini we have

$$\begin{aligned}\langle \mu_1 \otimes \mu_2, F \rangle &= \int_{X_1 \times X_2} F(x_1, x_2) \, d(\mu_1 \otimes \mu_2)(x_1, x_2) \\ &= \int_{X_2} \left(\int_{X_1} F(x_1, x_2) \, d\mu_1(x_1) \right) d\mu_2(x_2) = 0,\end{aligned}$$

since $F(-, y) \colon X \to \mathbb{K}$ vanishes outside U for each y and U is a μ-null set. Thus the claim is established. In particular, this applies to $U = X_1 \setminus \operatorname{supp}(\mu_1)$. So

$$X_1 \times X_2 \setminus \operatorname{supp}(\mu_1) \times \operatorname{supp}(\mu_2) = \bigl((X_1 \setminus \operatorname{supp}(\mu_1)) \times X_2\bigr) \cup \bigl(X_1 \times (X_2 \setminus \operatorname{supp}(\mu_2))\bigr)$$

is an open $\mu_1 \otimes \mu_2$-null set and thus $S \stackrel{\text{def}}{=} \operatorname{supp}(\mu_1 \otimes \mu_2) \subseteq \operatorname{supp}(\mu_1) \times \operatorname{supp}(\mu_2)$. By way of contradiction, suppose the containment is proper and pick $x_k \in \operatorname{supp}(\mu_k)$, $k = 1, 2$ with $(x_1, x_2) \notin S$. Find $f_k \geq 0$ in $C(X_k, \mathbb{K})$ with $f_k(x_k) > 0$ such that $\operatorname{supp}(f_1 \otimes f_2)$ does not meet the closed set S. Then $\int f_k d\mu_k > 0$, $k = 1, 2$ by Proposition 3.3(2). Thus $\int (f_1 \otimes f_2) \, d(\mu_1 \otimes \mu_2) = \langle \mu_1 \otimes \mu_2, f_1 \otimes f_2 \rangle = \langle \mu_1, f_1 \rangle \langle \mu_2, f_2 \rangle > 0$ on the one hand, and $\int (f_1 \otimes f_2) \, d(\mu_1 \otimes \mu_2) = 0$ on the other, since $\operatorname{supp}(f_1 \otimes f_2) \cap S = \emptyset$. This contradiction shows that $S = \operatorname{supp}(\mu_1) \times \operatorname{supp}(\mu_2)$ as asserted. □

4. Measures on Compact Semigroups and Groups

For the following discussion we fix a compact semigroup G, later to be specialized to a group.

4.1. Convolution. Let $m \colon G \times G \to G$ denote the multiplication of G. Then we have the operator
$$M(m) \colon M(G \times G) \to M(G).$$
For $\mu_1, \mu_2 \in M(G)$ we set
$$\mu_1 * \mu_2 = M(m)(\mu_1 \otimes \mu_2),$$

$$\begin{aligned}\langle \mu_1 * \mu_2, f \rangle = \langle \mu_1 \otimes \mu_2, C(m)(f) \rangle &= \int_{G \times G} f(g_1 g_2) \, d(\mu_1 \otimes \mu_2)(g_1, g_2) \\ &= \int_G \left(\int_G f(gh) \, d\mu_1(g) \right) d\mu_2(h) \\ &= \int_G \left(\int_G {}_h f \, d\mu_1 \right) d\mu_2(h).\end{aligned}$$

The product $\mu_1 * \mu_2$ is called the *convolution* of the measures μ_1 and μ_2.

For $f \in C(G, \mathbb{K})$ and $\mu \in M(G, \mathbb{K})$, the function $g \mapsto \mu({}_g f) \colon G \to \mathbb{K}$ is in $C(G, \mathbb{K})$.

EXERCISE 4.1. Prove directly that $g \mapsto \int_G f(xg) \, d\mu(x) \colon G \to \mathbb{K}$ is continuous. Hint. We do this for $\mathbb{K} = \mathbb{R}$ and derive the assertion for $\mathbb{K} = \mathbb{C}$ from this result. Note that ${}_g f$ is uniformly continuous on G (why?). If G is a compact group, this means that, given $\epsilon > 0$ there is an identity neighborhood U of G such that

$|_gf(xu) -{_g}f(x)| < \epsilon$ for all $u \in U$. Write the inequality in the form $_gf(x) - \epsilon < {_g}f(xu) < {_g}f(x) + \epsilon$. The positivity of μ yields

$$\mu(_gf) - \epsilon \cdot \mu(1) < \mu(_gf) < \int_G f(xgu)\, d\mu(x) < \mu(_gf) + \epsilon \cdot \mu(1).$$

Complete the proof.

In the more general case that G is a compact topological semigroup, we have to argue differently by using the fact, that on a compact space G, the neighborhoods of the diagonal in $G \times G$ for a unique uniform structure on G inducing the given topology on G. Indeed, for given $f \in C(G, \mathbb{K})$ and given $\epsilon > 0$, the set $\{(g,h) \in G \times G \mid (\forall x \in G)|f(xg) - f(xh)| < \epsilon\}$ is an open neighborhood of the diagonal in $G \times G$ (proof by contradiction: if the claim is false, then there is a $(g,h) \in G \times G$ such that $|f(xf) - f(xh)| < \epsilon$ for all $x \in X$, but that for ever neighborhood U of g and every neighborhood V of h the set $F_{UV} = \{x \in G \mid (\exists (u,v) \in U \times V)|f(xu) - f(xv)| \geq \epsilon\}$ is not empty. Due to the compactness of G and the continuity of the multiplication of G, the filterbasis $\{F_{UV} \mid U, V\}$ has a point of adherence x which would have to satisfy $|f(xg) - f(xh)| \geq \epsilon$). This means that $g \mapsto {_g}f : G \to C(G, \mathbb{K})$ is uniformly continuous and so, in particular, continuous.

We recall that $M(G, \mathbb{K})$ has a natural norm, the *dual norm*, that is, the *operator norm* on continuous functionals on $C(G, \mathbb{K})$ defined by $\|\mu\| = \sup_{\|f\| \leq 1} \|\mu(f)\|$. If G is a compact group, then inversion $\sigma \colon G \to G$, $\sigma(g) = g^{-1}$ is an involution that is, satisfies $\sigma(gh) = \sigma(h)\sigma(g)$ and $\sigma^2 = \mathrm{id}_G$. We shall abbreviate $M(\sigma)(\mu)$ by μ^*.

LEMMA 4.2. *Let G be a compact semigroup. Then convolution $(\mu, \nu) \mapsto \mu * \nu$ makes $M(G, \mathbb{K})$ into a Banach algebra with respect to the dual norm on $M(G, \mathbb{K})$. If G is a group, then $\mu^{**} = \mu$ and $(\mu * \nu)^* = \nu^* * \mu^*$, that is, $M(G, \mathbb{K})$ is an involutive Banach algebra.*

PROOF. This is the topic of the following exercise:

EXERCISE 4.3. *Hint. Show the bilinearity and associativity of $*$ and verify $\|\mu * \nu\| \leq \|\mu\| \cdot \|\nu\|$. Check the condition on the involution in the group case.* □

LEMMA 4.4.

(1) *For positive measures $\mu, \nu \in M(G, \mathbb{K})$ one has*

(†) $\quad\quad\quad\quad\quad\quad\quad \mathrm{supp}(\mu * \nu) = \mathrm{supp}(\mu)\mathrm{supp}(\nu).$

(2) *The support of an idempotent probability measure is a compact subsemigroup of G.*

PROOF. (1) The (semi)group multiplication $m \colon G \times G \to G$ induces an operator

$$M(m) \colon M(G \times G, \mathbb{K}) \to M(G, \mathbb{K}) \quad \text{satisfying} \quad M(m)(\mu \otimes \nu) = \mu * \nu.$$

Proposition 3.3(2) implies $\mathrm{supp}(\mu * \nu) = \mathrm{supp}\big(M(m)(\mu \otimes \nu)\big) = m\big(\mathrm{supp}(\mu \otimes \nu)\big)$; by Proposition 3.4 (**) we have $\mathrm{supp}(\mu \otimes \nu) = \mathrm{supp}(\mu) \times \mathrm{supp}(\nu)$. Thus $\mathrm{supp}(\mu * \nu) = m\big(\mathrm{supp}(\mu) \times \mathrm{supp}(\nu)\big) = \mathrm{supp}(\mu)\,\mathrm{supp}(\nu)$.

(2) This is immediate from (1). □

There are two significant topologies on $M(G, \mathbb{K})$: firstly, the dual norm topology, endowing $M(G, \mathbb{K})$ with the structure of a Banach algebra, and, secondly, the topology of pointwise convergence of functionals, that is, the topology induced from the inclusion $M(G, \mathbb{K}) \subseteq \mathbb{K}^{C(G,\mathbb{K})}$. This latter one is called the *weak $*$-topology*, endowing $M(G, \mathbb{K})$ with a locally convex algebra topology. It is the weak $*$-topology we are interested in here.

Recall that a measure $\mu \in M(G, \mathbb{K})$ is called a *probability measure* if it is positive and normalized. The topological space of all probability measures equipped with the weak $*$-topology will be denoted $P(G) \subseteq M(G, \mathbb{K})$.

LEMMA 4.5. *For a compact semigroup G, endowed with the weak $*$-topology, the space $P(G)$ has the following properties.*
  (1) $P(G)$ is a compact and convex subset of $M(G, \mathbb{R})$.
  (2) $P(G)$ is a compact topological semigroup with respect to convolution.
  (3) For $\mu_1, \mu_2, \nu \in P(G)$ and $0 \leq r_1, r_2$, $r_1 + r_2 = 1$ we have
  $$\nu * (r_1 \cdot \mu_1 + r_2 \cdot \mu_2) = r_1 \cdot (\nu * \mu_1) + r_2 \cdot (\nu * \mu_2),$$
  $$(r_1 \cdot \mu_1 + r_2 \cdot \mu_2) * \nu = r_1 \cdot (\mu_1 * \nu) + r_2 \cdot (\mu_2 * \nu).$$
  (4) *If G is a group, the involution $\mu \mapsto \mu^*$ leaves $P(G)$ invariant.*

PROOF. For (1) we have to show compactness, convexity, and for (2) that $P(G)$ is closed under convolution, and that the multiplication
$$(\mu, \nu) \mapsto \mu * \nu : P(G) \times P(G) \to P(G)$$
is weak $*$–continuous. Assertion (3) is due to the fact that multiplication on $P(G)$ is the restriction of an algebra multiplication to a convex subset. Claim (4) is straightforward. The details are left to the following exercise.. $\square$

EXERCISE 4.6. Prove details for Lemma 4.5.
Hint. (i) show convexity directly. Show that $P(G)$ is weak-$*$-closed and bounded; then apply the Theorem of Bourbaki-Banach-Alaoglu to prove compactness.
  (ii) For proving continuity of a function $\alpha \colon X \to P(G)$, recall that this amounts to showing that the functions $\alpha \mapsto \langle \alpha(x), f \rangle : X \to \mathbb{K}$ are continuous for all $f \in C(G, \mathbb{K})$.
  (iii) In a real algebra, multiplication is linear in each argument separately and this implies the assertion.

Recall from Definition 1.16 that an *affine semigroup* S satisfies
$$u(r_1 \cdot v + r_2 \cdot w) = r_1 \cdot uv + r_2 \cdot uw \quad \text{and} \quad (r_1 \cdot u + r_2 \cdot v)w = r_1 \cdot uw + r_2 \cdot vw,$$
for all $u, v, w \in S$ and real numbers $0 \leq r_1, r_2$, $r_1 + r_2 = 1$.

Recall from Definition 1.22 that an involutive semigroup S has an involution $s \mapsto s^*$ satisfying $s^{**} = s$ and $(st)^* = t^* s^*$. Accordingly, we can say that

COROLLARY 4.7. *For any compact semigroup G, respectively group, $P(G)$ is a locally convex, affine, compact, respectively involutive semigroup.*

Recall that the *point-mass* δ_g concentrated at $g \in G$ is the probability measure defined by $\delta_g(f) = f(g)$. The function $g \mapsto \delta_g \colon G \to P(G)$ is an injective morphism of compact topological semigroups. The element $\delta_\mathbf{1}$ is the identity of $P(G)$.

EXERCISE 4.8.
  (1) Discover the group $H(P(G))$ of elements of $P(G)$ which are invertible with respect to δ_1.
    Hint. Let $\mu, \nu \in P(G)$ be such that $\mu * \nu = \delta_1$. Then $\text{supp}(\mu) \text{supp}(\nu) = \text{supp}(\delta_1) = 1$. This implies that μ and ν are Dirac measures.

(2) Prove that $\mu_g = \mu * \delta_g$.

5. Semigroup Theoretical Characterization of Haar Measure

Recall from the introduction to the section on the definition of Haar measure that $\langle \mu_g, f \rangle = \langle \mu, {}_g f \rangle$; likewise we now define $\langle {}_g \mu, f \rangle = \langle \mu, f_g \rangle$ for ${}_g f(x) = f(xg)$ and $f_g(x) = f(gx)$ and call $\gamma_g = \gamma$ the *right invariance* and ${}_g \gamma = \gamma$ the *left invariance* of γ. We observe that

(+) $$\mu_g = \mu * \delta_g \quad \text{and} \quad {}_g \mu = \delta_g * \mu.$$

(Exercise.)

PROPOSITION 5.1. *For a probability measure $\gamma \in P(G)$ the following statements are equivalent:*

(1) *γ is a left- and right invariant measure of G.*
(2) *$\gamma * \mu = \mu * \gamma = \gamma$ for all $\mu \in P(G)$, that is, γ is the zero element of the compact semigroup $P(G)$.*

Moreover, if these conditions are satisfied, then $\mathrm{supp}(\gamma) = G$.

PROOF. This is the subject of the following exercise:

EXERCISE 5.2. Prove (+) and Proposition 5.1.

Hint. (+) is straightforward, e.g.

$$\langle \mu * \delta_g, f \rangle = \int_G \int_G f(xy) \, d\delta_g(y) d\mu(x) = \int_G f(xg) \, d\mu(x) = \langle \mu, {}_g f \rangle = \langle \mu_g, f \rangle.$$

(1)$\Rightarrow$(2). By definition of the convolution, $(\gamma * \mu)(f) = \int_G \gamma({}_g f) \, d\mu(g)$. Now $\gamma({}_g f) = \gamma(f)$ by right invariance of γ. Proceed.

Next $(\mu * \gamma)(f) = \int_G \mu({}_g f) \, d\gamma(f) = \int_G \left(\int_G f(xg) \, d\mu(x) \right) d\gamma(g)$. By the Fubini Theorem we can invert the order of integration:

$$(\mu * \gamma)(f) = \int_G \left(\int_G f(xg) \, d\gamma(g) \right) d\mu(x),$$

We have $\int_G f(xg) \, d\gamma(g) = \gamma(f)$. So $\mu * \gamma = \gamma$ follows.

(2)$\Rightarrow$(1). Via (+) this is straightforward by taking $\mu = \delta_g$.

Moreover, assume (2) satisfied. Then from Lemma 4.4 we derive $g \cdot \mathrm{supp}(\gamma) = \mathrm{supp}(\delta_g * \gamma) = \mathrm{supp}(\gamma)$ and since G acts transitively under left translation and $\mathrm{supp}(\gamma) \neq \emptyset$, we have $\mathrm{supp}(\gamma) = G$. □

Proposition 5.1 shows that there is at most one measure which is both left and right invariant, since a zero of a semigroup is unique. If an arbitrary compact space X is endowed with the semigroup multiplication $gh = g$ (the so-called *left zero multiplication*) then *every* measure is right invariant, but if X has at least two elements, a point measure δ_g is not left invariant. In fact, the proof of Proposition 5.1 shows that a right invariant measure on G is a left zero of $P(G)$.

After Proposition 5.1, the Existence and Uniqueness of a two-sided-invariant probability measure of a compact groups is equivalent to the assertion that

for a compact group G the compact topological semigroup $P(G)$ has a zero.

In particular, Haar measure is an idempotent in $P(G)$, if it exists. The element δ_1 is an idempotent. One needs to understand the idempotents of $P(G)$.

6. Idempotent Probability Measures on a Compact Group

We now return to our study of measures on a compact group.

PROPOSITION 6.1. *Assume that μ is an idempotent probability measure on a compact group G. Then the following conclusions hold:*

(1) *The support $\mathrm{supp}(\mu)$ is a closed subgroup of G.*
(2) *If $g \in \mathrm{supp}(\mu)$ then $\mu_g = \mu$, that is $\int {}_g f \, d\mu = \int f \, d\mu$ for all $f \in C(G, \mathbb{K})$.*
(3) *If $\nu \in P(G)$ and $\mathrm{supp}(\nu) \subseteq \mathrm{supp}(\mu)$, then $\mu * \nu = \nu * \mu = \mu$.*
(4) $\mu^* = \mu$.

PROOF. (1) Lemma 4.4(2) and Proposition 1.13 imply the claim.

(2) It is sufficient to prove (2) for positive f. So we assume that f is positive. We must show that the function $F \colon G \to \mathbb{K}$ defined by $F(g) = \langle \mu, {}_g f \rangle$ is constant on the compact subgroup $\mathrm{supp}(\mu)$. Since supp is compact and F is continuous, there is an $m \in \mathrm{supp}(\mu)$ such that $F(m) = \max F(\mathrm{supp}(\mu))$. Then ${}_m F$ attains its maximum on $\mathrm{supp}(\mu)$ in the identity 1, and it is no loss of generality if we replace F by ${}_m F$ and assume now that F attains its maximum on $\mathrm{supp}(\mu)$ in 1. Thus $F(1) - F$ is a continuous function on G which is nonnegative on $\mathrm{supp}(\mu)$. Then Proposition 3.3(1) allows us to conclude $\langle \mu, F(1) - F \rangle \geq 0$, that is $\langle \mu, F \rangle \leq \langle \mu, F(1) \rangle = F(1)$. Now we calculate

$$\begin{aligned} F(1) &= \langle \mu, f \rangle = \langle \mu * \mu, f \rangle = \int_G \int_G f(xy) \, d\mu(x) d\mu(y) \\ &= \int_G \left(\int_G {}_y f(x) \, d\mu(x) \right) d\mu(y) = \int_G F(y) \, d\mu(y) = \langle \mu, F \rangle \leq F(1). \end{aligned}$$

Thus equality holds and therefore we have, for the continuous function $F(1) - F \colon G \to \mathbb{R}$, which takes nonnegative values on $\mathrm{supp}(\mu)$, the relation $\int_G (F(1) - F) \, d\mu = 0$. Now Proposition 3.3(2) implies $(F(1) - F)|\mathrm{supp}(\mu) \equiv 0$ and this proves that F is constant on $\mathrm{supp}(\mu)$.

(3) We compute $(\mu * \nu)(f) = \int_G \mu({}_g f) \, d\nu(g)$. Now $\mu({}_g f) = \mu(f)$ for $g \in \mathrm{supp}(\mu)$ and so certainly for $g \in \mathrm{supp}(\nu)$. So Proposition 3.3(3) implies $(\mu * \nu)(f) = \int_G \mu({}_g f) \, d\nu = \int_G \langle \mu, f \rangle \, d\nu = \mu(f)$. The proof of the relation $\nu * \mu = \mu$ is similar.

(4) By Proposition 3.2(3) applied with $\phi(g) = g^{-1}$ we have $\mathrm{supp}(\mu) = \mathrm{supp}(\mu^*)$ and therefore $\mu * \mu^* = \mu$ by (3). But we also have $\mu^* * \mu^* = \mu^*$ and thus we may apply the results of (3) to μ^* and find $\mu * \mu^* = \mu^*$. Thus $\mu^* = \mu$ follows. □

COROLLARY 6.2. $P(G)$ *has a zero.*

PROOF. By Proposition 6.1, $P(G)$ is a compact affine involutive semigroup in which all idempotents are involutive. The assertion then follows from Corollary 1.23. □

In view of Proposition 5.1 this implies the existence and uniqueness of Haar measure on a compact group. Thus Theorem 2.4 is proved.

In the process we gained considerable insight into the structure of the convolution semigroup $P(G)$ of probability measures on a compact group G. In particular, each idempotent probability measure μ determines a compact subgroup $H = \mathrm{supp}(\mu)$ and is measure supported by H and invariant under translation by elements of H. Conversely: Each compact subgroup H has its own Haar measure $\gamma^H \in M(H, \mathbb{K})$. We recall that the inclusion $j: H \to G$ yields a measure $(\gamma^H)_G \stackrel{\text{def}}{=} M(j)(\gamma^H)$ defined by $\langle (\gamma^H)_G, f \rangle = \langle \gamma^H, f|H \rangle$ for each $f \in C(G, \mathbb{K})$.

PROPOSITION 6.3. *Let H be a compact subgroup of a compact group G and let $\gamma^H \in M(H, \mathbb{K})$ be Haar measure on H. Then $\mu \stackrel{\text{def}}{=} (\gamma^H)_G$ is an idempotent measure in $P(G)$ with support $\operatorname{supp}(\mu) = H$ such that $\mu_h = \mu$ for all $h \in H$, and each idempotent measure on G arises in this fashion.*

PROOF. We outline to proof in the following exercise.

EXERCISE 6.4. *Hint. (i) If H and G are compact groups and $\phi: H \to G$ is a morphism of compact groups then the linear operator $M(\phi): M(H) \to M(G)$ is a morphism of Banach algebras preserving the involution mapping $P(H)$ into $P(G)$.*

(ii) Now, apply this to the inclusion $j: H \to G$ and $\mu = M(j)(\gamma^H)$. Thus the fact that γ^H is an idempotent of $P(H)$ by Proposition 5.1 implies that μ is an idempotent in $P(G)$. By Proposition 3.3 we have $\operatorname{supp}(\mu) = j(\operatorname{supp}(\gamma^H)) = j(H) = H$. Let $h \in H$. Then $\gamma^H = \gamma_h^H = \gamma^H * \delta_h^H$ and so $\mu = M(j)(\gamma^H) = M(j)(\gamma^H) * M(j)(\delta_h^H) = \mu * \delta_h^G = \mu_h$. The preceding discussion shows that each idempotent measure on G arises in this fashion □

We have seen that for $\mu \in E\bigl(P(G)\bigr)$ the support $\operatorname{supp}(\mu)$ is a compact subgroup of G. On $E\bigl(P(G)\bigr)$ we have the partial order defined in Definition 1.14.

EXERCISE 6.5. *$\mu, \nu \in E\bigl(P(G)\bigr)$ be idempotent probability measures such that $\mu \leq \nu$. Then $\operatorname{supp}(\nu) \subseteq \operatorname{supp}(\mu)$.*

7. The Category of Compact Affine Monoids

The knowledge of the functional analytical background of the compact affine measure semigroup of a compact semigroup now allows us to complete our theory of locally convex compact affine semigroups by elucidating the universal property of the compact affine monoid $P(S)$ among all locally convex compact affine monoids. This will compel us to discuss some functorial aspects of the objects we have considered individually so far.

A probability measure μ on a compact space is called *simple* if it is a convex combination of a finite set of point measures, that is, if it of the form $\mu = \sum_i r_i \delta_{t_i}$ for a finite set of points $t_i \in S$, where $0 \leq r_i$ and $\sum_i r_i = 1$. The simple probability measures are precisely those that have finite support. The following is easily verified.

LEMMA 7.1. *The subset $P_s(S) \subseteq P(S)$ of simple probability measures is an affine dense subsemigroup of $P(S)$.*

THEOREM 7.2.
  (1) *If S is a compact semigroup, then the mapping $\eta_S: S \to P_s(S)$ defined by $\eta_S(s) = \delta_s$ is a morphism of compact semigroups, as well a homeomorphism onto the family of point masses.*
  (2) *Furthermore, if T is a locally convex compact affine semigroup, then there is a morphism of compact affine semigroups $\epsilon_T: P(T) \to T$ satisfying $\epsilon_T(\sum_i r_i \delta_{t_i}) = \sum_i r_i t_i$ for each simple measure $\sum_i r_i \delta_{t_i} \in P_s(T)$.*
  (3) *A locally convex, compact affine semigroup T is a homomorphic retract of $P(T)$ via $\epsilon_T \circ \eta_T = 1_T$.*

PROOF. For (1), the map η_S is continuous because $P(S)$ inherits its topology from the weak $*$-topology on the unit ball of $M(S)$, and for $s, s' \in S$ and $f \in C(S, \mathbb{R})$, we have

$$(\eta_S(s) * \eta_S(s'))(f) = \int f\, d(\delta_s * \delta_{s'}) = \iint (f \circ m_S)\, d\delta_s\, d\delta_{s'} = f(ss') = \eta_S(ss')(f),$$

so η_S is a semigroup homomorphism and a bijection onto the set of point masses. Since η_S is a continuous map whose domain is compact and whose range is Hausdorff, η_S is a homeomorphism onto its image.

For (2), the mapping $\epsilon \colon P_s(T) \to T$ is clearly well-defined on simple measures. Moreover, a routine calculation shows ϵ is a homomorphism with respect to convolution that also preserves the identity, as well as an affine map on $P_s(T)$. So, we need to show that ϵ extends to a function $\epsilon_T \colon P(T) \to T$ that is a continuous affine monoid homomorphism.

If $\mu \in P(T)$, then since $P_s(T)$ is weak $*$-dense in $P(T)$, we can find a net of simple measures $\{\mu_\alpha\} \subseteq P_s(T)$ with $\lim \mu_\alpha = \mu$. Since T is compact, there is a subnet $(\nu_\beta)_\beta$ of $(\mu_\alpha)_\alpha$ such that $\epsilon(\nu_\beta)$ converges to some $t_\mu \in T$. Then $\lim_\beta \nu_\beta = \mu$ and $\lim_\beta \epsilon(\mu_\beta) = t_\mu$. The definition of the weak $*$-topology implies that $\int f\, d\mu = \lim_\beta \int f\, d\nu_\beta$, and $\lim_\beta f(\epsilon(\nu_\beta)) = f(t_\mu)$ for each $f \in C(T, \mathbb{R})$. Now, if f is affine, then $\int f\, d\nu_\beta = f(\epsilon(\nu_\beta))$, and since the affine maps separate the points, we can conclude that t_μ is the unique limit point of $\epsilon(\mu_\alpha)$. Hence $\epsilon_T \colon P(T) \to T$ is well-defined and continuous.

Part (3) is now obvious. $\square$

REMARK 7.3. The mapping $\epsilon_T \colon P(T) \to T$ defined in part (2) of the Theorem is called the *barycenter map*. It exists in case T is a locally convex, compact affine semigroup, and sends each probability measure to its "center of gravity."

Let CS denote the category of compact semigroups and continuous, semigroup homomorphisms, and let CAS denote the category of locally convex, compact affine semigroups and continuous, affine semigroup homomorphisms. We then have

COROLLARY 7.4. *The assignment $S \to P(S)$ extends to a functor* CS $\to$ CAS *which is left adjoint to the forgetful functor $T \mapsto |T|$, with unit $\eta_S \colon S \to |P(S)|$ and counit $\epsilon_T \colon T \to P(|T|)$.*

PROOF. If $\phi \colon S \to T$ is a morphism of compact semigroups, then a morphism of compact affine spaces $P(\phi) \colon P(S) \to P(T)$ is defined, for any $f \colon T \to \mathbb{R}$, by $P(\phi)(\mu)(f)\langle P(\phi)(\mu), f\rangle = \langle \mu, f \circ \phi\rangle = \int (f \circ \phi)\, d\mu$. Moreover, for $\mu, \nu \in P(S)$ and $f \in C(T, \mathbb{R})$, we have

$$\begin{aligned}
P(\mu *_S \nu)(f) = \int_S (f \circ \phi)\, d(\mu *_S \nu) &= \iint_{S\,S} f \circ \phi \circ m_S\, d\mu d\nu \\
&= \iint_{S\,S} f \circ m_T \circ (\phi \times \phi)\, d\mu d\nu \\
&= (P(\phi)(\mu) *_T P(\phi)(\nu))(f),
\end{aligned}$$

where $m_S \colon S \times S \to S$ and $m_T \colon T \times T \to T$ are the semigroup operations and $*_S, *_T$ denote convolution. Thus $P \colon (P(S), *_S) \to (P(T), *_T)$ is s semigroup homomorphism.

The unit of the adjunction is $\eta_S\colon S \to P(S)$ defined in Theorem 7.2, and the second part of Theorem 7.2 shows that $\epsilon_T\colon P(T) \to T$ serves as a co-unit, a fact we now use.

If T is a compact, affine semigroup and $\phi\colon S \to T$ is a morphism of compact semigroups, then we define $\widehat{\phi}\colon P(S) \to T$ by $\widehat{\phi} = \epsilon_T \circ P(\phi)$. Theorem 7.2 and the first part of the proof show that this is a well-defined morphism of compact, affine monoids, Moreover,

$$\widehat{\phi} \circ \eta_S(s) = \widehat{\phi}(\delta_s) = \epsilon_T \circ P(\phi)(\delta_s) = \epsilon_T(\delta_{\phi(s)}) = \phi(s),$$

as required. The uniqueness of $\widehat{\phi}$ follows from the fact that any morphism of compact affine semigroups $g\colon P(S) \to T$ satisfying $g \circ \eta_S = \phi$ agrees with $\widehat{\phi}$ on the simple measures, and they are dense in $P(S)$. □

This same adjunction cuts down to an adjunction between the category CM of compact monoids and continuous, monoid homomorphisms, and the category CAM of locally convex, compact affine monoids and continuous, affine monoid homomorphisms. It's interesting to note that since $M(S)$ is locally convex for any compact semigroup S, the semigroup $P(S)$ is a compact, convex subset of a locally convex vector space. Hence, Krein-Milman implies that $P(S)$ is the closed convex hull of its set of extreme points. Moreover, a basic result in functional analysis is the $\eta_X(X)$ is the set of extreme points of $P(X)$, for any compact Hausdorff space X (cf. [11, 21]). Of course, this implies that, for a compact affine semigroup T, the unit $\eta_T\colon T \to P(T)$ is affine iff T is degenerate, since all point masses are extreme points in $P(T)$.

Let CG denote the category of compact groups and group homomorphisms, which is a full subcategory of CM. It's tempting to try to restrict the adjunction $P\colon \mathsf{CM} \rightleftarrows \mathsf{CAM}\colon |\ |$ to CG, but there's a hitch: $P(G)$ is not a group unless G is degenerate. To define such an adjunction, we must refine the forgetful functor from CAM to CM.

THEOREM 7.5. *The functor $P\colon \mathsf{CG} \to \mathsf{CAM}$ is left adjoint to the functor $H\colon \mathsf{CAM} \to \mathsf{CG}$ which associates to a locally convex, compact affine monoid T its group of units $H_T(1)$, and associates to a morphism of compact monoids its restriction to the group of units.*

PROOF. Surely $H\colon \mathsf{CM} \to \mathsf{CG}$ is well-defined, since the group of units of a compact monoid is closed, and hence compact, and since morphisms of compact monoids preserve the identity, and hence all units as well. It's then a triviality that the inclusion $\mathsf{CG} \hookrightarrow \mathsf{CM}$ is left adjoint to H, so the result follows, since adjoints compose. □

8. The Monoid of Stochastic Matrices

We next consider a primary example of compact affine monoids, the $n \times n$-stochastic matrices. This material is not new – one finds much of it in the original sources for results about compact affine semigroups, namely the papers of Cohen and Collins [8, 9].

DEFINITION 8.1. An $n \times n$-*stochastic matrix* is an $n \times n$ matrix $M = (x_{ij})_{1 \le i,j \le n}$ with non-negative, real entries satisfying $\sum_j x_{ij} = 1$ for each $i \le n$. We let $\mathsf{ST}(n)$ denote the family of $n \times n$-stochastic matrices.

Here are some basic results about $\mathsf{ST}(n)$:

LEMMA 8.2. *If $X = \{1, \ldots, n\}$, then*
(1) *$P(X) = \{\sum_{i \leq n} r_i \delta_i \mid r_i \in [0,1] \land \sum_i r_i = 1\}$ is a compact, convex subset of $\mathbb{R}^{|X|}$ whose set of extreme points is $\{\delta_i \mid i \in X\}$.*
(2) *$\mathsf{ST}(n)$ defines a family of continuous, affine maps of $P(X)$.*

PROOF. For (1), the fact that X is finite implies that all probability measures on X are simple. It then follows that the extreme points of $P(X)$ are the point masses.

For (2), first note that $P(X)$ inherits its topology from $\mathbb{R}^{|X|}$; recall that this agrees with the weak $*$-topology on $P(X)$, which is the same as the sup-norm topology (cf. [**2**]). Now, if $M \in \mathsf{ST}(n)$, then $M(i) = (x_{i1}, x_{i2}, \ldots, x_{in})$, the ith row of M, forms the coefficients of a probability distribution on X, by definition of $\mathsf{ST}(n)$. If $i \in X$, then define $M(\delta_i) = \sum_{j \leq n} x_{ij} \delta_j$, and extend M to $P(X)$ by:

$$M(m) = \sum_{i \in X} r_i \left(\sum_{j \leq n} x_{ij} \delta_j \right) = \sum_{1 \leq i,j \leq n} r_i x_{ij} \delta_j,$$

where $m = \sum_{i \in X} r_i \delta_i \in P(X)$. This mapping is affine, by definition. $\square$

THEOREM 8.3. *For each $n > 0$, the following hold:*
(1) *$\mathsf{ST}(n)$ is an $n(n-1)$-dimensional compact, affine monoid whose identity is I_n, the $n \times n$ identity matrix.*
(2) *The extreme points of $\mathsf{ST}(n)$ are the matrices having all rows representing point masses – i.e., one entry is 1 and all other entries are 0.*
(3) *The group of units $H(I_n)$ in $\mathsf{ST}(n)$ is the family of $n \times n$-permutation matrices. $H(I_n)$ is contained in the extreme points of $\mathsf{ST}(n)$.*
(4) *The minimal ideal $\mathcal{M}(\mathsf{ST}(n))$ is the set of right zeroes of $\mathsf{ST}(n)$,*

$$\mathcal{M}(\mathsf{ST}(n)) = \{M \in \mathsf{ST}(n) \mid (\forall N \in \mathsf{ST}(n)) \; NM = M\}$$

In fact, the right zeroes of $\mathsf{ST}(n)$ are the matrices having all rows identical.
(5) *Since $\mathcal{M}(\mathsf{ST}(n))$ is a compact convex subspace of $\mathbb{R}^{n^2}$, it is the convex hull of its extreme points, which are precisely the matrices O_k, $k = 1, \ldots, n$, where O_k is the matrix all of whose rows have 1 in the kth column, and all other entries 0. The O_k are extreme points of $\mathsf{ST}(n)$ as well. It follows that $\mathcal{M}(\mathsf{ST}(n)) \subseteq \mathbb{R}^{n^2}$ is a compact convex semigroup of dimension $n-1$.*

PROOF. (1): Since $\mathsf{ST}(n) \subseteq \mathbb{R}^{n^2}$ is a closed subset of $[0,1]^{n^2}$, it is compact and Hausdorff in the inherited topology. The product of stochastic matrices is another such, and $\mathsf{ST}(n)$ contains the identity matrix, so it is a compact monoid. Since convex combinations of stochastic matrices are stochastic, $\mathsf{ST}(n)$ is also an affine monoid.

For $m \in \{1, \ldots, n\}$, let L_m be the linear functional on the n^2-dimensional vector space of all matrices $(x_{jk})_{j,k=1,\ldots,n}$ defined by $L_m((x_{jk})_{j,k=1,\ldots,n}) = x_{m1} + \cdots + x_{mn}$. Then $V = \bigcap_{m=1}^{n} L_m(1)$ is an $n^2 - n = n(n-1)$ dimensional affine variety, and $\mathsf{ST}(n)$ is the intersection of V with the nonnegative hyperquadrant $Q = \{(x_{jk})_{j,k=1,\ldots,n} \mid x_{jk} \geq 0\}$ having $O \in \mathsf{ST}(n)$ defined by $x_{jk} = 1/n$, $j,k = 1, \ldots, n$ in its interior. It follows that $\mathsf{ST}(n)$ contains interior points of V and therefore has dimension $n(n-1)$.

(2): The standard unit vectors in $\mathbb{R}^n$, whose entries are all 0s, except for one 1, are extreme points of their closed, convex hull, and it follows that each matrix in $\mathsf{ST}(n)$ whose rows consist of all such vectors, are extreme points among the stochastic matrices. It's a triviality that any stochastic matrix is a convex combination of these matrices, so the matrices whose rows are the unit vectors comprise the extreme points of $\mathsf{ST}(n)$.

(3). Clearly the identity matrix is the identity for $\mathsf{ST}(n)$, and since all the $n\times n$-permutation matrices are in $\mathsf{ST}(n)$, the group of units contains these matrices. For the converse, Lemma 8.2(2) implies $\mathsf{ST}(n)$ is a monoid of affine selfmaps of the space $P(X)$, where $X = \{1,\dots,n\}$. Since each mapping $M \in \mathsf{ST}(n)$ is affine, it maps extreme points in $P(X)$ to extreme points in $M(P(X))$. If $M \in \mathsf{ST}(n)$ is invertible, then $M(P(X)) = P(X)$. But, Lemma 8.2(1) implies $\eta_S(X)$ is the set of extreme points of $P(X)$, so M maps point masses to point masses, and it is a bijection. Thus, M is a permutation matrix.

(4): Any matrix in $\mathsf{ST}(n)$ whose rows are identical is a right zero, as is easily checked. Such matrices form a closed left ideal $I \subseteq \mathsf{ST}(n)$. On the other hand, if $M \in \mathsf{ST}(n)$ is arbitrary and $N \in ST(n)$ has identical rows, then so does NM, which implies I is an ideal of $\mathsf{ST}(n)$. Since $MN = N$ for any $M \in \mathsf{ST}(n)$ and any $N \in I$, it follows that $I = \mathcal{M}(\mathsf{ST}(n))$.

(5): Since the convex combination of matrices each having identical rows also has identical rows, $\mathcal{M}(\mathsf{ST}(n))$ is a convex space. It's clear that each matrix O_k has identical rows, with each row all 0s except for a 1 in the kth column, is in $\mathcal{M}(\mathsf{ST}(n))$, and it's clear that any matrix having identical rows is a convex combination of the family $\{O_k \mid k \leq n\}$. The matrices O_k are extreme points by (2).

The final statement is now clear, since $\{O_k\}$ is a family of n linearly independent vectors in $\mathbb{R}^{n^2}$, whose convex hull is $n-1$-dimensional. $\square$

DEFINITION 8.4. A matrix $M = (x_{ij})_{1\leq i,j\leq n} \in \mathsf{ST}(n)$ is *doubly stochastic* if $\sum_i x_{ij} = 1$ for each j. We denote the family of doubly stochastic matrices by $\mathsf{DT}(n)$.

COROLLARY 8.5. *The set $\mathsf{DT}(n)$ is an $(n-1)^2$-dimensional compact affine submonoid of $\mathsf{ST}(n)$. The group $H(I_n)$ of units of $\mathsf{ST}(n)$ is contained in $\mathsf{DT}(n)$ and $\mathcal{M}(\mathsf{DT}(n)) = \{O\}$ contains only the zero O of $\mathsf{DT}(n)$, all of whose entries are $\frac{1}{n}$. The compact affine space $\mathsf{DT}(n)$ is the intersection of $\mathsf{ST}(n)$ with an affine subvariety of dimension $n-1$ of the vector space of all $n\times n$ matrices.*

Moreover, $\mathsf{DT}(n)$ is the closed convex hull $\overline{\mathrm{conv}}(H(I_n))$ of the set of permutation matrices.

PROOF. First we determine $\dim \mathsf{DT}(n)$.

We consider the vector space $\mathrm{M}_n(\mathbb{R})$ of all $n\times n$ real matrices and recall from the proof of Theorem 8.3 the linear forms $L_m \in \mathrm{Hom}(\mathrm{M}_n(\mathbb{R}),\mathbb{R})$, $L_m((x_{jk})_{j.k=1,\dots,n}) = x_{1m} + \cdots + x_{nm}$. The vector space S_1 spanned by the L_m, $m = 1,\dots, n$ contains the the linear form $L = L_1 + \cdots + L_n$, $L((x_{jk})_{j.k=1,\dots,n}) = \sum_{j,k=1,\dots,n} x_{jk}$.

Now let L^m, $m = 1,\dots,n$, be the linear functional on $\mathrm{M}_n(\mathbb{R})$ defined by $L^m((x_{jk})_{j,k=1,\dots,n}) = x_{1m} + \cdots + x_{nm}$ and let S_2 be the vector space spanned by the L^m. Then $L \in S_2$ as well, and we claim that $\dim(S_1 \cap S_2) = 1$.

We write $U = \mathbb{R}^n$ with the standard basis $e_j = (\delta_{jk})_{k=1,\dots,n}$ for the Kronecker deltas. Set $e = e_1 + \cdots + e_n$. We may identify $\mathrm{Hom}(\mathrm{M}_n(\mathbb{R}),\mathbb{R})$ with $U \otimes U$ in such a fashion that

$$\begin{aligned}
L_m &= e_m \otimes e_1 + \cdots + e_m \otimes e_n &= e_m &\otimes e, \\
L^m &= e_1 \otimes e_1 + \cdots + e_n \otimes e_m &= e &\otimes e_m, \\
L &= \textstyle\sum_{j,k=1}^n e_j \otimes e_k &= e &\otimes e.
\end{aligned}$$

The vectors $e_1, \cdots, e_{n-1}, e$ form a basis of U and thus the set of vectors

$$B \stackrel{\text{def}}{=} \{e_m \otimes e, \quad m = 1, \ldots, n-1; \quad e \otimes e; \quad e \otimes e_m, \quad m = 1, \ldots n-1\}$$

is linearly independent in $U \otimes U$. We have

$$\begin{aligned}
S_1 &= \mathbb{R}{\cdot}(e_1 \otimes e) + \cdots + \mathbb{R}{\cdot}(e_{n-1} \otimes e) + \mathbb{R}{\cdot}(e \otimes e), \\
S_2 &= \mathbb{R}{\cdot}(e \otimes e_1) + \cdots + \mathbb{R}{\cdot}(e \otimes e_{n-1}) + \mathbb{R}{\cdot}(e \otimes e), \\
S_1 + S_2 &= \textstyle\sum_{m=1}^n \mathbb{R}{\cdot}(e_m \otimes e) + \mathbb{R}{\cdot}(e \otimes e) + \sum_{m=1}^n \mathbb{R}{\cdot}(e \otimes e_m).
\end{aligned}$$

Thus $S_1 + S_2 = \operatorname{span} B$ and $S_1 \cap S_2 = \mathbb{R}{\cdot}(e \otimes e)$ and so

$$\dim(S_1 + S_2) = \dim S_1 + \dim S_2 - \dim(S_1 \cap S_2) = 2n - 1$$

The annihilator $(S_1 + S_2)^\perp$ of $S_1 + S_2 \subseteq \operatorname{Hom}(\mathrm{M}_n(\mathbb{R}), \mathbb{R})$ in $\mathrm{M}_n(\mathbb{R})$ thus has the dimension $n^2 - (2n-1) = (n-1)^2$.

Set $W = \bigcap_{m=1}^n (L^m)^{-1}(1)$. Then $V \cap W = (S_1 + S_2)^\perp + E_n$, where E_n is the unit matrix is an $(n-1)^2$-dimensional affine variety, and $\mathsf{DT}(n)$ is the intersection of $V \cap W$ with Q containing O (as in the proof of Theorem 8.3(2) in its interior. Hence $\mathsf{DT}(n)$ is $(n-1)^2$-dimensional.

Clearly the family of doubly stochastic matrices is a closed subset of $\mathsf{ST}(n)$, hence it is compact. Moreover, a convex combination of doubly stochastic matrices is clearly another such, so $\mathsf{DT}(n)$ is a convex subset of $\mathsf{ST}(n)$. It's routine to show that the product of doubly stochastic matrices is again doubly stochastic, so $\mathsf{DT}(n)$ is a subsemigroup of $\mathsf{ST}(n)$ that clearly contains the permutation matrices, which are $H(I_n)$. Hence $\mathsf{DT}(n)$ is a compact, affine submonoid of $\mathsf{DT}(n)$. The claim about $\mathcal{M}(\mathsf{DT}(n))$ follows from the fact that the only right zero in $\mathsf{ST}(n)$—i.e., a stochastic matrix all of whose rows are identical—that is doubly stochastic must have all entries $\frac{1}{n}$.

For the final claim, it's clear that $\overline{\operatorname{conv}}(H(I_n)) \subseteq \mathsf{DT}(n)$, since the latter is compact and convex. The reverse containment is the content of the Birkhoff-von Neumann Theorem (cf. [**19**]). □

The next result, which is somewhat technical, will be needed in Section 11.

PROPOSITION 8.6. *If $M \in \mathsf{ST}(n)$, then*

$$\mathcal{M}(\mathsf{ST}(n)M) = \mathsf{ST}(n)M \cap \mathcal{M}(\mathsf{ST}(n)) = \mathcal{M}(\mathsf{ST}(n))M.$$

PROOF. First, $\mathcal{M}(\mathsf{ST}(n)) \subseteq \mathsf{ST}(n) \cap \mathcal{M}(\mathsf{ST}(n))$ follows from $\mathcal{M}(\mathsf{ST}(n)M) \subseteq \mathcal{M}(\mathsf{ST}(n))$, which is obvious, and from the fact that $\mathsf{ST}(n)M \cap \mathcal{M}(\mathsf{ST}(n))$ is an ideal of $\mathsf{ST}(n)M$. For the reverse inclusion, observe that every element of $\mathsf{ST}(n)M \cap \mathcal{M}(\mathsf{ST}(n))$ is a right zero of $\mathsf{ST}(n)$, using Theorem 8.3(4), hence an element of $\mathcal{M}(\mathsf{ST}(n))M$. Next, $\mathcal{M}(\mathsf{ST}(n))M$ is an ideal of $\mathsf{ST}(n)M$, and so it contains $\mathcal{M}(\mathsf{ST}(n)M)$; conversely, $\mathcal{M}(\mathsf{ST}(n))M \subseteq \mathsf{ST}(n))M \cap \mathcal{M}(\mathsf{ST}(n)) = \mathcal{M}(\mathsf{ST}(n)M)$. This shows the two equalities in the assertion. □

The technical nature of the last result can be simplified if we recall that $\mathsf{ST}(n)$ is a transformation semigroup.

DEFINITION 8.7. Let X be a set. The *full transformation semigroup over X* is the family of selfmaps of X. This semigroup is denoted $T(X)$.

Clearly, $\mathsf{ST}(n)$ is a transformation semigroup.

PROPOSITION 8.8. *Let S be a transformation semigroup on the set X. Then*
(1) $\mathcal{M}(T(X)) = \{\rho_x \mid x \in X\}$, where $\rho_x \colon X \to X$ is $\rho_x(y) = x$.
(2) *If $S \cap \mathcal{M}(T(X)) \neq \emptyset$, then $\mathcal{M}(S) = S \cap \mathcal{M}(T(X))$.*
(3) *If $S \cap \mathcal{M}(T(X)) \neq \emptyset$, then $\mathcal{M}(S)s \subseteq \{\rho_x \mid x \in s(X)\}$ ($\forall s \in S$).*
(4) *If $\mathcal{M}(S) = \mathcal{M}(T(X))$, then $\mathcal{M}(S)s = \{\rho_x \mid x \in s(X)\}$ ($\forall s \in S$).*

PROOF. For (1), if $f \in T(X)$ and $x \in X$, then $f\rho_x = \rho_x$, while $\rho_x f = \rho_{f(x)}$. It follows that $\{\rho_x \mid x \in X\}$ is an ideal of $T(X)$, and since each ρ_x is a right zero in $T(X)$, we conclude that $\mathcal{M}(T(X)) = \{\rho_x \mid x \in X\}$.

For (2), that $\mathcal{M}(T(X)) \cap S$ is an ideal of S is clear, and from (1) we know $\mathcal{M}(T(X)) \cap S$ is a family of right zeros, and hence is contained in any ideal if S.

For (3), note that $\mathcal{M}(S)s = (\mathcal{M}(T(X)) \cap S)s$ is a family of right zeroes. If $\rho_x \in S$, then $x = \rho_x(y) \in \{s(y) \mid s \in S\}$.

Finally we prove (4). Suppose $\mathcal{M}(S) = \mathcal{M}(T(X))$, and let $s \in S$ and $x \in s(X)$. Then $x = s(y)$ for some $y \in X$. Since $\mathcal{M}(S) = \mathcal{M}(T(X))$, $\rho_y \in S$, and then $\rho_y s = \rho_x$, so $\rho_x = \rho_y s \in Ss$. But $\rho_y \in \mathcal{M}(S)$ by (2), so $\rho_x = \rho_z s \in \mathcal{M}(S)s$. Hence $\mathcal{M}(S)s \supseteq \{\rho_x \mid x \in s(X)\}$, and the reverse containment then follows from (3). □

COROLLARY 8.9. *For each $n > 0$, if $M \in \mathsf{ST}(n)$, then*
$$\mathcal{M}(\mathsf{ST}(n)M) \stackrel{\phi}{\simeq} \overline{\mathrm{conv}}(M(1), \ldots, M(n)) = M(P(n)),$$
where $\phi(\sum_{1 \leq i \leq n} r_i M(i)) = M(\sum_{1 \leq i \leq n} r_i \delta_i)$ for $i \in \underline{n}$.

PROOF. Theorem 8.3(4) implies the members of $\mathcal{M}(\mathsf{ST}(n))$ are those matrices all of whose rows are the same. For $p \in P(n)$, the constant selfmap $\rho_p \colon P(n) \to P(n)$ is represented by the stochastic matrix M_p all of whose rows are p, and then Proposition 8.8(4) implies $\mathcal{M}(\mathsf{ST}(n))M = \{\rho_p \mid p \in M(P(n))\} = \{M_{M(p)} \mid p \in P(n)\}$. Then the isomorphism $\phi \colon \mathcal{M}(\mathsf{ST}(n))M \to M(P(n))$ is given by $\phi(M_{M(p)}) = M(p)$, from which it follows that
$$\phi(M_{\sum_{1 \leq i \leq n} r_i M(i)}) = \sum_{1 \leq i \leq n} r_i M(i) = M\left(\sum_{1 \leq i \leq n} r_i \delta_i\right),$$
the last equality following from the fact that M acts affinely on $P(n)$. □

We conclude the discussion of $\mathsf{ST}(n)$ with an example relating to the matrices O_k described in Theorem 8.3(5). By a *channel*, we simply mean a mechanism that takes inputs and transforms them to outputs, usually introducing some noise in the process. These objects are the focus of study in information theory, a theme we explore in the next part of the paper.

A *Z-channel* is a 2×2-stochastic matrix of the form
$$Z = \begin{pmatrix} (1-p) & p \\ 0 & 1 \end{pmatrix}.$$

Z can be interpreted as a channel that sends 1 to 1 with probability 1, but that sends 0 to 1 with probability p. This probability – p – is called the *crossover probability of Z*. Such a channel is a convex combination

$$Z = \begin{pmatrix} (1-p) & p \\ 0 & 1 \end{pmatrix} = (1-p)\begin{pmatrix} 1 & 0 \\ 0 & 1 \end{pmatrix} + p\begin{pmatrix} 0 & 1 \\ 0 & 1 \end{pmatrix}$$

of the identity I_2 and O_2. The family of all such convex combinations, $(1-p) \cdot I_2 + p \cdot O_2$ forms a *one-parameter semigroup* from I_2 to O_2; that is, the mapping

$$p \mapsto (1-p) \cdot I_2 + p \cdot O_2 \;:\; ([0,1], \cdot) \to \mathsf{ST}(2)$$

is a monoid homomorphism, where we endow $[0,1]$ with the multiplication $(r,s) \mapsto r + s - rs$. More generally,

DEFINITION 8.10. Given $n > 0$, we define a *Z-channel* in $\mathsf{ST}(n)$ to be a convex combination $X = (1-p)I_n + pO_k$, where $p \in [0,1]$ and $0 \leq k \leq n$. We denote that family of Z-channels in $\mathsf{ST}(n)$ by $Z(n)$; i.e.,

$$Z(n) = \{(1-p)I_n + pO_k \mid p \in [0,1] \;\&\; k \leq n\},$$

and we let $\mathsf{ZT}(n) = \overline{\mathrm{conv}}(Z(n))$ denote the closed, convex hull of $Z(n)$.

DEFINITION 8.11. Let S be a semigroup, and let $X \subseteq S$. The *normalizer of X* is $N(X) = \{s \in S \mid sX = Xs\}$.

PROPOSITION 8.12. *For each $n > 0$, $\mathsf{ZT}(n)$, the closed convex hull of $Z(n)$ satisfies:*
 (1) *$\mathsf{ZT}(n)$ is a compact affine submonoid of $\mathsf{ST}(n)$ whose only unit is I_n.*
 (2) *The minimal ideal of $\mathsf{ZT}(n)$ is $\mathcal{M}(\mathsf{ST}(n))$.*
 (3) *$\mathsf{ZT}(n)$ is a union of line segments running from I_n to elements in $\mathcal{M}(\mathsf{ST}(n))$. Each of these line segments is a one-parameter semigroup.*
 (4) *$H(I_n)$ is contained in the normalizer of $\mathsf{ZT}(n)$.*

PROOF. $\mathsf{ZT}(n)$ is compact and convex by definition. By Theorem 8.3(5), $\mathcal{M}(\mathsf{ST}(n)) = \overline{\mathrm{conv}}(\{O_k \mid k \leq n\})$, so $\mathcal{M}(\mathsf{ST}(n))$ is also compact and convex, and then we can realize $\mathsf{ZT}(n) = \overline{\mathrm{conv}}(\{I_n\} \cup \mathcal{M}(\mathsf{ST}(n)))$, which implies $\mathsf{ZT}(n)$ is the union of line segments from I_n to some element in $\mathcal{M}(\mathsf{ST}(n))$ by Proposition 1.34. That each of these is a one-parameter semigroup follows from Lemma 1.30 and the fact that each element of $\mathcal{M}(\mathsf{ST}(n))$ is an idempotent which lies below the identity. Finally, it's clear that $H \cdot I_n = I_n \cdot H$, and that $H \cdot \mathcal{M}(\mathsf{ST}(n)) = \mathcal{M}(\mathsf{ST}(n)) \cdot H$ for each $H \in H(I_n)$, and since $\mathsf{ST}(n)$ is an affine semigroup, it follows that $H \cdot \overline{\mathrm{conv}}(\{I_n\} \cup \mathcal{M}(\mathsf{ST}(n))) = \overline{\mathrm{conv}}(\{I_n\} \cup \mathcal{M}(\mathsf{ST}(n))) \cdot H$, for each $H \in H(I_n)$. But $\mathsf{ZT}(n) = \overline{\mathrm{conv}}(\{I_n\} \cup \mathcal{M}(\mathsf{ST}(n)))$. □

Quite generally, due to Theorem 8.3, the submonoid $\mathsf{ST}(n)$ of stochastic matrices in the monoid of all $n \times n$ real matrices is a locally convex compact affine monoid with a right zero minimal ideal, and thus everything we said in Sections 1.1 – 1.3 (namely, Definition 1.16–Corollary 1.33) applies to this finite-dimensional monoid.

8.1. Postscript. The proof of the existence and uniqueness of Haar measure on a compact group that we present here is a modern version of a proof due to JAMES WENDEL (see [40]). This proof was one of the earliest applications to harmonic analysis of the theory of compact topological semigroups, which was in the process of being developed at the time Wendel's proof was published. It is an appropriate choice in the context of our presentation of compact groups; it provides information on compact groups beyond the existence of Haar measure as such, and it is compatible with the spirit of topological algebra, providing a good deal of information on the semigroup structure of the compact affine monoid of probability measures $P(G)$ on G. Wendel's proof, on the other hand, does not yield a proof of the existence of Haar measure on locally compact, noncompact groups. Our presentation here is new in the sense that, first, it puts forward more clearly what kind of measure theory is needed (namely, the rudiments of BOURBAKI's Radon measure theory on compact spaces to the extent they are presented here from first principles), and, second, that it separates cleanly the basic theory of compact semigroups and brings the two together in the last minute to obtain the result as a blend of analysis and topological algebra.

Our proof also varies from Wendel's original in the proof that a compact, affine group is a point (Theorem 1.20). The proof known at the time of Wendel's paper relied on the Krein-Milman Theorem, which in turn required assuming the group was a subspace of a locally convex vector space. Our approach not only removes this assumption, it also applies in the setting of a much more abstract notion of an affine semigroup (Definition 1.16).

Our presentation of the universal properties of the compact affine monoid $P(G)$ emphasizes on a general level the significance of this monoid beyond the existence and uniqueness proof of Haar measure.

Finally, the monoid of stochastic matrices has been the subject of considerable attention over the years, notably in the work of Mukherjea and Tserpes [29], and later Högnäs and Mukherjea [19]. In anticipation of work in the next section, we have focused our discussion on the structure of $\mathcal{M}(\mathsf{ST}(n))$ and the example above on the example of Z-channels, which arise in information theory. An example of the application of compact monoid theory is a natural generalization of Z-channels from the binary case, and that they form a compact convex submonoid spanned by one-parameter submonoids. We return to this example at the end of the paper.

Part 3. Domains and Information

Information theory arose from the seminal work of Claude Shannon [35] in the 1940s. Shannon's theory gives a precise meaning to *information* and provides a mechanism to evaluate the information in a *channel* – a mapping from inputs to outputs. The tools Shannon's theory provides involve the concepts of *entropy, mutual information,* and *capacity,* the last of which gives a precise measure of the rate at which information can be pushed through the channel without information loss. Much of the work in information theory focuses on individual channels and their characteristics, using statistical analyses to understand the capabilities of a channel, the potential for information leakage, etc. In the seminal work [25], a new approach to understanding channels was put forward, one in which the focus is on the family of all channels from a set I of inputs to a set O of outputs. The point of departure for [25] is that the family of channels for the case of *binary channels* ($I = O = \{0, 1\}$) forms a compact monoid whose structure offers insights into the information-theoretic properties of binary channels. In establishing their results, the authors also apply *domain theory* [1, 14], a tool that arises in the theory of computation.

Here, we elaborate on the results in [**25**], broadening the range of examples to which the general approach applies from binary to n−ary channels. This means generalizing from 2×2 stochastic matrices – which represent binary channels – to $\mathsf{ST}(n)$, the $n \times n$-stochastic matrices, which have long been known to form a compact affine monoid (cf., e.g., [**19**]) – what's new in [**25**] is the application to Shannon information theory. Our goal is to extend the results of Martin, et al., and to clarify the role of domain theory in this setting, showing how domains that can be used to analyze channel behavior arise in this setting; in particular, we show that *capacity* induces a corresponding map on a naturally-defined quotient of $\mathsf{ST}(n)$ which is a domain (but *not* a quotient monoid).

9. Domains and Compact Monoids

Our next goal is to relate the results about the monoid of stochastic matrices to their representation of channels in information theory. To accomplish this, we first need to recall some facts from domain theory. Domain theory has its origins in the theory of computation, where it was initially used to provide mathematical models for abstract programming languages. More recently, its use has spread to a number of areas, owing to the fact that domains capture the essence of a number of phenomena that arise in many different areas. The use of domains in physics, and the adaptation of "classical domain theory" to accommodate the needs of this and other areas, was pioneered by Martin [**23**] and by Coecke and Martin [**7**]. We briefly review the relevant aspects here, but the reader is advised to check these references for more details about the constructions involved.

We begin with some terminology. A non-empty subset $D \subseteq P$ of a partially ordered set is *directed* if any pair of elements in D has an upper bound *in D*. P is *directed complete* if every directed subset of P has a least upper bound in P. If $x, y \in P$, then we write $x \ll y$ iff for each directed subset $D \subseteq P$, if $y \leq \sup D$, then $(\exists d \in D)\ x \leq d$. If this holds, we say, "x is *way-below* y". We let $\downarrow\!y \stackrel{\text{def}}{=} \{x \in P \mid x \ll y\}$ denote the set of elements of P that are way-below y. A mapping $f \colon P \to Q$ between partial orders is *Scott continuous* if f is monotone and $f(\sup D) = \sup f(D)$ for all $D \subseteq P$ directed.

DEFINITION 9.1. A *domain* is a directed complete partial order P in which $\downarrow\!y$ is directed and $y = \sup \downarrow\!y$, for all $y \in P$.

A simple example of a domain is $([0,1], \leq)$, where $x \ll y$ iff $x = 0$ or $x < y$. There are a number of natural topologies that arise on a domain. The *Scott topology* has for its open sets those $U \subseteq P$ satisfying:

- $U = \uparrow\!U \stackrel{\text{def}}{=} \{x \in P \mid (\exists u \in U)\ u \leq x\}$, and
- If $\sup D \in U$ for a directed set D, then $D \cap U \neq \emptyset$.

The Scott topology on a poset is always T_0; it is *sober* (cf. [**1, 14**]) if P is a domain. On a domain, the family $\{\Uparrow\!y \mid y \in P\}$ is a base for the Scott topology. A mapping between partial orders is Scott continuous in the sense defined above, if and only if it is monotone and continuous with respect to the Scott topologies on its domain and range.

A refinement of the Scott topology can be obtained by adding as open sets those sets of the form $P \setminus \uparrow\!F$, where $F \subseteq P$ is finite, and $\uparrow\!F = \{x \in P \mid (\exists y \in F)\ y \leq x\}$. This is called the *Lawson topology* on P; it is always Hausdorff on a domain, and when it is compact, P is called a *coherent domain*. For example, the Scott topology on $[0,1]$ is the topology of lower semi-continuity: the family $\{(r,1] \mid r < 1\} \cup \{[0,1]\}$ is the set of Scott open sets. The Lawson topology on $[0,1]$ is just the usual topology.

We also want to consider Lawson continuous monotone maps between coherent domains. Recall that a map $f\colon X \to Y$ between topological spaces is *proper* (also called *perfect*) if

- $\downarrow f(A)$ is closed in Y if $A \subseteq X$ is closed, where $\downarrow f(A) = \{y \in Y \mid y \in f(x) \text{ for some } x \in A\}$, and
- $f^{-1}(C) \subseteq X$ is compact for each compact saturated subset $C \subseteq Y$, where a set is *saturated* if it is the intersection of open sets..

If X and Y are domains, and f is Scott continuous, then these two properties reduce to the first one (cf. Lemma VI-6.21, [14]). We also say f is *proper at* $x \in X$ if $f^{-1}(\uparrow y) \cap \downarrow x$ is Scott compact in $\downarrow x$, for each $y \in Y$.

We collect some properties of coherent domains. The proofs can be found in [14]: part (1) is contained in Proposition III-1.6 and its proof, while the first statement in part (2) is a direct consequence of the definition of the Lawson topology, and the last part follows from Lemma VI-6.1 and Proposition VI-6.24.

PROPOSITION 9.2. *Let P and Q be coherent domains, and let $f\colon P \to Q$ be Scott continuous.*

(1) *If $X \subseteq P$ is Scott closed, then the inherited Scott (respectively, Lawson) topology on X is the same as the intrinsic Scott (respectively, Lawson) topology on X.*

(2) *f is Lawson continuous iff, for all $y \in Q$, $f^{-1}(\uparrow y) \subseteq P$ is closed in the Lawson topology, iff $f^{-1}(\uparrow y) \subseteq P$ is compact in the Scott topology, iff f is proper.*

From (2), *it follows that the monotone Lawson-continuous maps between coherent domains are exactly the Scott-continuous proper maps.* □

An example of importance to our work is the following. Let X be a compact Hausdorff space. The family $\mathsf{CL}(X)$ of non-empty closed subsets of X is a domain under reverse inclusion. Indeed, a filter basis of compact non-empty subsets of a Hausdorff space has a non-empty compact intersection (this result is well-known in topology; it has a generalization to sober spaces that is a corollary of the Hofmann-Mislove Theorem (cf. [14])). In this partial order, $C \ll D$ iff $D \subseteq C^\circ$, the interior of C. The domain $(\mathsf{CL}(X), \supseteq)$ also is coherent.

REMARK 9.3. Topologists are familiar with the hyperspace $\mathsf{CL}(X)$ of closed subsets of X in the *Vietoris topology*, which has for a sub basis sets of the form $W(U,V) = \{C \in \mathsf{CL}(X) \mid A \subseteq U \wedge A \cap V \neq \emptyset\}$, where $U, V \subseteq X$ are open. This hyperspace is again compact and Hausdorff. The Lawson topology on $\mathsf{CL}(X)$ agrees with the Vietoris topology: indeed, sets of the form $\{C \in \mathsf{CL}(X) \mid C \subseteq U\}$ are Scott open, and those of the form $\{C \in \mathsf{CL}(X) \mid C \cap V \neq \emptyset\}$ correspond to the open lower sets (cf. see Example VI-3.8 of [14]). Thus domain theory gives us an order-theoretical approach to analyzing this hyperspace.

If X is also a convex subset of a locally convex vector space, then the family $\mathsf{Conv}(X) = \{C \subseteq X \mid C = \overline{\mathrm{conv}}(C)\}$ of (non-empty) closed convex subsets of X is a retract of $\mathsf{CL}(X)$ under the mapping $C \mapsto \overline{\mathrm{conv}}(C)$; it follows that $(\mathsf{Conv}(X), \supseteq)$ also is a domain, where again, $C \ll D$ iff $D \subseteq C^\circ$. The mapping $C \mapsto \overline{\mathrm{conv}}(C)$

is an example of a *kernel operator:* a selfmap $k\colon P \to P$ of a domain satisfying $k(x) \leq x$ for each $x \in P$. In this case, the kernel operator is Scott continuous, which assures that its image is also a domain. Details of these results are presented in Proposition 10.6.

Another concept that will be useful is that of a *measurement* on a domain. A measurement is a Scott-continuous function $\mu\colon P \to [0,\infty)^{op}$ from P into the non-negative reals in the *dual order:* $r \sqsubseteq s$ iff $s \leq r$ in the usual order.

DEFINITION 9.4. Let P be a dcpo.
(1) A Scott continuous map $\mu\colon P \to [0,\infty)^{op}$ *measures the content at* x if, for each Scott-open set $U \subseteq P$, $x \in U$ implies

$$(\exists \epsilon > 0)\ x \in \mu_\epsilon(x) \stackrel{\text{def}}{=} \{y \in P \mid y \leq x\ \&\ |\mu(y) - \mu(x)| < \epsilon\}$$
$$= \{y \in P \mid y \leq x\ \&\ \mu(y) < \mu(x) + \epsilon\} \subseteq U,$$

the last equality following from the monotonicity of μ.
(2) $\mu\colon P \to [0,\infty)^{op}$ is a *measurement* if μ measures the content at x, for each $x \in \ker(\mu) \stackrel{\text{def}}{=} \{x \in P \mid \mu(x) = 0\}$.
(3) If $\mu\colon P \to [0,\infty)^{op}$ measures the content at x for every $x \in P$, then we say μ *measures* P.

EXAMPLE 9.5. Consider the domain $\mathbf{I}([0,1]) = (\{[a,b] \mid 0 \leq a \leq b \leq 1\}, \supseteq)$ of non-empty subintervals of the unit interval, ordered by reverse inclusion. Then $[a,b] \ll [c,d]$ iff $[c,d] \subseteq [a,b]^\circ$, and the maximal elements of $\mathbf{I}([0,1])$ are the singleton sets. An example of a measurement is the mapping $\mu\colon \mathbf{I}([0,1]) \to [0,\infty)^{op}$ by $\mu([a,b]) = b - a$, the length of the interval. This mapping is clearly Scott continuous ($\inf\{\mu([b,a]) \mid [b,a] \in \mathcal{F}\} = \mu(\cap \mathcal{F})$ for a filter basis of intervals), and if $U \subseteq \mathbf{I}([0,1])$ is Scott open, then $[a,b] \in U$ implies there is some $\epsilon > 0$ with $\uparrow([a-\frac{\epsilon}{2}, b+\frac{\epsilon}{2}] \cap [0,1]) \subseteq U$. Then $[c,d] \in \mu_\epsilon([a,b])$ implies $[c,d] \supseteq [a,b]$ and $(d-c) - (b-a) < \epsilon$, so $[c,d] \in \uparrow([a-2\epsilon, b+2\epsilon] \cap [0,1]) \subseteq U$, as required. Thus length measures $\mathbf{I}([0,1])$.

One important property of measurements on domains is that they allow proofs that *non-monotone* maps have fixed points. In [23], several examples are given of computational processes that do not give rise to Scott-continuous maps, but for which appropriate measurements on their domains provide a method to show the processes have unique fixed points, thus allowing a precise definition of such processes. While Martin's thesis [23] gives examples of fixed point theorems, such as Brouwer's Fixed Point Theorem, that can be proved using measurements, the results have been substantially extended by the work of Waszkiewicz [38] who showed that domains whose Scott topology is partially metrizable are exactly those that admit measurements. This opened the door to results such as those by Valero [36] who proved generalizations of the Banach Fixed Point Theorem for partial metric spaces.

DEFINITION 9.6. Let $f\colon P \to Q$ be a monotone mapping between partially ordered sets. Recall that f is *strictly monotone* if $x < y \in P$ implies $f(x) < f(y) \in Q$.

PROPOSITION 9.7. *Let P be a domain and let $m\colon P \to [0,\infty)^{op}$ be Scott continuous. If m is proper at $x \in P$, then the following are equivalent:*
(1) *m measures the content at x.*

(2) m is strictly monotone at $x \in P$: i.e., $y \leq x$ & $m(y) = m(x) \Rightarrow y = x$. In particular, a Scott-continuous, proper map $m \colon P \to [0,\infty)^{op}$ is a measurement iff m is strictly monotone at each $x \in \ker m$.

PROOF. (1) implies (2): Suppose m measures the content at $x \in P$ and let $y \leq x$ with $m(y) = m(x)$. Then, for each Scott-open set $U \subseteq P$ with $x \in U$, there is some $\epsilon > 0$ with $x \in m_\epsilon(x) \subseteq U$. But then $y \in m_\epsilon(x) \stackrel{\text{def}}{=} \{z \leq x \mid |m(z) - m(x)| < \epsilon\}$. Since m is Scott continuous, $\cap \{m_\epsilon(x) \mid \epsilon > 0\} = \uparrow x$, so $y \in \uparrow x$, which implies $y = x$, as claimed.

(2) implies (1): Let $x \in P$ and let $U \subseteq P$ be Scott open with $x \in U$. Let $y \leq x$ with $m(y) = m(x)$, and let $\mathcal{F} = \{m^{-1}(\uparrow(m(x) + \epsilon)) \mid \epsilon > 0\}$ (note that $\uparrow r = \{s \in [0, \infty) \mid s \leq r\}$ since we are using the dual order). Then $y \in X \cap \downarrow x$ for each $X \in \mathcal{F}$. Since m is Scott continuous and proper at x, $\{X \cap \downarrow x \mid X \in \mathcal{F}\}$ is a filter basis of Scott-compact sets in the dcpo $\downarrow x$. Moreover, $z \in \cap \{X \cap \downarrow x \mid X \in \mathcal{F}\}$ implies $z \leq x$ and $m(z) \leq m(x) + \epsilon$ for all $\epsilon > 0$, so $m(z) = m(x)$. Since m is strictly monotone, it follows that $z = x$. Hence $\cap \{X \cap \downarrow x \mid X \in \mathcal{F}\} = \{x\} \subseteq U$, and since U is Scott open, there is some $X \in \mathcal{F}$ with $X \cap \downarrow x \subseteq U$. By definition of $\mathcal{F}$, $X = m^{-1}(\uparrow(m(x) + \epsilon))$ for some $\epsilon > 0$, and so $\{y \in \downarrow x \mid m(y) < m(x) + \epsilon\} \subseteq U$. □

The following will be useful in the next section.

COROLLARY 9.8. Let P be a coherent domain and let $m \colon P \to [0,\infty)^{op}$ be Lawson continuous and monotone. Then m measures P iff m is strictly monotone at each $x \in P$. □

PROOF. This follows from the Proposition since a monotone Lawson-continuous map is proper at x for each $x \in P$ by Proposition 9.2(2). □

REMARK 9.9. Proposition 9.7 and Corollary 9.8 can be derived using the principal results in [**27**], but the characterization using proper maps is not stated there.

10. Some Results from Information Theory

We next recall some concepts from Information Theory; a standard reference is [**10**], but the results also can be found in the original source [**35**]. Given a discrete probability space (X, p_X), Shannon's theory [**35**] defines the (base 2) *entropy* of X as

$$\mathcal{H}(X) = \sum_{x \in X} p_X(x) \log_2 \frac{1}{p_X(x)}.$$

(While it is traditional to use H to denote entropy, we use $\mathcal{H}$ because we have already used H to denote the maximal group of an idempotent in a semigroup.) The quantity $\mathcal{H}(X)$ represents the average information in X. Given a second discrete probability space $(Y, p(\cdot \mid x))$ for each $x \in X$, the *conditional entropy in Y given X* is

$$(10.1) \quad \mathcal{H}(Y|X) = \sum_{x \in X} p_X(x) \mathcal{H}(Y|X = x) = \sum_{x \in X} p_X(x) \sum_{y \in Y} p(y|x) \log_2 \frac{1}{p(y|x)},$$

and the *mutual information in X and Y* is

$$\mathcal{I}(Y, X) = \mathcal{H}(Y) - \mathcal{H}(Y|X).$$

A *channel* from inputs X to outputs Y is a mapping $C \colon X \to P(Y)$ that maps each input to a distribution on Y. The idea is that a channel may introduce noise into the transmission of data, but we assume the channel is lossless. In other words, $C = (p(y|x))$ is an $X \times Y$-matrix whose (x, y)-entry is the *conditional probability* of output y occurring, given that the input was x. Note that each distribution p on the inputs X then produces a corresponding distribution on Y, the outputs, which we denote $p \cdot C$ (here we adopt the dot product notation to denote the dot product of the vector p and the matrix C). Then, letting $[P(X) \to P(Y)]$ denote the functions from distributions on X to distributions on Y, the capacity of a channel is given by

$$\mathsf{Cap}\colon [P(X) \to P(Y)] \to [0,1] \quad \text{by} \quad \mathsf{Cap}(C) = \sup_{p \in Pr(X)} \mathcal{H}(p \cdot C) - \mathcal{H}(p \cdot C \mid p),$$

(where $p \cdot C = M(C)(p)$, regarding C as a mapping $M(C)$ on measures) i.e., $\mathsf{Cap}(C)$ is the supremum of the possible mutual information values $\mathcal{I}(p \cdot C, p)$ as p ranges over the distributions on X, the set of inputs.

EXAMPLE 10.1 (Binary Channels). The prototypical example is that of a *binary channel*: Let $X = Y = \{0, 1\}$. Then a channel is a 2×2-matrix

$$C = \begin{pmatrix} a & 1-a \\ b & 1-b \end{pmatrix},$$

where $0 \leq a, b \leq 1$. In the binary setting, a distribution on $X = \{0, 1\}$ is a pair $p = [x \mid 1 - x]$ which assigns probability x to 0, and probability $1 - x$ to 1; note that it is traditional to denote the entropy of p by $\mathcal{H}(x)$.

Given a distribution $p = [x \mid 1-x]$ on $X = \{0, 1\}$, the corresponding distribution on $Y = \{0, 1\}$ under the channel C is then

$$p \cdot C = [xa + (1-x)b \mid x(1-a) + (1-x)(1-b)],$$

and the capacity of C is

$$\begin{aligned}\mathsf{Cap}(C) &= \sup_{p \in Pr(X)} \mathcal{H}(p \cdot C) - \mathcal{H}(p \cdot C \mid p) \\ &= \sup_{r \in [0,1]} \mathcal{H}(ra + (1-r)b) - (r\mathcal{H}(a) + (1-r)\mathcal{H}(b)),\end{aligned}$$

where the second equality follows from Equation (10.1). Assuming $b < a$, we can visualize this as follows: The curve $y = \mathcal{H}(x)$ from $x = b$ to $x = a$ corresponds

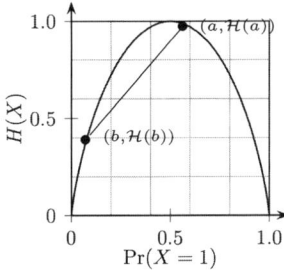

FIGURE 1. Visualizing $\mathsf{Cap}(C)$.

to $\mathcal{H}(ra + (1-r)b)$, while the line joining $(b, \mathcal{H}(b))$ to $(a, \mathcal{H}(a))$ corresponds to $r\mathcal{H}(a) + (1-r)\mathcal{H}(b)$, both as r runs from 0 to 1. Thus,

$$r \mapsto \mathcal{H}(ra + (1-r)b) - (r\mathcal{H}(a) + (1-r)\mathcal{H}(b))\colon [0,1] \to [0,1]$$

gives the vertical distance between the point $(r, \mathcal{H}(ra + (1 - r)b))$ and point $(r, r\mathcal{H}(a) + (1 - r)\mathcal{H}(b))$; this is a continuous function from $[0, 1]$ to $[0, 1]$, being the difference of two such. This mapping extends to a continuous selfmap of $\mathbf{I}([0, 1])$, sending each subinterval $[b, a]$ to the range of distances between the curve $y = \mathcal{H}(x)$ and the straight line joining $(b, \mathcal{H}(b))$ and $(a, \mathcal{H}(a))$ over the subinterval $[b, a]$. Following this by $\max \colon \mathbf{I}([0, 1]) \to [0, 1]$, which is also continuous, yields the function **Cap**. Notice that, as a function from subintervals $[b, a] \subseteq [0, 1]$ to $[0, 1]$, **Cap** also is monotone with respect to inclusion, and $\mathsf{Cap} \colon \mathbf{I}([0, 1]) \to [0, 1]^{op}$ is continuous with respect to reverse inclusion.

The rest of this section is devoted to giving a firm foundation for this analysis in the binary case, and to extending it to higher dimensions.

10.1. The domain of compact convex sets. We begin our analysis with the following.

DEFINITION 10.2. Let $K \subseteq \mathbb{R}^n$ be a convex set. A function $f \colon K \to \mathbb{R}$ is *strictly concave* if

$$f(rk + (1 - r)k') > rf(k) + (1 - r)f(k')$$

for all $r \in (0, 1)$ and all $k, k' \in K$.

Recall *Jensen's Inequality* [**15**]: If $f \colon K \to \mathbb{R}$ is a convex function defined on a convex subset K of a vector space V, then $E(f(X)) \geq f(E(X))$ for a finite random variable $X \colon \Omega \to K$, where E denotes expectation. Moreover, if f is strictly convex, then $E(f(X)) = f(E(X))$ implies X is constant.

Jensen's inequality is a fundamental result of information theory; for example, it is crucial for proving the non-negativity of mutual information, that the mutual information in a pair of random variables is 0 iff the random variables are independent, and in proving that entropy is strictly concave (cf. [**10**], Chapter 2). Using the duality of concavity and convexity, the following generalizes this by strengthening the result in case f is strictly convex.

LEMMA 10.3. *If $K \subseteq \mathbb{R}^n$ is compact and convex, and if $f \colon K \to \mathbb{R}$, then the following are equivalent:*

(1) *f is strictly concave.*
(2) *For all $r_1, \ldots, r_m \in (0, 1)$ and all $k_1, \ldots, k_m \in K$,*

$$\sum_{i \leq m} r_i = 1 \quad \Rightarrow \quad f\left(\sum_{i \leq m} r_i k_i\right) > \sum_{i \leq m} r_i f(k_i).$$

PROOF. (1) $\Rightarrow$ (2) is obvious. For the reverse direction, we proceed by induction on m. The base case, $m = 2$, is just the definition of strict concavity. So suppose (2) holds for some m, and consider a family $r_1, \ldots, r_{m+1} \in (0, 1)$ and

$k_1, \ldots, k_{m+1} \in K$. Since $r_i \in (0,1)$ for each i,

$$f\left(\sum_{i \leq m+1} r_i k_i\right)$$
$$= f\left(\sum_{i \leq m-1} r_i k_i + (r_m + r_{m+1})(\frac{r_m}{r_m + r_{m+1}} k_m + \frac{r_{m+1}}{r_m + r_{m+1}} k_{m+1})\right)$$
$$> \sum_{i \leq m-1} r_i f(k_i) + (r_m + r_{m+1}) f(\frac{r_m}{r_m + r_{m+1}} k_m + \frac{r_{m+1}}{r_m + r_{m+1}} k_{m+1})$$
$$> \sum_{i \leq m-1} r_i f(k_i) + r_m f(k_m) + r_{m+1} f(k_{m+1}) = \sum_{i \leq m+1} r_i f(k_i). \qquad \square$$

NOTATION 10.4. If K be a compact convex set, then $\mathsf{Conv}_n(K)$ denotes the family of convex polytopes $\overline{\mathrm{conv}}(\{x_1, \ldots, x_k\})$ generated by finite subsets $\{x_1, \ldots, x_k\} \subseteq K$, where $k \leq n$.

Also note that $P(n) \stackrel{\text{def}}{=} \{x \in [0,1]^n \mid \sum_i x_i = 1\}$ is a compact, convex subset of $[0,1]^n$, which we identify with the family $P\{1, \ldots, n\}$ of probability distributions on $\{1, \ldots, n\}$.

We already commented following Remark 9.3 that the compact convex subsets of a locally convex topological vector space form a domain under reverse inclusion. Because this is crucial to our results, we now provide a more detailed account of this fact. We begin by collecting some facts about locally convex topological vector spaces and their compact convex subsets. These results can all be found in any introductory text on functional analysis, for example, cf. [**33**].

PROPOSITION 10.5. *Let V be a locally convex real topological vector space. Then*

(1) *If $O, O' \subseteq V$ are open, then $O + O'$ and $O - O'$ are open.*
(2) *If $O \subseteq V$ is open and $r \neq 0$, then $rO = \{rx \mid x \in O\}$ is open.*
(3) *If $A, B \subseteq V$ are convex, then*

$$\langle A \cup B \rangle \equiv \{ra + (1-r)b \mid a \in A, b \in B, r \in [0,1]\}$$

is the convex hull of $A \cup B$.
(4) *If V is locally convex and $K \subseteq V$ is compact and convex, then*

$$K = \bigcap \{C \mid K \subseteq C^\circ \subseteq C, \text{ with } C^\circ \text{ and } C \text{ convex and } C \text{ compact}\},$$

where the interior C° is taken in K, and the family of such sets C is filtered.

PROOF. Part (1) follows from the fact that $(V, +)$ is a topological group, while part (2) follows from the fact that scalar multiplication by a non-zero number, r, has an inverse, multiplication by $\frac{1}{r}$.

For part (3), it is clear that $\langle A \cup B \rangle$ is contained in the convex hull of $A \cup B$. Conversely, we show that $\langle A \cup B \rangle$ is convex. Indeed, suppose that

$$r_1 a_1 + (1 - r_1) b_1, r_2 a_2 + (1 - r_2) b_2 \in \langle A \cup B \rangle,$$

and assume at least one of r_1, r_2 is neither 0 nor 1. Then, given $r \in (0,1)$, we have

$$r(r_1a_1 + (1-r_1)b_1) + (1-r)(r_2a_2 + (1-r_2)b_2)$$
$$= (rr_1a_1 + (1-r)r_2a_2) + (r(1-r_1)b_1 + (1-r)(1-r_2)b_2)$$
$$= (rr_1 + (1-r)r_2)\left(\frac{rr_1}{rr_1 + (1-r)r_2}a_1 + \frac{(1-r)r_2}{rr_1 + (1-r)r_2}a_2\right)$$
$$+ (r(1-r_1) + (1-r)(1-r_2))\left(\frac{r(1-r_1)}{r(1-r_1) + (1-r)(1-r_2)}b_1\right.$$
$$\left.+ \frac{(1-r)(1-r_2)}{r(1-r_1) + (1-r)(1-r_2)}b_2\right),$$

and it is easy to check that the coefficients all add to 1, as required

For part (4), the Hahn-Banach Separation Theorem implies that any compact convex subset of a locally convex real vector space is the intersection of open half spaces that contain it. But we are asking for a bit more. First note that if $A \subseteq V$ is convex, then the continuity of scalar multiplication implies that $\overline{A}$ also is convex: if $x, y \in \overline{A}$ and $0 < r < 1$, then choosing nets $\{x_i\}_{i \in I}, \{y_i\}_{i \in I} \subseteq A$ with $x = \lim_i x_i$ and $y = \lim_i y_i$ implies that

$$rx + (1-r)y = r(\lim_i x_i) + (1-r)(\lim_i y_i) = \lim_i(rx_i + (1-r)y_i) \in \overline{A},$$

the containment following from the convexity of A.

Now, given $K \subseteq V$ compact and convex, and $H \subseteq V$ is an open half space containing K, then for each $k \in K$, we can find a convex open set O_k with compact closure and satisfying $k \in O_k \subseteq \overline{O_k} \subseteq H$. Since K is compact, finitely many of the O_ks cover K, say $O_1, \ldots, O_m$, and the convex hull of their union is

$$\langle \bigcup_{i \leq m} O_i \rangle = \{\sum_{1 \leq m} r_i x_i \mid x_i \in O_i, r_i \geq 0, \sum_{i \leq m} r_i = 1\},$$

by part (3) and induction on m. Clearly $\overline{\langle \bigcup_i O_i \rangle} = \overline{\mathrm{conv}}(\bigcup_i \overline{O_i}) \subseteq H$, the containment following from the convexity of H. Since Hahn-Banach implies K is the intersection of open half spaces, it follows that K is the intersection of compact, convex neighborhoods, and by taking finite intersections, we can conclude that K is the filtered intersection of compact, convex neighborhoods. □

PROPOSITION 10.6. *Let K be a compact, convex set and let $\Gamma(K)$ denote the family of non-empty compact subsets of K, under reverse containment. The following are equivalent:*

(1) *K is locally convex.*
(2) *The mapping*

$$C \mapsto \overline{\mathrm{conv}}(C) : \Gamma(K) \to \Gamma(K), \quad \mathrm{im}(\overline{\mathrm{conv}}) = \mathsf{Conv}(K)$$

is Scott continuous.

In this case, $\overline{\mathrm{conv}}$ is order theoretically a kernel operator, and topologically a continuous retraction onto the space of non-empty compact convex subsets of K, so $\mathsf{Conv}(K)$ is a coherent domain.

Moreover, if the equivalent conditions listed above hold, then the subfamily $\mathsf{Conv}_n(K) = \{\overline{\mathrm{conv}}(F) \mid \emptyset \neq F \subseteq K, |F| \leq n\}$ of convex hulls of non-empty finite sets of cardinality at most n is a coherent subdomain of $\mathsf{Conv}(K)$.

PROOF. Clearly the mapping $C \mapsto \overline{\text{conv}}(C) \colon \Gamma(K) \to \Gamma(K)$ is a kernel operator whose image is the family $\mathsf{Conv}(K)$ of non-empty compact convex subsets of K. If K is locally convex, then $\overline{\text{conv}}$ is Scott continuous (wrt reverse containment): indeed, monotonicity is clear, and if $\{C_i\}_{i \in I}$ is a filter basis of compact sets with $C = \bigcap_i C_i \in \Gamma(K)$, then part (4) of the previous Proposition implies $\overline{\text{conv}}(C)$ has a basis of convex compact neighborhoods in K. If U is one such, then $C \subseteq \overline{\text{conv}}(C) \subseteq U$ implies $C_i \subseteq U$ for some index i. Then $\overline{\text{conv}}(C_i) \subseteq U$, and so $\bigcap_i \overline{\text{conv}}(C_i) \subseteq U$. Hence $\bigcap_i \overline{\text{conv}}(C_i) = \overline{\text{conv}}(C)$. This shows (1) implies (2).

Conversely, suppose the K is not locally convex. Then there is a point $k \in K$ and an open set $U \subseteq K$ containing k that contains no compact convex neighborhood of k in K. Since the topology on K is compact Hausdorff, the family $\mathcal{N}(k)$ of compact neighborhoods of k is a filter basis of compact sets whose intersection is $\{k\}$. But the fact that U contains no compact convex neighborhood of k implies that $\bigcap\{\overline{\text{conv}}(N) \mid N \in \mathcal{N}(k)\} \not\subseteq U$, since otherwise the fact that U is open would imply $N \subseteq U$ for some $N \in \mathcal{N}(k)$. Thus $\overline{\text{conv}}(\{k\}) = \{k\} \neq \bigcap_{N \in \mathcal{N}(k)} \overline{\text{conv}}(N)$, which is the supremum of $\mathcal{N}(k)$ in $(\mathsf{Conv}(K), \supseteq)$. Thus, $\overline{\text{conv}}$ is not Scott continuous.

Assuming now that conditions (1) and (2) hold, the mapping $\overline{\text{conv}}$ clearly maps $\Gamma(K)$ onto $\mathsf{Conv}(K)$ and the fact that $\overline{\text{conv}}$ is a Scott-continuous kernel operator implies this image is a domain (cf. Corollary IV-1.7 of [14]). The mapping also is proper: if $K' \subseteq K$ is compact and convex, then

$$\overline{\text{conv}}^{-1}(\uparrow_{\mathsf{Conv}(K)} K') = \{C \in \Gamma(K) \mid \overline{\text{conv}}(C) \subseteq K'\} = \uparrow_{\Gamma(K)} K',$$

which is a Scott-compact upper set, hence a Lawson-compact subset of $\Gamma(K)$. This implies $\mathsf{Conv}(K)$ is a coherent domain and that $\overline{\text{conv}} \colon \Gamma(K) \to \mathsf{Conv}(K)$ is Lawson continuous.

We next show $\Gamma_n(K) = \{F \mid \emptyset \neq F \subseteq K, |F| \leq n\} \subseteq \Gamma(K)$ is closed in the Lawson topology: Since the family $\Gamma(K)$ is compact in the Lawson topology, any net $\{F_i\}_{i \in I}$ of finite sets has a cluster point; by taking a subnet, we can assume convergence. If $\lim_i F_i = C \subseteq K$, then C is closed, and if C had more than n points, then we could choose a family of pairwise disjoint open sets $U_1, \ldots, U_{n+1}$ with $U_k \cap C \neq \emptyset$ for each k. (First choose $n+1$ points from C, and then choose open sets U_k which are pairwise disjoint so that each contains exactly one of the points from C.) Since $C = \lim_i F_i$, for each k there is some index $j_k \in I$ with $F_i \cap U_k \neq \emptyset$ for each $i \geq j_k$. Since I is directed, choosing an index i with $i \geq j_k$ for each k shows $F_i \cap U_k \neq \emptyset$, but then F_i has at least $n+1$ points since the U_k are pairwise disjoint. This is a contradiction to F_i having at most n points for each i. This shows the family $\Gamma_n(K)$ of finite non-empty subsets of cardinality at most $n < \infty$ is Lawson-closed in $\Gamma(K)$.

Now, since $\Gamma_n(K)$ is Lawson-closed in $\Gamma(K)$, its image under $\overline{\text{conv}}$, namely $\mathsf{Conv}_n(K)$, is a compact subspace of $\mathsf{Conv}(K)$. So $\mathsf{Conv}_n(K)$ it is closed under filtered intersections, which means it is directed complete. Next, $\mathsf{Conv}_n(K)$ is also a domain: if $F, G \in \Gamma_n(K)$ and $G \subseteq \overline{\text{conv}}(F)^\circ$, then local convexity of K implies there is a convex neighborhood of G that is contained in $\overline{\text{conv}}(F)$, and then $\overline{\text{conv}}(G) \subseteq \overline{\text{conv}}(F)^\circ$, so $\overline{\text{conv}}(F) \ll \overline{\text{conv}}(G)$. This also shows the way-below relation $\ll$ on $\mathsf{Conv}_n(K)$ is the one it inherits from $\mathsf{Conv}(K)$, and since $\mathsf{Conv}_n(K)$ is Lawson closed in the coherent domain $\mathsf{Conv}(K)$, it then follows that $\mathsf{Conv}_n(K)$ is a coherent domain in the inherited order. $\square$

10.2. Real-valued functions on compact convex subsets of a compact convex space.
Throughout this subsection, K is a compact convex subspace of a locally convex topological vector space, but most arguments would apply to a compact locally convex affine space in the sense of Definition 1.27. We also let X be a compact space; valid examples are $X = |K|$, the underlying compact space of K, or, a finite set with the discrete topology. The elements of the space $C(X,K)$ of continuuous functions $X \to K$ will often be denoted ξ, ξ_1 etc., and we shall consider a given real-valued function $f \colon K \to \mathbb{R}$, often taking values in the unit interval $\mathbb{I} = [0,1]$. This yields real-valued functions $f \circ \xi \colon X \to \mathbb{R}$ from each of which we shall provide a real number $g(\xi)$ using intermediate procedures and integration, giving us a function $g \colon X \to \mathbb{R}$.

On $C(X,K)$ we define a binary relation $\preceq$ as follows: Recall $\overline{\mathrm{conv}} \colon \Gamma(K) \to \mathsf{Conv}(K)$ from the previous subsection.

DEFINITION 10.7. For each $\xi_1, \xi_2 \in C(X,K)$, define
$$\xi_1 \preceq \xi_2 \quad \text{if and only if} \quad \xi_1(X) \subseteq \overline{\mathrm{conv}}(\xi_2(X)).$$

Note that $\xi_1 \preceq \xi_2$ iff $\overline{\mathrm{conv}}(\xi_1) \subseteq \overline{\mathrm{conv}}(\xi_2)$, and so $\preceq$ is a transitive and reflexive binary relation yielding an equivalence relation $\sim$ by $\xi_1 \sim \xi_2$ to hold if and only if $\xi_1 \preceq \xi_2$ and $\xi_2 \preceq \xi_1$, that is,
$$\xi_1 \sim \xi_2 \Leftrightarrow \overline{\mathrm{conv}}(\xi_1) = \overline{\mathrm{conv}}(\xi_2).$$
The relation $\preceq$ induces on the quotient $C(X,K)/\sim$ a partial order $\leq$ such that $\mathrm{class}_\sim(\xi_1) \leq \mathrm{class}_\sim(\xi_2)$ unambiguously iff $\xi_1 \preceq \xi_2$. Now let
$$\mathsf{Conv}_X(K) = \{\overline{\mathrm{conv}}(\xi(X)) \mid \xi \in C(X,K)\}$$
and define $\gamma \colon C(X,K) \to \mathsf{Conv}_X(K)$ by $\gamma(\xi) = \overline{\mathrm{conv}}(\xi(X))$. Then $\overline{\mathrm{conv}}$ factors uniquely through the quotient map $q \colon C(X,K) \to C(X,K)/\sim$ with an order preserving bijection $\beta \colon (C(X,K)/\sim) \to \mathsf{Conv}_X(K)$:

$$\begin{array}{ccc} (C(X,K),\preceq) & \xrightarrow{\gamma} & (\mathsf{Conv}_X(K), \subseteq) \\ {\scriptstyle q}\downarrow & & \downarrow{\scriptstyle \mathrm{id}} \\ (C(X,K)/\sim, \leq) & \xrightarrow{\beta} & (\mathsf{Conv}_X(K), \subseteq). \end{array}$$

LEMMA 10.8. *Suppose the function* $g \colon C(X,K) \to \mathbb{R}$ *satisfies* $g(\xi_1) \leq g(\xi_2)$ *whenever* $\xi_1 \preceq \xi_2$. *Then there is a unique order preserving function*

(10.2) $\quad \widehat{f} \colon (\mathsf{Conv}_X(K), \subseteq) \to (\mathbb{R}, \leq), \ \widehat{f}(C) = g(\xi) \ \textit{for} \ C = \overline{\mathrm{conv}}(\xi(X)).$

PROOF. Since $C \in \mathsf{Conv}_X(K)$ there is at least one $\xi \colon X \to K$ such that $\gamma(\xi) = \overline{\mathrm{conv}}(\xi(X)) = C$. If $\xi_1 \in C(X,K)$ also satisfies $\gamma(\xi_1) = C$, then $\xi \sim \xi_1$ and so $g(\xi) = g(\xi_1)$ and we can unambiguously define $\widehat{f}(C) = g(\xi)$. That $\widehat{f}$ is order preserving follows from the information in the commuting diagram above. $\square$

The conditions under which the mapping $\widehat{f}$ is continuous occupy the remainder of this section. We begin with the following simple observation.

LEMMA 10.9. *Assume the following conditions:*
  (1) *g is continuous.*
  (2) $\overline{\mathrm{conv}} \colon C(X,K) \to \mathsf{Conv}_X(K)$ *is a closed function.*
Then $\widehat{f} \colon \mathsf{Conv}_X(K) \to \mathbb{R}$ *is continuous.*

PROOF. By Lemma 10.8 we have $g = \widehat{f} \circ \gamma$. For any set $S \subseteq \mathbb{R}$ we have $\widehat{f}^{-1}(S) = \gamma(g^{-1}(S))$. Now if S is closed, then $g^{-1}(S)$ is closed by (1) and then $\widehat{f}^{-1}(S)$ is closed by (2). $\square$

LEMMA 10.10. *The function* $\gamma \colon C(X,K) \to \mathsf{Conv}_X(K)$, $\gamma(\xi) = \overline{\mathrm{conv}}(\xi(X))$ *is continuous.*

PROOF. We prove the claim in two steps: Firstly, the function $\xi \mapsto \xi(X) : C(X,K) \to \Gamma(K)$ where $\Gamma(K)$ is the space of closed subsets of K in the Vietoris topology. The Vietoris topology on K is a uniform topology defined as follows: Let $\mathfrak{U}$ be a symmetric entourage of the uniform structure of the compact space K, that is, $\mathfrak{U}$ is a symmetric neighborhood of the diagonal in $K \times K$. Then ξ_1 and ξ_2 are $\mathfrak{U}$-close iff $(\forall x \in X)\, (\xi_1(x), \xi_2(x)) \in \mathfrak{U}$. This implies $\xi_1(X) \subseteq \mathfrak{U}(\xi_2(X))$ (where $A \subseteq \mathfrak{U}(B)$ iff $(\forall a \in A)(\exists b)\, a \in \mathfrak{U}(b)$ (i.e., $(b,a) \in \mathfrak{U}$)) and likewise $\xi_2(X) \subseteq \mathfrak{U}(\xi_1(X))$. This means that $\xi_1(X)$ and $\xi_2(X)$ are $\mathfrak{U}$-close in the uniformity of the compact space $\Gamma(K)$. Thus $\xi \mapsto \xi(X) : C(X,K) \to \Gamma(K)$ is in fact uniformly continuous.

Secondly we recall from Proposition 10.6, that $C \mapsto \overline{\mathrm{conv}}(C) : \Gamma(K) \to \mathsf{Conv}(K)$ is continuous. $\square$

COROLLARY 10.11. *If g is continuous and X is finite, then $\widehat{f}$ is continuous.*

PROOF. If X is finite, then $C(X,K) = K^X$ is compact. By Lemma 10.10 $\gamma \colon K^X \to \mathsf{Conv}_X(K)$ is a continuous surjective function between compact spaces and is, therefore, a closed map. $\square$

10.3. Concavity at work. The developments of Lemma 7.1, Theorem 7.2 and Remark 7.3 depend on the assumption that S is a compact semigroup only in as much as the continuous functions η_S and ϵ_T are shown to be morphisms of compact semigroups. In the context of compact spaces and compact affine spaces we know that for each continuous function $\xi \colon X \to K$ there is a unique continuous affine map $\xi' \colon P(X) \to K$, namely, $\xi'(\mu) = \int_X \xi\, d\mu$ such that $\xi = |\xi'| \circ \eta$, $\eta(x) = \delta_x$, the point mass at x. The notation $|\bullet|$ means forgetting the affine structure. Diagrammatically we have:

$$\begin{array}{ccccc} X & \xrightarrow{\eta} & |P(X)| & & P(X) \\ \xi \downarrow & & \downarrow |\xi'| & & \downarrow \xi' \\ |K| & \xrightarrow{\mathrm{id}} & |K| & & K \end{array}$$

LEMMA 10.12.

(1) *The function*
$$(\xi, \mu) \mapsto \int_X \xi\, d\mu : C(X,K) \times P(X) \to \mathbb{R}$$
is continuous.

(2) *For $\xi \in C(X,K)$ we have*
$$\overline{\mathrm{conv}}(\xi(X)) = \left\{ \int_X \xi\, d\mu \mid \mu \in P(X) \right\}.$$

PROOF. (1): By definition of $\int_X \xi d\mu$, given any continuous affine function $\omega: K \to \mathbb{R}$ we have
$$\omega\left(\int_X \xi d\mu\right) = \int_X (\omega \circ \xi) d\mu = \langle \omega \circ \xi, \mu \rangle.$$
Then for each fixed ω, the function $(\xi, \mu) \mapsto \langle \omega \circ \xi, \mu \rangle : C(X, K) \times P(X) \to \mathbb{R}$ is continuous. Since $\mathsf{Aff}(K, \mathbb{R})$ separates the points of K and X is compact, a function $\alpha: Y \to K$ is continuous if $\omega \circ \alpha: Y \to \mathbb{R}$ is continuous for all $\omega \in \mathsf{Aff}(K, \mathbb{R})$. Hence $(\xi, \mu) \mapsto \int_X \xi d\mu : C(X, K) \times P(X) \to \mathbb{R}$ is continuous.

(2): The function $\mu \mapsto \int_X \xi_2 d\mu : P(X) \to K$ is continuous and affine. Hence its image is a compact convex subset of K contained in $\overline{\mathrm{conv}}(\xi_2(X))$. Since it contains $\xi_2(x) = \int_X \xi_2 d\delta_x$ for all x, it does contain $\overline{\mathrm{conv}}(\xi_2(X))$ and the assertion follows. □

Assume now that we have a continuous function $f: |K| \to |\mathbb{I}|$, $\mathbb{I} = [0, 1]$ giving rise to two continuous functions
(1) $f \circ |\xi'| : |P(X)| \to |\mathbb{I}|$, $(f \circ |\xi'|)(\mu) = f(\int_{x \in X} \xi(x) \, d\mu(x))$, and
(2) $(f \circ \xi)' : P(X) \to \mathbb{I}$, $(f \circ \xi)'(\mu) = \int_{x \in X} (f(\xi(x))) \, d\mu(x)$.
If f is concave, the continuous functions (1) and (2) compare as follows:

LEMMA 10.13. *A continuous function $f: K \to \mathbb{R}$ is concave if (and only if) for all compact spaces X, all $\xi \in C(X, K)$ and all $\mu \in P(X)$ the following inequality holds:*

(†) $$\int_X (f \circ \xi) d\mu \leq f\left(\int_X \xi d\mu\right)$$

PROOF. First assume that f is concave. For given X and $\xi \in C(X, K)$, the inequality (†) holds for simple measures by Jensen's Inequality (cf. the discussion following Definition 10.2). By Lemma 7.1, the space $P_s(X)$ of simple measures is dense in $P(X)$. The continuity of f and the continuity of $(\xi, \mu) \mapsto \int_X \xi d\mu$ according to Lemma 10.12(1) then shows (†) in general.

Conversely, if (†) is satisfied for all X, all $\xi \in C(X, K)$, and all $\mu \in P(X)$, then it holds, in particular, for $X = K$, $\xi = \mathrm{id}_K$ and all simple $\mu \in P_s(K)$ and thus yields the definition of a concave function. □

Now we measure how far the two real numbers in (†) are apart:

DEFINITION 10.14. For each $\xi \subset C(X, K)$, define

(∗) $\underline{f}: X \times P(X) \to \mathbb{R}$, $\underline{f}(\xi, \mu) = f\left(\int_{x \in X} \xi(x) \, d\mu(x)\right) - \int_{x \in X} (f(\xi(x))) \, d\mu(x)$

If $f(K) \subseteq \mathbb{I}$, then $\underline{f}(X \times P(X)) \subseteq \mathbb{I}$.

LEMMA 10.15. *Let A be a topological and B a compact space, and assume that $\phi: A \times B \to \mathbb{R}$ be a continuous function. Then the function $g: A \to \mathbb{R}$ defined by $g(a) = \max \phi(\{a\} \times B)$ is continuous.*

PROOF. Since a continuous real-valued function on a compact space attains its maximum there is a function $m: A \to B$ such that $\phi(a, m(a)) = \max \phi(\{a\} \times B)$. (In general, the existence of m requires the Axiom of Choice.)

Let $g_b: A \to \mathbb{R}$ be defined by $g_b(a) = \phi(a, b)$. The continuity of ϕ implies that of g_b for all $b \in B$. Then $g = \sup_{b \in B} g_b$ is a pointwise supremum of continuous functions and is, therefore, lower semicontinuous, so we have to show that it is also upper semicontinuous; that is, for all $a \in A$ and all $t \in \mathbb{R}$ such that $g(a) < t$ there is a neighborhood U of a from the neighborhood filter $\mathcal{U}(a)$ such that $g(u) < t$ for all $u \in U$.

Now suppose that this is not the case. Then there are an $a \in A$ and a $t \in \mathbb{R}$ with $g(a) < t$ such that for all $U \in \mathcal{U}(a)$ there is some $u \in U$ such that $g(u) = \max_{b \in B} \phi(u,b) \geq t$, and so the set $B_U \stackrel{\text{def}}{=} \{b \in B \mid (\exists u \in U) \, \phi(u,b) \geq t\}$ is not empty. Now $\{B_U \mid U \in \mathcal{U}(a)\}$ is a filter basis in the compact space B. Let b a point of adherence of this filterbasis, which exists because of the compactness of B. If $\mathcal{U}(b)$ is the neighborhood filter of b in B, then $\{B_U \cap V \mid (U,V) \in \mathcal{U}(a) \times \mathcal{U}(b)\}$ is a filter basis converging to b and thus by the continuity of ϕ the filter basis

$$\{\phi(U \times (B_U \cap V)) \mid (U,V) \in \mathcal{U}(a) \times \mathcal{U}(b)\}$$

in $\mathbb{R}$ converges to a number $\phi(a,b) \geq t > g(a) = \max \phi(\{a\} \times B)$. This is a contradiction which proves that g is upper semicontinuous, hence is continuous. $\square$

The proof of the lemma shows, by the way, that separation axioms are not needed for either A or B.

Before we go on we observe that $C(X,K)$ is an affine space via pointwise convex combinations.

LEMMA 10.16. *For $\underline{f}$ we have the following conclusions:*
(1) *$\underline{f}$ is continuous.*
(2) *For each $\xi \in C(X,K)$ the set $\underline{f}(\{\xi\} \times P(X)) \subseteq \mathbb{R}$ is bounded, and the function $g \colon C(X,K) \to \mathbb{R}$ defined by $g(\xi) = \sup \underline{f}(\{\xi\} \times P(X))$ is continuous.*
(3) *For each $\xi \in C(X,K)$ there is some $\mu_\xi \in P(X)$ such that $g(\xi) = \underline{f}(\xi, \mu_\xi)$.*

PROOF. (1) The function $\underline{f}$ defined in $(*)$ is continuous as the difference of two continuous functions.

(2) Since $P(X)$ is compact and $\underline{f}$ is continuous by (1), $\underline{f}(\{\xi\} \times P(X))$ is bounded in $\mathbb{R}$ and thus g is well defined. We now apply Lemma 10.15 with $A = C(X,K)$, $B = P(X)$, and $\phi = \underline{f}$, and the assertion follows.

(3) Since every real valued continuous function on a compact space attains a maximum, the element μ_ξ exists. $\square$

By Lemma 10.12(2), $\xi_1 \preceq \xi_2$ implies

$$(\#) \qquad (\forall x \in X)(\exists \mu^x \in P(X)) \; \xi_1(x) = \int_X \xi_2 \, d\mu^x.$$

In the sequel we shall use some notation as follows:

DEFINITION 10.17. *Let μ_ξ be as in Lemma 10.16(3) and μ^x as in $(\#)$. If $\xi_1 \preceq \xi_2$ we say that Hypothesis (H) is satisfied if and only if the function $x \mapsto \mu^x \colon X \to P(X)$ is μ_{ξ_1}-integrable, that is,*

$$(\text{H}) \qquad (\exists \nu \in P(X)) \, \nu = \int_{x \in X} \mu^x \, d\mu_{\xi_1}(x).$$

We note that hypothesis (H) may be satisfied in a variety of contexts, but it certainly holds if X is finite, which is still a relevant situation.

LEMMA 10.18. *If $\xi_1 \preceq \xi_2$ and if Hypothesis (H) is satisfied, then $g(\xi_1) \leq g(\xi_2)$.*

PROOF. Assume $\xi_1 \preceq \xi_2$. Then by (#), for each $x \in X$ we have some $\mu^x \in P(X)$ such that
$$\xi_1(x) = \int_X \xi_2 d\mu^x.$$
Therefore,
$$\begin{aligned}
g(\xi_1) &= f\left(\int_{x \in X} \xi_1(x) d\mu_{\xi_1}(x)\right) - \int_{x \in X} f(\xi_1(x)) d\mu_{\xi_1}(x) \\
&= f\left(\int_{x \in X} d\mu_{\xi_1}(x) \left(\int_{y \in X} \xi_2(y) d\mu^x(y)\right)\right) \\
&\quad - \int_{x \in X} f\left(\int_{y \in X} \xi_2(y) d\mu^x(y)\right) d\mu_{\xi_1}(x) \\
&\leq f\left(\int_{x \in X} d\mu_{\xi_1}(x) \left(\int_{y \in X} \xi_2(y) d\mu^x(y)\right)\right) \\
&\quad - \int_{x \in X} d\mu_{\xi_1}(x) \int_{y \in Y} f(\xi_2(y)) d\mu^x(y)
\end{aligned}$$

by Lemma 10.13. Now by Hypothesis (H) we have a measure $\nu \in P(X)$ defined by $\nu = \int_{x \in X} \mu^x d\mu_{\xi_1}(x)$. By reordering integration, the right side then equals

$$\begin{aligned}
f\left(\int_X \xi_2 d\nu\right) - \int_X (f \circ \xi_2) d\nu &= \underline{f}(\xi_2, \nu) \\
&\leq \sup_{\mu \in P(X)} \underline{f}(\xi_2, \mu) \\
&= g(\xi_2),
\end{aligned}$$

the final equality following from the definition of g in Lemma 10.16(2). □

Recall that the space $C(X, K)$ is given the topology of uniform convergence which is the compact open topology as X and K are compact. Taking the information from Lemma 10.8 and Lemma 10.18 above we obtain the following summary of what we have up to this point:

PROPOSITION 10.19. *Let K be a compact convex set in a locally convex vector space and X a compact space. Assume that $f\colon K \to \mathbb{I}$ is a continuous concave function and that hypothesis* (H) *is satisfied. Define*

$$g\colon C(X, K) \to \mathbb{I}, \quad g(\xi) = \sup_{\mu \in P(X)} \left\{ f\left(\int_X \xi \, d\mu\right) - \int_X (f \circ \xi) \, d\mu \right\}.$$

Then g is continuous and there is a unique order preserving function

$$\widehat{f}\colon (\mathsf{Conv}_X(K), \subseteq) \to (\mathbb{I}, \leq) \text{ such that } \widehat{f}(\overline{\mathrm{conv}}(\xi(X)) = g(\xi). \quad \square$$

As a corollary, we obtain the following result, for which we observe that for a finite set X, say, $\{1, 2, \ldots, n\}$ the space $C(X, K)$ is K^n whose elements ξ are n-tuples $(x_1, \ldots, x_n)$, while $P(X)$ is the simplex $\mathcal{R}_n = \{(r_1, \ldots, r_n) \in \mathbb{I}^n \mid \sum_{k=1}^n r_k = 1\}$, and that $\int_X \xi \, d\mu = \sum_{k=1}^n r_k x_k$. In place of $\mathsf{Conv}_X(K)$ we write $\mathsf{Conv}_n(K)$.

THEOREM 10.20. *Let n be a natural number and K a compact convex set in a locally convex vector space. Assume that $f\colon K \to \mathbb{I}$ is a continuous concave function* and define $g: K^n \to \mathbb{I}$,

$$g(x_1,\ldots,x_n) = \sup_{(r_1,\ldots,r_n)\in\mathcal{R}_n}\left\{f\left(\sum_{k=1}^n r_k x_k\right) - \sum_{k=1}^n r_k f(x_k)\right\}.$$

Then g is continuous and there is a unique order preserving and continuous *function*
$\widehat{f}: (\mathsf{Conv}_n(K), \subseteq) \to (\mathbb{I}, \leq)$ such that $\widehat{f}(\overline{\mathrm{conv}}(\{x_1,\ldots,x_n\})) = g(x_1,\ldots,x_n)$.

PROOF. For a proof of this theorem we apply Proposition 10.19 and Corollary 10.11. □

11. The Topological Nature of Capacity

In this section we show that capacity as a function on the monoid $\mathsf{ST}(n)$ factors through the family of finite polytopes over $P(n) = \{(x_1,\ldots,x_n) \in \mathbb{I}^n \mid \sum_i x_i = 1\}$: that is, the capacity of a channel $M \in \mathsf{ST}(n)$ is the value of an induced function at the finite polytope $M(P(n))$ that is the image of $P(n)$ under the channel M. The point here is that the topology of the image set – $M(P(n))$ – determines the capacity of M, and two channels having the same image have equal capacity. Moreover, the induced function is continuous and monotone from the family of finite polytopes endowed with the Vietoris topology, to the reals in the usual topology and order.

To set the stage, we review the binary case derived by Martin, et al. in [**25**]. Figure 2 illustrates that $\mathsf{ST}(2) \simeq [0,1]^2$ under $\begin{pmatrix} a & 1-a \\ b & 1-b \end{pmatrix} \mapsto \begin{pmatrix} a \\ b \end{pmatrix}$ and $\mathcal{M}(\mathsf{ST}(2))$ corresponds to the diagonal.

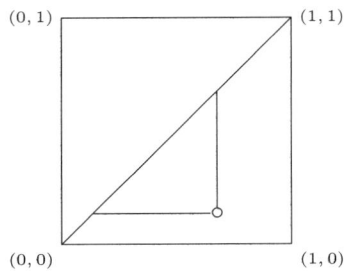

FIGURE 2. $\mathsf{ST}(2) \cdot C$ is the region within the triangle with vertex ○

The following lists the principal results from [**25**]:
(1) If $C \in \mathsf{ST}(2)$, then the principal left ideal C generates is $\mathsf{ST}(2)C$, and this compact subsemigroup satisfies $\mathcal{M}(\mathsf{ST}(2)C) = \mathcal{M}(\mathsf{ST}(2)) \cap (\mathsf{ST}(2)C)$.
(2) Using the fact that $\mathsf{ST}(2) \simeq [0,1]^2$, if $C = \begin{pmatrix} a & 1-a \\ b & 1-b \end{pmatrix}$, then $\mathcal{M}(\mathsf{ST}(2)C) = [b, a] \subseteq [0, 1]$, assuming $b \leq a$.
(3) There is a pre-order – i.e., a reflexive, transitive relation — on binary channels defined by reverse containment on intervals:
$$C \leq C' \quad \Leftrightarrow \quad \mathcal{M}(\mathsf{ST}(2)C) \supseteq \mathcal{M}(\mathsf{ST}(2)C').$$
If we denote the relation $\leq \cap \geq$ by $\equiv$, then $\mathsf{ST}(2)/\equiv$ has a corresponding partial order $\leq/\equiv$.

(4) The mapping $\pi_2\colon (\mathsf{ST}(2), \leq) \to (\mathsf{ST}(2)/\!\equiv, \leq/\!\equiv)$ by $\pi_2(C) = \mathcal{M}(\mathsf{ST}(2))C$ is continuous and monotone, and the quotient structure, $\mathsf{ST}(2)/\!\equiv$ is isomorphic as an ordered space to the interval domain $\mathbf{I}([0,1])$ over the unit interval.

(5) The capacity function on $\mathsf{ST}(2)$ factors through the quotient space $\mathsf{ST}(2)/\!\equiv$, so that $\mathsf{Cap}\colon \mathsf{ST}(2) \to [0,1]^{op}$ satisfies $\mathsf{Cap}(C) = \mathsf{cap}([b,a])$, where $C = \begin{pmatrix} a & 1-a \\ b & 1-b \end{pmatrix}$, and $\mathsf{cap}\colon \mathbf{I}([0,1]) \to \mathbb{I}$ by

$$\mathsf{cap}([b,a]) = \sup_{x \in [0,1]} \mathcal{H}(xb + (1-x)a) - x\mathcal{H}(b) - (1-x)\mathcal{H}(b)$$

measures $\mathbf{I}([0,1])$, in the sense of Definition 9.4.

(6) In fact, if

$$\mathsf{ST}(2)_+ = \left\{ \begin{pmatrix} a & 1-a \\ b & 1-b \end{pmatrix} \mid b \leq a \in [0,1] \right\},$$

then $\mathsf{ST}(2)_+$ is a submonoid of $\mathsf{ST}(2)$, the pre-order on $\mathsf{ST}(2)_+$ is a partial order, and the quotient mapping π_2 restricted to $\mathsf{ST}(2)_+$ is an order-isomorphism, so the monoid structure on $\mathsf{ST}(2)_+$ can be transferred to $\mathsf{ST}(2)/\!\equiv$.

To summarize, there is a pre-order on binary channels defined by comparing the minimal ideal of the principal left ideal each generates, and this pre-order induces an ordered quotient of $\mathsf{ST}(2)$. Moreover, the quotient map preserves capacity, the "capacity map" cap measures the quotient *qua* domain, when given the quotient order induced by the pre-order on $\mathsf{ST}(2)$, and this quotient structure is isomorphic — as a monoid and as a ordered space — to a submonoid of $\mathsf{ST}(2)$.

In higher dimensions, things become more complicated, but the same approach yields almost the same results. The idea is to replace subintervals of $[0,1]$ with finite polytopes in $P(n) \subseteq [0,1]^{n-1}$, for each n. To obtain this analog for $\mathsf{ST}(n)$, we recall Proposition 8.6 and Corollary 8.9 which characterize intersections of principal left ideals in $\mathsf{ST}(n)$ with its minimal ideal: If $M \in \mathsf{ST}(n)$, then

$$\mathcal{M}(\mathsf{ST}(n)M) = \mathsf{ST}(n)M \cap \mathcal{M}(\mathsf{ST}(n)) = \overline{\mathrm{conv}}(\{M(1), \ldots, M(n)\}) = M(P(n)),$$

the image of $P(n)$ under the affine mapping M. This result tells us something new about $\mathsf{ST}(2)$ — that the equivalence defined above is simply that channels have the same image:

$$C \equiv C' \Leftrightarrow \mathcal{M}(\mathsf{ST}(2))C = \mathcal{M}(\mathsf{ST}(2))C' \Leftrightarrow C(P(2)) = C'(P(2)).$$

Concentrating on $\mathsf{ST}(n)$, we have the following. If we formalize the pre-order in Corollary 8.9 by defining

$$(\forall M, M' \in \mathsf{ST}(n))\ M \leq M' \Leftrightarrow M(P(n)) \supseteq M'(P(n)),$$

then:

PROPOSITION 11.1. *Consider the compact affine monoid* $\mathsf{ST}(n)$.

(1) *The mapping* $M \mapsto \mathcal{M}(\mathsf{ST}(n)M)\colon (\mathsf{ST}(n), \leq) \to (\mathsf{Conv}_n(\mathcal{M}(\mathsf{ST}(n))), \supseteq)$ *is continuous, where we endow* $\mathsf{Conv}_n(\mathcal{M}(\mathsf{ST}(n)))$ *with the Vietoris topology.*

(2) The family $\mathsf{Conv}_n(\mathcal{M}(\mathsf{ST}(n)))$ is isomorphic to the family of finitely-generated, closed convex left ideals of $\mathcal{M}(\mathsf{ST}(n))$, when both are ordered by reverse inclusion.

PROOF. The mapping $(C, M) \mapsto CM \colon \mathsf{ST}(n) \times \mathsf{ST}(n) \to \mathsf{ST}(n)$ is continuous, and lifts to a continuous mapping $(X, M) \mapsto XM \colon \Gamma(\mathsf{ST}(n)) \times \mathsf{ST}(n) \to \Gamma(\mathsf{ST}(n))$. Restricting the first component to the fixed set $\mathcal{M}(\mathsf{ST}(n)) \in \Gamma(\mathsf{ST}(n))$ shows the mapping $M \mapsto \mathcal{M}(\mathsf{ST}(n))M \colon \mathsf{ST}(n) \to \Gamma(\mathsf{ST}(n))$ is continuous. But $\mathcal{M}(\mathsf{ST}(n))M = \mathcal{M}(\mathsf{ST}(n)M)$ by Proposition 8.6. Corollary 8.9 then implies $\mathcal{M}(\mathsf{ST}(n)M) = \overline{\mathrm{conv}}(\{M(1), \ldots, M(n)\}) \in \mathsf{Conv}_n(\mathcal{M}(\mathsf{ST}(n)))$. Finally, $M \leq M'$ iff $M(P(n)) \supseteq M'(P(n))$ implies the mapping is monotone, which shows (1).

For (2), any finite set $F \subseteq \mathcal{M}(\mathsf{ST}(n))$ generates a closed left ideal $\mathcal{M}(\mathsf{ST}(n))F$, and the convex hull $\overline{\mathrm{conv}}(F)$ is the left ideal F generates in $\mathcal{M}(\mathsf{ST}(n))$. $\square$

The final piece of the puzzle is contained in the following result. We recall from our Notation 10.4 that $P(n)$ is the simplex of all points $X = (x_1, \ldots, x_n)$ satisfying $\sum_i x_i = 1$, considered as probability distributions on $\{1, \ldots, n\}$.

We define the mapping $\mathsf{cap} \colon (\mathsf{Conv}_n(P(n)), \subseteq) \to \mathbb{R}_+$ as follows: For an n-tuple of vectors $(v_1, \ldots, v_n) \in P(n)$, and for probability distributions $p = (r_1, \ldots, r_n)$ ranging through $P(n)$ we set

$$\mathsf{cap}(\overline{\mathrm{conv}}(\{v_1, \ldots, v_n\})) = \sup\left\{\mathcal{H}(\sum_{i=1}^n r_i \cdot v_i) - (\sum_{i=1}^n r_i \mathcal{H}(v_i)) \mid p \in P(n)\right\}.$$

Now we have:

PROPOSITION 11.2. *The poset* $\mathsf{Conv}_n(P(n)))$ *is a domain, and* cap *is a Vietoris continuous and monotone map.*

PROOF. That $\mathsf{Conv}_n(P(n))$ is a domain follows from Proposition 10.6. It is well-known that entropy $\mathcal{H}$ is (strictly) concave (cf. [**10**], Chapter 2), and so taking $\mathsf{cap} = \widehat{\mathcal{H}}$ in Theorem 10.20 implies the result. $\square$

We now give some examples of how our results about compact monoids and capacity can be used to understand classical channels. One consequence of our results is the ability to guarantee capacity decreases along ordered line segments, in particular along one-parameter semigroups. For example, when we analyzed the subsemigroup $\mathsf{ZT}(n)$ in Proposition 8.12 we noted that $\mathsf{ZT}(n)$ consists of one-parameter semigroups running from I_n to $\mathcal{M}(\mathsf{ST}(n))$ each of which is a line segment of the form $\{(1 - r) \cdot I_n + r \cdot Z \mid 0 \leq r \leq 1\}$ for a fixed $Z \in \mathcal{M}(\mathsf{ST}(n))$. For $Z \in \mathcal{M}(\mathsf{ST}(n))$, the one-parameter semigroup running from I_n to Z is indexed by the parameter $r \in [0, 1]$, and fixing one such, the associated channel $X_r = (x_{ij})$ has entries

$$x_{i,j} = \begin{cases} (1 - r) + rz_{ii} & \text{if } i = j, \\ rz_{ij} & \text{if } i \neq j, \end{cases}$$

for each $r \in [0, 1]$, where $X_r = (1 - r)I_n + rZ$. Since Z has identical rows (cf. Theorem 8.3(4)), it follows that the mapping $\phi \colon [0, 1] \to \mathsf{Conv}_n(\mathcal{M}(\mathsf{ST}(n)))$ by $\phi(r) = \overline{\mathrm{conv}}(\{X_r(1) \ldots, X_r(n)\})$ is monotone and continuous. Since cap is montone and continuous on $(\mathsf{Conv}_n(\mathcal{M}(\mathsf{ST}(n))), \supseteq)$ by Proposition 11.2, we conclude that $\mathsf{cap} \circ \phi$ is monotone and continuous as well. Hence capacity is decreasing along the one-parameter semigroup $r \mapsto (1 - r)I_n + rZ$ for each $Z \in \mathcal{M}(\mathsf{ST}(n))$.

The same argument would apply, for example, starting with an arbitrary channel $C \in \mathsf{ST}(n)$ and considering the line segment $r \mapsto (1-r)C + rZ_C$, where $Z_C \in \mathcal{M}(\mathsf{ST}(n))$ is the channel with identical rows, all of which are of the form $\vec{p} \cdot C$, where $\vec{p} \in P(\{1,\ldots,n\})$ is an optimal input distribution for which C achieves capacity. Thus, capacity is decreasing along this straight line in $\mathsf{ST}(n)$. Moreover, if we consider the one-parameter semigroup $(1-r)I_n + rZ_C$, then this straight line is the family $\{C((1-r)I_n + rZ_C) \mid 0 \le r \le 1\}$, which realizes the line segment as a translate of a one-parameter semigroup. So, to achieve a specified reduction in capacity (for example, to lower the chance of a covert channel being used in conjunction with the channel C), we can simply use the appropriate channel $(1-r_0)C + r_0 Z_C$, which can be viewed as following C with the channel $(1-r_0)I_n + rZ_C$ under composition.

11.1. Summary. The aim of our development in this section was to generalize the results of Martin, et al. in [**25**]. When we consider the list of the results from that paper we enumerated at the beginning of this section, we have fulfilled our goal for items (1) – (4). Moreover, we also have completed most of (5) — the exception being that we have not shown that cap measures $(\mathsf{Conv}_n(\mathcal{M}(\mathsf{ST}(n))), \supseteq)$ — and (6). For the first of these, recalling Corollary 9.8 and noting Proposition 11.2, what is needed is a proof that cap is strictly monotone, but there is some evidence that this may well be false. For the second — generalizing (6) — what is needed is to identify a suitable submonoid S of $\mathsf{ST}(n)$ so that the restriction $\pi_n|_S$ of the quotient map to S is an order-isomorphism onto $\mathsf{Conv}_n(P(n))$. The result in the case of $\mathsf{ST}(2)$ is due fundamentally to the fact that the identity is the only idempotent outside $\mathcal{M}(\mathsf{ST}(2))$, but things are much more complicated in the case of $\mathsf{ST}(n)$ for larger n.

Acknowledgment

We would like to thank the participants of the 2007–2008 Clifford Lectures, the Workshops on Informatic Phenomena, WIP 2008 and 2009, all of which took place at Tulane University, as well as the participants of the Seminar on Semantics of Information, Schloß Dagstuhl, Germany, June, 2010, for their interest in the rôle of compact monoids in information theory. The results in Part II of this paper were presented at the Dagstuhl Seminar by the first author, and a preliminary version of the results in Part III were part of an invited address by the second author at MFPS 25, which took place in Oxford, UK in 2009.

The second author also wishes to thank KEYE MARTIN for many stimulating conversations on classical and quantum channels, and the potential rôle that compact affine monoids play in that area.

Finally, we owe to two anonymous and exceptionally thorough referees a great debt of thanks for their very detailed examination of earlier drafts of this paper which now is greatly improved by the many detailed and perceptive comments and suggestions that were included in the referees' reports.

References

[1] Abramsky, S., and A. Jung, "Domain theory," in: S. Abramsky, D. M. Gabbay, and T. S. E. Maibaum, editors, Handbook of Logic in Computer Science, III. Oxford University Press. ISBN 0-19-853762-X. http://www.cs.bham.ac.uk/~axj/pub/papers/handy1.pdf.

[2] Bourbaki, N., "Espaces vectoriels topologiques", Chap. 1 à 5, Masson Paris etc., 1981, vii+368 pp.
[3] Carruth, J. H., J. Hildebrant, and R. J. Koch, "Introduction to Topological Semigroups I and II," Marcel Dekker, New York 1983, respectively, 1986.
[4] Chow, H. L., *On compact affine semigroups*, Semigroup Forum **11** (1975), 146–152.
[5] Clifford, A. H., and D. D. Miller, *Union and symmetry preserving endomorphisms of the semigroup of all binary relations on a set*, Czechoslovak Mathematical Journal, vol. 20 (1970), issue 2, pp. 303-314
[6] Clifford, A. H., and G. B. Preston, "The Algebraic Theory of Semigroups, I," Math. Surveys **7**, Amer. Math. Soc., Providence, R. I., 1962, 224 pp.
[7] Coecke, B., and K. Martin, *Partiality in physics*, 2003, http://arxiv.org/abs/quant-ph/0312044
[8] Cohen, H., and H. S. Collins, *Affine semigroups*, Trans. Amer. Math. Soc. **93** (1959), 97–113.
[9] Collins, H. S., *Remarks on affine semigroups*, Pac. J. Math. **14** (1964), 449–455.
[10] Cover, T., and J. Thomas, "Elements of Information Theory", Wiley (2006), xiii + 748 pp.
[11] Conway, J. B., "A Course in Functional Analysis", Springer-Verlag, (1994).
[12] Ellis, R., *Distal transformation group*, Pac. J. Math. **8** (1958), 401–405.
[13] Furstenberg, H., and Y. Katznelson, *Idempotents in compact Semigroups and Ramsey Theory*, Israel J. Math. **68** (1989), 257–270.
[14] Gierz, G., K. H. Hofmann, K. Keimel, J. Lawson, M. Mislove and D. Scott, "Continuous Lattices and Domains", Cambridge University Press (2003).
[15] Hardy, G. H., J. E. Littlewood and G. Pólya, "Inequalities", 2nd ed., Cambridge University Press, Cambridge, UK, pp. 83-84, 1988.
[16] Hofmann, K. H., *An illustration of the power of structure theory*, Applied Categorical Structures **8** (2000), 145–160.
[17] Hofmann, K. H., and P. S. Mostert, "Elements of Compact Semigroups," Charles E. Merrill, Columbus (Ohio), 1966, xiii + 384 pp.
[18] Hofmann, K. H., and S. A. Morris, "The Structure of Compact Groups," Verlag Walter De Gruyter Berlin, 1998, xvii+834pp. Second Revised and Augmented Edition 2006, xviii+858pp.
[19] Högnäs, G., and A. Mukherjea, "Probability Measures on Semigroups," (Second Edition of [**29**]), Springer-Verlag, 2011, 430 pp.
[20] Kelley, J. L., "General Topology," Van Nostrand, New York (1955); Republished in 1975 by Springer-Verlag.
[21] Knapp, A., "Advanced Real Analysis", Birkhäuser (2005). xxv + 465pp.
[22] Lawson, J. D., *Embeddings of compact convex sets and locally compact cones*, Pac. J. Math. **66** (1976), 443–453.
[23] Martin, K., "A Foundation for Computation," PhD Thesis, Department of Mathematics, Tulane University, New Orleans, LA, 2000
[24] Martin, K., *The scope of a quantum channel*, submitted.
[25] Martin, K., G. Allwein and I. Moskowitz, *Algebraic information theory for binary channels*, Theor. Comput. Sci. **411** (2010), 1918–1927.
[26] Martin, K., J. Feng and S. Krishnan, *A free object in quantum information theory*, ENTCS **265** (2010), 35–47.
[27] Martin, K., and P. Panangaden, A technique for verifying measurements, Electronic Notes Theoretical Computer Science **218** (2008), pp 261-273.
[28] Michael, A. A., G., On three classical results about compact groups, Scientiae Mathematicae Japonicae Online **e-2011** (2011), 271–273.
[29] Mukherjea, A., and N. A. Tserpes, "Probability Measures on Semigroups: Convolution Products, Random Walks, and Random Matrices," Lecture Notes in Math., Springer-Verlag 1976.
[30] von Neumann, J., On complete topological spaces, Trans. Amer. Math. Soc. vol. 37 (1935) pp. 1-20.
[31] Paalman–de Miranda, A. B., "Topological Semigroups," Mathematical Centre Tracts 11, Mathematisch Centrum Amsterdam, 1964, 169 pp.
[32] Roberts, J. W., *A compact convex set with no extreme points*, Studia Math. **60** (1977), 255–266.
[33] Rudin, Walter, "Functional Analysis" (2nd ed.), McGraw-Hill Science/Engineering/Math, 1991.

[34] Ruppert, W., "Compact Semitopological Semigroups: An Intrinsic Theory," Lecture Notes in Math. **1079** (1984), Springer-Verlag, iii+260 pp.
[35] Shannon, C. E., *A mathematical theory of communication*, Bell Systems Technical Journal **27** (1948), 379–423, 623–656.
[36] Valero, O., *On Banach fixed point theorems for partial metric spaces,* Applied General Topology **6** (2005), pp 229–240.
[37] Wallace, A. D., *On the structure of topological semigroups*, Bull. Amer. Math. Soc. **61** (1955), 95–112.
[38] Waszkiewicz, P., *Distance and measurement in domain theory,* Electronic Notes in Theoretical Computer Science **45** (2001), 15pp.
[39] Wehausen, J. V., Transformations in linear topological spaces, Duke Math. J. vol. 4 (1938) pp. 157-169.
[40] Wendel, J. G., *Haar measure and the semigroup of measures on a compact group,* Proc. Amer. Math. Soc. **5** (1954), 923–929.

FACHBEREICH MATHEMATIK, TECHNISCHE UNIVERSITÄT DARMSTADT, GERMANY AND DEPARTMENT OF MATHEMATICS, TULANE UNIVERSITY, NEW ORLEANS, LA
E-mail address: `hofmann@mathematik.tu-darmstadt.de`

DEPARTMENT OF MATHEMATICS, TULANE UNIVERSITY, NEW ORLEANS, LA
E-mail address: `mislove@tulane.edu`

The Scope of a Quantum Channel

Keye Martin

ABSTRACT. The capacity of a classical channel is a single number that to a large extent captures its ability to transmit information. Though analogous notions exist for quantum channels, the use of a single number is not particularly informative. For instance, any basis of the state space can be used to represent classical bits, and each representation leads to a classical channel with a capacity all its own. So the ability of a quantum channel to transmit information should minimally depend on all of the different ways classical bits can be represented: instead of a single number, we should measure it with a *set* of numbers. We call this set, which consists of the range of achievable classical capacities, the *scope*.

For unital channels on qubits, we establish that scope is in fact a *compact interval*, provide an exact characterization of it and then show how to systematically calculate it. These results, which rely extensively on the algebraic structure of quantum channels, are then used to design an *adaptive* scheme for communication in which the participants can maximize the information transmitted after first determining the state of the environment (which we show how to do) and then performing a scope calculation. When this technique is applied to quantum cryptography, it becomes possible to minimize the error rate over any time interval where the environment remains stable. For familiar forms of noise, like bit flipping, the error rate is cut in half.

1. Introduction

We understand communication as being the following process: a sender takes some information, represents it in a certain way, sends this representation to a receiver, who then performs some operation on the representation to extract the information. To measure the amount of information being transmitted, one needs some way of measuring the correlation between the sent and the received. Many factors can affect the amount of information that gets transmitted, the most well-known being noise in the environment. This paper is about how our ability to transmit classical information through a noisy quantum channel depends on how that information is represented.

Given a quantum channel f, each basis of the state space defines a particular way to represent classical information. For each such representation, there is an

---

2010 Mathematics Subject Classification. Primary 94A17; Secondary 81Q99.
Keywords. Capacity, scope, channel, information theory, quantum communication.
Research supported by the Naval Research Laboratory and the Office of Naval Research.

associated classical channel having a capacity all its own. The range of capacities achieved as we vary over all possible representations is called the *scope* of the channel f. It is denoted by $s(f)$. Intuitively, the largest value of scope tells us how we should represent information so that we *maximize* the amount transmitted, while the smallest value could be used to measure the optimal performance of a method designed to interrupt communication or remove a steganographic message. If all representations are not available to us, then we may want to choose the *best available*, which might mean considering a value in between the maximum and minimum.

In order to calculate scope, we need a simpler representation of quantum channels, and for this we turn in the next section to the Bloch representation. We give a self-contained presentation of many well-known results whose proofs are difficult to find in the literature, as well as some more obscure results that are at times taken for granted. After studying the Bloch representation, we obtain an elementary characterization of the channels that are our principal concern in this paper: the *unital* qubit channels, which occur naturally in communication as conservative models of noise. It turns out that such a channel can always be canonically represented by a convex sum of rotations in $\mathbb{R}^3$. We consider some of their fascinating algebraic properties and use them to establish that qubit unitality is the quantum analogue of a binary symmetric channel.

After our study of unitality, we have the simple language needed to not only characterize scope precisely as the solution to an optimization problem, but also to give a systematic method for solving it: each unital channel can be replaced by a symmetric unital channel with the same scope – but the scope of a symmetric channel can be calculated from its eigenvalues! This is very surprising given that we are then able to show that each symmetric channel is a convex sum of *four* involutive rotations which collectively comprise a copy of the Klein-four group. We give scope calculations for projective measurements, bit flipping, bit-phase flipping, phase flipping, depolarization, the two-Pauli channel, the intercept-resend attack in quantum cryptography and general unitary evolution.

Finally, we focus on the significance of scope to quantum information itself. For symmetric channels, we show that the largest value in the scope coincides with the Holevo capacity and that this is also true for unitary channels. We show how the ability to perform scope calculations can be used to minimizing the error rate in protocols like QKD, leading to a new idea we call *adaptive quantum communication*. Along the way, it becomes clear that scope can be used to classify physical effects according to the degree with which they disturb a system.

2. The Bloch representation

Let $\mathcal{H}^2$ denote a two dimensional complex Hilbert space with specified inner product $\langle \cdot | \cdot \rangle$.

DEFINITION 2.1. A *quantum state* is a density operator $\rho : \mathcal{H}^2 \to \mathcal{H}^2$, i.e., a self-adjoint, positive, linear operator with $\mathrm{tr}(\rho) = 1$. The quantum states on $\mathcal{H}^2$ are denoted Ω^2.

We also sometimes call density operators *mixed states*.

DEFINITION 2.2. A quantum state ρ on $\mathcal{H}^2$ is *pure* if $\mathrm{spec}(\rho) \subseteq \{0, 1\}$.

Pure states always have the form $f : \mathcal{H}^2 \to \mathcal{H}^2$ given by

$$f(u) = \langle \psi | u \rangle \psi$$

for some unit vector $\psi \in \mathcal{H}^2$. However, because unit vectors in quantum mechanics are normally written $|\psi\rangle$, the pure state associated to a unit vector is normally denoted $|\psi\rangle\langle\psi|$. Such operators define *projections* on $\mathcal{H}^2$. If $|\psi\rangle$ and $|\phi\rangle$ are orthogonal, then $|\psi\rangle\langle\psi| + |\phi\rangle\langle\phi| = I$.

DEFINITION 2.3. The *spin operators* are denoted $\{I, \sigma_x, \sigma_y, \sigma_z\}$ and given by

$$I = \begin{pmatrix} 1 & 0 \\ 0 & 1 \end{pmatrix} \quad \sigma_x = \begin{pmatrix} 0 & 1 \\ 1 & 0 \end{pmatrix} \quad \sigma_y = \begin{pmatrix} 0 & -i \\ i & 0 \end{pmatrix} \quad \sigma_z = \begin{pmatrix} 1 & 0 \\ 0 & -1 \end{pmatrix}$$

Each spin operator is self adjoint and unitary. They have many wonderful and valuable properties, including:
- $\sigma_i^2 = I$ (involutivity),
- $\sigma_i \sigma_j + \sigma_j \sigma_i = 0$ (anticommutativity),

for distinct $i, j \in \{x, y, z\}$. Notice that

(2.1) $$(\forall i \in \{x, y, z\}) \; \text{tr}(\sigma_i) = 0$$

And since $\text{tr}(ab) = \text{tr}(ba)$ for any two matrices, linearity of the trace gives

(2.2) $$\text{tr}(\sigma_i \sigma_j) = \text{tr}(\sigma_j \sigma_i) = 0$$

for distinct $i, j \in \{x, y, z\}$. Since the spin operators anticommute and are all self adjoint,

(2.3) $$\langle \sigma_i(v) | \sigma_j(v) \rangle = -\langle \sigma_j(v) | \sigma_i(v) \rangle$$

for distinct $i, j \in \{x, y, z\}$. These facts will be important to us later.

There is a 1-1 correspondence between density operators on a two dimensional state space and points on the unit ball $\mathbb{B}^3 = \{x \in \mathbb{R}^3 : |x| \leq 1\}$: each density operator $\rho : \mathcal{H}^2 \to \mathcal{H}^2$ can be written uniquely as

$$\rho = \frac{1}{2}(I + r_x \sigma_x + r_y \sigma_y + r_z \sigma_z) := \frac{1}{2}(I + r \cdot \vec{\sigma})$$

where $r = (r_x, r_y, r_z) \in \mathbb{R}^3$ satisfies $|r| = \sqrt{r_x^2 + r_y^2 + r_z^2} \leq 1$ and $\vec{\sigma} = (\sigma_x, \sigma_y, \sigma_z)$ is the vector of spin operators. The vector r is called the *Bloch vector* associated to ρ. We sometimes write $r = [\![\rho]\!]$ and denote the *bijection* between Ω^2 and $\mathbb{B}^3$ as $[\![\cdot]\!] : \Omega^2 \to \mathbb{B}^3$. For the sake of completeness:

LEMMA 2.4. $[\![\cdot]\!] : \Omega^2 \to \mathbb{B}^3$ *is a bijection.*

Proof. The spin operators form a basis for the vector space $M_{2\times 2}(\mathbb{C})$ of 2×2 matrices with complex entries. Thus, any $A \in M_{2\times 2}(\mathbb{C})$ can be written as

$$A = t \cdot I + x \cdot \sigma_x + y \cdot \sigma_y + z \cdot \sigma_z$$

where t, x, y and z are all complex. The following are all standard well-known facts:
- A is self-adjoint iff $(t, x, y, z) \in \mathbb{R}^4$
- A is positive iff $|(x, y, z)| \leq t$
- $\text{tr}(A) = 1$ iff $t = 1/2$

Thus, for a density operator ρ, we have $t = 1/2$ and $|(x, y, z)| \leq 1/2$. If we set $r = 2(x, y, z)$, then
$$\rho = \frac{1}{2}(I + r_x \sigma_x + r_y \sigma_y + r_z \sigma_z) := \frac{1}{2}(I + r \cdot \vec{\sigma})$$
with $r \in \mathbb{B}^3$. That such an r is unique follows from the fact that spin matrices form a basis for $M_{2 \times 2}(\mathbb{C})$. This proves that $[\![\cdot]\!]$ is an injective function. To see that it is surjective, we simply take $r \in \mathbb{B}^3$ and form the matrix $\rho = \frac{1}{2}(I + r \cdot \vec{\sigma})$, for which we have $[\![\rho]\!] = r$. □

PROPOSITION 2.5. *Let ρ and σ be density operators with respective Bloch vectors r and s.*

(i) *The eigenvalues of ρ are $(1 \pm |r|)/2$,*
(ii) *The base two von Neumann entropy of ρ is $S\rho = H((1+|r|)/2) = H((1-|r|)/2)$, where $H : [0,1] \to [0,1]$ is the base two Shannon entropy,*
(iii) *If ρ and σ are pure states and $r + s = 0$, then ρ and σ are orthogonal, and thus form a basis for the state space. Conversely, the Bloch vectors associated to a pair of orthogonal pure states form antipodal points on the sphere.*
(iv) *The Bloch vector for a convex sum of mixed states is the convex sum of the Bloch vectors.*
(v) *The Bloch vector for the completely mixed state $I/2$ is $0 = (0, 0, 0)$.*

Proof. For (i), we simply use the fact that
$$\rho = \frac{1}{2}(I + r \cdot \vec{\sigma}) = \frac{1}{2}\begin{pmatrix} 1 + r_z & r_x - ir_y \\ r_x + ir_y & 1 - r_z \end{pmatrix}$$
where r is the Bloch vector for ρ. Assertion (ii) is then immediate since $S\rho$ is equal to the Shannon entropy of the eigenvalues of ρ.

For (iii), since ρ and σ are pure states, there are unit vectors $|\psi\rangle$ and $|\phi\rangle$ such that $\rho = |\psi\rangle\langle\psi|$ and $\sigma = |\phi\rangle\langle\phi|$. Because $r + s = 0$, $\rho + \sigma = I$, so applying this operator to $|\phi\rangle$ gives
$$|\psi\rangle \cdot \langle\psi|\phi\rangle + |\phi\rangle \cdot 1 = |\phi\rangle \implies |\psi\rangle \cdot \langle\psi|\phi\rangle = 0$$
But $|\psi\rangle$ is a unit vector, so this means $\langle\psi|\phi\rangle = 0$, proving that ρ and σ are orthogonal pure states. Conversely, suppose that $|\psi\rangle$ and $|\phi\rangle$ are orthogonal pure states with Bloch vectors r and s. Then their associated density operators are
$$\rho = |\psi\rangle\langle\psi| = \frac{1}{2}(I + r \cdot \vec{\sigma}) \quad \& \quad \sigma = |\phi\rangle\langle\phi| = \frac{1}{2}(I + s \cdot \vec{\sigma})$$
where $\vec{\sigma} = (\sigma_x, \sigma_y, \sigma_z)$ is the vector of spin operators. We then see that $\rho + \sigma = I + ((r + s) \cdot \vec{\sigma})/2$. Applying this operator to $|\psi\rangle$ gives $v := ((r + s) \cdot \vec{\sigma})|\psi\rangle = 0$. Taking the inner product of v with itself, applying linearity and conjugate linearity of the inner product, and repeatedly applying equation (2.3) gives $(r_1 + s_1)^2 + (r_2 + s_2)^2 + (r_3 + s_3)^2 = 0$, which implies that $r + s = 0$, proving that r and s are antipodal points on the sphere.

For (iv), let ρ and σ be density operators and $x \in [0, 1]$. Then
$$x \cdot \rho + (1 - x) \cdot \sigma = \frac{x}{2} \cdot (I + r \cdot \vec{\sigma}) + \frac{1 - x}{2} \cdot (I + s \cdot \vec{\sigma})$$
$$= \frac{1}{2}(I + (xr + (1 - x)s) \cdot \vec{\sigma})$$

which proves that the Bloch vector of a convex sum is the convex sum of the Bloch vectors.

(v) follows from (iii) and (iv): the completely mixed state can be written

$$I/2 = \frac{1}{2}|0\rangle\langle 0| + \frac{1}{2}|1\rangle\langle 1|$$

where $|0\rangle$ and $|1\rangle$ are orthogonal pure states. Since $I/2$ is a convex sum of density operators, its Bloch vector r is the convex sum of the associated Bloch vectors by (iv). However, by (iii), the sum of these Bloch vectors has to be zero since the vectors used to define them are orthogonal. □

The fact that antipodal points on the Bloch sphere correspond to orthogonal pure states is something we will make repeated use of in this paper.

DEFINITION 2.6. A *qubit channel* is a trace preserving function $\varepsilon : \Omega^2 \to \Omega^2$ that is convex linear and completely positive.

Each qubit channel $\varepsilon : \Omega^2 \to \Omega^2$ can be represented by a function $f_\varepsilon : \mathbb{B}^3 \to \mathbb{B}^3$ on the *Bloch sphere* $\mathbb{B}^3$. The map f_ε is called the *Bloch representation* of ε and is the unique map which makes the following diagram commute:

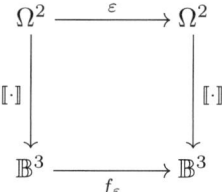

Formally, f_ε is defined by $f_\varepsilon(r) = [\![\varepsilon([\![r]\!]^{-1})]\!]$ and satisfies $[\![\varepsilon(\rho)]\!] = f_\varepsilon([\![\rho]\!])$ for all $\rho \in \Omega^2$.

PROPOSITION 2.7. *Let ε be a qubit channel and let f_ε be its Bloch representation.*
  (i) *The function f_ε is convex linear.*
 (ii) *Composition of qubit channels corresponds to composition of Bloch representations: for channels $\varepsilon_1, \varepsilon_2$, we have $f_{\varepsilon_1 \circ \varepsilon_2} = f_{\varepsilon_1} \circ f_{\varepsilon_2}$.*
(iii) *Convex sum of qubit channels corresponds to convex sum of Bloch representations: for channels $\varepsilon_1, \varepsilon_2$ and $x \in [0,1]$, we have $f_{x\varepsilon_1 + \bar{x}\varepsilon_2} = xf_{\varepsilon_1} + \bar{x}f_{\varepsilon_2}$.*

Proof. Throughout this proof, $x \in [0,1]$ and ρ and σ are density operators with Bloch vectors r and s respectively. Let us also recall in general that a qubit channel ε is related to its Bloch representation f_ε by $[\![\varepsilon\rho]\!] = f_\varepsilon([\![\rho]\!])$.

(i) The function f_ε is convex linear:

$$\begin{aligned}
f_\varepsilon(xr + (1-x)s) &= f_\varepsilon(x[\![\rho]\!] + (1-x)[\![\sigma]\!]) \\
&= f_\varepsilon([\![x\rho + (1-x)\sigma]\!]) \qquad \text{(Prop. 2.5(iv))} \\
&= [\![\varepsilon(x\rho + (1-x)\sigma)]\!] \\
&= [\![x\varepsilon(\rho) + (1-x)\varepsilon(\sigma)]\!] \\
&= x[\![\varepsilon(\rho)]\!] + (1-x)[\![\varepsilon(\sigma)]\!] \\
&= xf_\varepsilon([\![\rho]\!]) + (1-x)f_\varepsilon([\![\sigma]\!]) \\
&= xf_\varepsilon(r) + (1-x)f_\varepsilon(s)
\end{aligned}$$

(ii) The representation of a composition is the composition of the representations:
$$f_{\varepsilon_1 \varepsilon_2}(r) = f_{\varepsilon_1 \varepsilon_2}(\llbracket \rho \rrbracket) = \llbracket \varepsilon_1(\varepsilon_2(\rho)) \rrbracket = f_{\varepsilon_1}(\llbracket \varepsilon_2(\rho) \rrbracket) = f_{\varepsilon_1}(f_{\varepsilon_2}(\llbracket \rho \rrbracket)) = f_{\varepsilon_1}(f_{\varepsilon_2}(r)).$$

(iii) The representation of a convex sum is the convex sum of the representations:
$$\begin{aligned}
f_{x\varepsilon_1 + (1-x)\varepsilon_2}(r) &= f_{x\varepsilon_1 + (1-x)\varepsilon_2}(\llbracket \rho \rrbracket) \\
&= \llbracket x\varepsilon_1(\rho) + (1-x)\varepsilon_2(\rho) \rrbracket \\
&= x\llbracket \varepsilon_1(\rho) \rrbracket + (1-x)\llbracket \varepsilon_2(\rho) \rrbracket \qquad \text{(Prop. 2.5(iv))} \\
&= x f_{\varepsilon_1}(\llbracket \rho \rrbracket) + (1-x) f_{\varepsilon_2}(\llbracket \rho \rrbracket) \\
&= x f_{\varepsilon_1}(r) + (1-x) f_{\varepsilon_2}(r)
\end{aligned}$$

$\square$

If we now denote f_ε by $\llbracket \varepsilon \rrbracket$, we see that the following desirable "semantic properties" all hold:

- $\llbracket \bot \rrbracket = 0$, where $\bot$ is the constant qubit channel $\rho \mapsto I/2$,
- $\llbracket I \rrbracket = I$
- $\llbracket f \circ g \rrbracket = \llbracket f \rrbracket \circ \llbracket g \rrbracket$
- $\llbracket xf + (1-x)g \rrbracket = x\llbracket f \rrbracket + (1-x)\llbracket g \rrbracket$

These properties allow one to calculate Bloch representations of channels that can be written as compositions and convex sums of simpler channels. The predominant example of a 'simple' channel is a *unitary channel*.

DEFINITION 2.8. A qubit channel ε is *unitary* if there is a unitary operator $U : \mathbb{C}^2 \to \mathbb{C}^2$ such that $\varepsilon(\rho) = U\rho U^\dagger$ for all $\rho \in \Omega^2$.

We will now show how to calculate the Bloch representation of a unitary channel.

DEFINITION 2.9. Let $r_x(\theta), r_y(\theta), r_z(\theta) : \mathbb{B}^3 \to \mathbb{B}^3$ denote the rotations about the x, y and z axes:

$$r_x(\theta) = \begin{pmatrix} 1 & 0 & 0 \\ 0 & \cos\theta & -\sin\theta \\ 0 & \sin\theta & \cos\theta \end{pmatrix} \quad r_y(\theta) = \begin{pmatrix} \cos\theta & 0 & \sin\theta \\ 0 & 1 & 0 \\ -\sin\theta & 0 & \cos\theta \end{pmatrix} \quad r_z(\theta) = \begin{pmatrix} \cos\theta & -\sin\theta & 0 \\ \sin\theta & \cos\theta & 0 \\ 0 & 0 & 1 \end{pmatrix}$$

An orthogonal matrix M is an invertible matrix such that $M^{-1} = M^t$. An orthogonal matrix M with $\det(M) = +1$ is called a *rotation*.

Every rotation M on $\mathbb{R}^3$ can be written in the form $M = r_x(\alpha) \cdot r_y(\beta) \cdot r_z(\theta)$ for angles α, β, θ. Though we are not aware of any proofs in the literature, the relationship between unitary qubit channels and rotations is well-known and is as follows:

PROPOSITION 2.10.
  (i) If ε is a unitary qubit channel, then f_ε is a rotation.
  (ii) For every rotation M on $\mathbb{R}^3$, there is a unitary qubit channel ε such that $f_\varepsilon = M$.

Proof. (i) By exponentiating the spin operators σ_y and σ_z, we obtain unitary operators

$$\hat{r}_y(\theta) = e^{-i\theta \sigma_y/2} = \begin{pmatrix} \cos\theta/2 & -\sin\theta/2 \\ \sin\theta/2 & \cos\theta/2 \end{pmatrix} \quad \text{and} \quad \hat{r}_z(\theta) = e^{-i\theta\sigma_z/2} = \begin{pmatrix} e^{-i\theta/2} & 0 \\ 0 & e^{i\theta/2} \end{pmatrix}$$

A standard fact [6] is that any unitary operator U can be written as
$$U = e^{i\lambda}\,\hat{r}_z(\alpha)\,\hat{r}_y(\beta)\,\hat{r}_z(\theta)$$
where $\lambda, \alpha, \beta, \theta$ are real numbers. The unitary channel $\varepsilon(\rho) = U\rho U^\dagger$ defined by U is the same as the one defined by the unitary operator $e^{-i\lambda}U$ since $(e^{i\lambda}U)^\dagger = e^{-i\lambda}U$. Since $\varepsilon = \varepsilon_\alpha \circ \varepsilon_\beta \circ \varepsilon_\theta$, where ε_α, ε_β and ε_θ are the unitary channels associated to $\hat{r}_z(\alpha)$, $\hat{r}_y(\beta)$ and $\hat{r}_z(\theta)$, respectively, Prop. 2.7(iii) tells us that its Bloch representation is
$$f_\varepsilon = f_{\varepsilon_\alpha} \circ f_{\varepsilon_\beta} \circ f_{\varepsilon_\theta}$$
However, calculations you wouldn't wish on your worst enemy reveal that for any density operator ρ written in terms of its Bloch vector $r = (r_x, r_y, r_z)$,
$$\varepsilon_\beta(\rho) = \frac{1}{2}\begin{pmatrix} 1 + (r_z\cos\beta - r_x\sin\beta) & (r_z\sin\beta + r_x\cos\beta) - ir_y \\ (r_z\sin\beta + r_x\cos\beta) + ir_y & 1 - (r_z\cos\beta - r_x\sin\beta) \end{pmatrix}$$
and
$$\varepsilon_\theta(\rho) = \frac{1}{2}\begin{pmatrix} 1 + r_z & (r_x\cos\theta - r_y\sin\theta) - i(r_x\sin\theta + r_y\cos\theta) \\ (r_x\cos\theta - r_y\sin\theta) + i(r_x\sin\theta + r_y\cos\theta) & 1 - r_z \end{pmatrix}$$
from which it follows that
$$f_{\varepsilon_\alpha} = r_z(\alpha) \quad \& \quad f_{\varepsilon_\beta} = r_y(\beta) \quad \& \quad f_{\varepsilon_\theta} = r_z(\theta)$$
and hence that f_ε is a composition of rotations.

(ii) Let M be a rotation on $\mathbb{R}^3$. Then there are real numbers α, β, θ such that $M = r_x(\alpha)r_y(\beta)r_z(\theta)$. By exponentiating the spin matrix σ_x, we obtain a unitary operator
$$\hat{r}_x(\theta) = e^{-i\theta\sigma_x/2} = \begin{pmatrix} \cos\theta/2 & -i\sin\theta/2 \\ -i\sin\theta/2 & \cos\theta/2 \end{pmatrix}$$
As in the previous proof, $r_x(\theta)$ is the Bloch representation of the unitary channel associated to $\hat{r}_x(\theta)$. Using the angles in the decomposition of M into basic rotations, we define a unitary operator $U = \hat{r}_x(\alpha)\hat{r}_y(\beta)\hat{r}_z(\theta)$ and let ε be its associated unitary channel. Since $\varepsilon = \varepsilon_\alpha \circ \varepsilon_\beta \circ \varepsilon_\theta$, where ε_α, ε_β and ε_θ are the unitary channels associated to $\hat{r}_x(\alpha)$, $\hat{r}_y(\beta)$ and $\hat{r}_z(\theta)$, respectively, Prop. 2.7(iii) tells us that its Bloch representation is
$$f_\varepsilon = f_{\varepsilon_\alpha} \circ f_{\varepsilon_\beta} \circ f_{\varepsilon_\theta} = r_x(\alpha)r_y(\beta)r_z(\theta) = M$$
which finishes the proof. □

The proof of the last result gives a way to calculate the Bloch representation of a unitary channel: if we can write U in the form
$$U = e^{i\lambda}\,\hat{r}_x(\alpha)\,\hat{r}_y(\beta)\,\hat{r}_z(\theta)$$
then its Bloch representation is the rotation $r_x(\alpha)r_y(\beta)r_z(\theta)$. Let us now apply this technique to a very important class of unitary channels, the *spin channels*.

LEMMA 2.11. *Let* ε_x, ε_y, ε_z *denote the unitary channels associated to the spin matrices* σ_x, σ_y, σ_z. *Then*

$$[\![\varepsilon_x]\!] = r_x(\pi) = \begin{pmatrix} 1 & 0 & 0 \\ 0 & -1 & 0 \\ 0 & 0 & -1 \end{pmatrix}, \quad [\![\varepsilon_y]\!] = r_y(\pi) = \begin{pmatrix} -1 & 0 & 0 \\ 0 & 1 & 0 \\ 0 & 0 & -1 \end{pmatrix},$$

$$[\![\varepsilon_z]\!] = r_z(\pi) = \begin{pmatrix} -1 & 0 & 0 \\ 0 & -1 & 0 \\ 0 & 0 & 1 \end{pmatrix}.$$

Proof. We have

$$\hat{r}_x(\pi) = e^{-i\pi\sigma_x/2} = -i\sigma_x, \quad \hat{r}_y(\pi) = e^{-i\pi\sigma_y/2} = -i\sigma_y, \quad \hat{r}_z(\pi) = e^{-i\pi\sigma_z/2} = -i\sigma_z.$$

The unitary channels associated to $\hat{r}_x(\pi)$, $\hat{r}_y(\pi)$ and $\hat{r}_z(\pi)$ are the spin channels ε_x, ε_y and ε_z. Thus, the Bloch representations of the spin channels are $r_x(\pi)$, $r_y(\pi)$ and $r_z(\pi)$. □

The matrices in the last lemma are all involutions: an *involution* on a set X is a function $f : X \to X$ for which $f \circ f = 1_X$ i.e. a function that is its own inverse. They are very important:

DEFINITION 2.12. The Bloch representations of the spin channels are denoted s_x, s_y, s_z:

$$s_x := r_x(\pi) = \begin{pmatrix} 1 & 0 & 0 \\ 0 & -1 & 0 \\ 0 & 0 & -1 \end{pmatrix}, \quad s_y := r_y(\pi) = \begin{pmatrix} -1 & 0 & 0 \\ 0 & 1 & 0 \\ 0 & 0 & -1 \end{pmatrix},$$

$$s_z := r_z(\pi) = \begin{pmatrix} -1 & 0 & 0 \\ 0 & -1 & 0 \\ 0 & 0 & 1 \end{pmatrix}.$$

Let us now use s_x, s_y and s_z along with Prop. 2.7 to calculate the representations of some very important channels:

EXAMPLE 2.13. Let $I : \Omega^2 \to \Omega^2$ denote the identity channel, $\bot$ denote the constant channel $\rho \mapsto I/2$ and let $p \in [0,1]$ be the probability that some form of noise affects a state.

(i) The Bloch representation of the *bit flipping channel* $\varepsilon = (1-p)I + p\,\varepsilon_x$ is

$$[\![\varepsilon]\!] = (1-p)[\![I]\!] + p[\![\varepsilon_x]\!] = (1-p)I + ps_x$$

(ii) The Bloch representation of the *phase flipping* channel $\varepsilon = (1-p)I + p\,\varepsilon_z$ is

$$[\![\varepsilon]\!] = (1-p)[\![I]\!] + p[\![\varepsilon_z]\!] = (1-p)I + ps_z$$

(iii) The Bloch representation of the *bit-phase flip* channel $\varepsilon = (1-p)I + p\,\varepsilon_y$ is

$$[\![\varepsilon]\!] = (1-p)[\![I]\!] + p[\![\varepsilon_y]\!] = (1-p)I + ps_y$$

(iv) The Bloch representation of the *depolarization channel* $d = p \cdot \bot + (1-p)I$ is

$$[\![d]\!] = p[\![\bot]\!] + (1-p)[\![I]\!] = p \cdot 0 + (1-p)I = (1-p)I$$

Because of the close correspondence between qubit channels and their Bloch representations, from here on we shall refer to both $\varepsilon : \Omega^2 \to \Omega^2$ and its Bloch representation $[\![\varepsilon]\!] = f_\varepsilon : \mathbb{B}^3 \to \mathbb{B}^3$ as being *qubit channels*.

DEFINITION 2.14. The Bloch representation $f_\varepsilon : \mathbb{B}^3 \to \mathbb{B}^3$ of a qubit channel $\varepsilon : \Omega^2 \to \Omega^2$ is also referred to as a *qubit channel*.

For instance, ε_x, ε_y, ε_z are spin channels, as are s_x, s_y and s_z. We now turn to the class of qubit channels that will be our principal concern in this paper.

3. Unitality

The channels in Example 2.13 are among the most important examples of noise. Each arises as a convex sum of unitary channels – a nondeterministic choice between rotations. Such channels are examples of what are called *unital channels*. In this section, we use a simple characterization of unital channels to first establish some of their crucial properties and then to reconcile them with what we feel are their classical counterparts, the *binary symmetric channels*.

DEFINITION 3.1. A qubit channel ε is *unital* if $\varepsilon(I/2) = I/2$. Equivalently, when $f_\varepsilon(0) = 0$.

In addition to the unital channels given in Example 2.13, any *projective measurement*
$$\varepsilon(\rho) = P_0 \rho P_0 + P_1 \rho P_1$$
for projections $P_0 + P_1 = I$ defines a unital channel. Another example was mentioned above: any convex sum of unitary channels is unital. In fact, it turns out that this is the only way to construct a unital qubit channel: the set of unital qubit channels is both compact and convex and its set of extreme points is exactly the set of unitary channels [3]. We thus obtain an inductive characterization of unital qubit channels in familiar and elementary terms:

THEOREM 3.2. *The class of unital channels $\mathcal{U}$ is the smallest class of functions of type $\mathbb{B}^3 \to \mathbb{B}^3$ such that*
- *For each angle θ, $r_x(\theta), r_y(\theta), r_z(\theta) \in \mathcal{U}$,*
- *If $f, g \in \mathcal{U}$, then $f \circ g \in \mathcal{U}$, and*
- *If $f, g \in \mathcal{U}$ and $p \in [0, 1]$, then $pf + (1-p)g \in \mathcal{U}$.*

Proof. Let $\mathcal{U}$ denote the smallest set of functions satisfying the three closure properties given in the statement of the theorem. This collection is a subset of the class of all unital channels since all maps in $\mathcal{U}$ take 0 to 0.

For the converse, all rotations on $\mathbb{R}^3$ belong to $\mathcal{U}$ since each can be written as the composition $r_x(\alpha) \cdot r_y(\beta) \cdot r_z(\theta)$. Thus, by Prop. 2.10, the Bloch representations of unitary channels belong to $\mathcal{U}$. Finally, the unitaries are the set of extreme points in the class of unital quantum operations [3], so every unital channel is a convex sum of rotations and thus a member of $\mathcal{U}$. □

COROLLARY 3.3. *If $f_i \in \mathcal{U}$ and $x_i \in [0,1]$ for $1 \leq i \leq n$ with $\sum_{i=1}^{n} x_i = 1$, then $\sum_{i=1}^{n} x_i f_i \in \mathcal{U}$.*

Proof. This is just the usual inductive observation about finite probability distributions:
$$\sum_{i=1}^{n} x_i f_i = x_1 f_1 + (1 - x_1) \left(\sum_{k=2}^{n} \frac{x_k f_k}{\sum_{i=2}^{n} x_i} \right)$$

when $x_1 < 1$. □

COROLLARY 3.4. *If $f \in \mathcal{U}$, then $|f(x)| \leq |x|$, for all $x \in \mathbb{B}^3$.*

Proof. The proof is a straightforward induction on the set $\mathcal{U}$ of unital channels. □

We now consider some valuable closure properties possessed by the class of unital channels.

PROPOSITION 3.5.
- *The maps $\mathrm{id}(x) = x$ and $z(x) = 0$ belong to $\mathcal{U}$,*
- *If $f \in \mathcal{U}$ and $p \in [0,1]$, then $pf \in \mathcal{U}$,*
- *If $f \in \mathcal{U}$, then $f^t \in \mathcal{U}$.*

Proof. The identity map $\mathrm{id}(x) = x$ can be written as $\mathrm{id} = r_x(0)$, so it is unital. The zero map z belongs to $\mathcal{U}$ since it can be written as a convex sum of the spin channels
$$z = \frac{1}{4}\mathrm{id} + \frac{1}{4}s_x + \frac{1}{4}s_y + \frac{1}{4}s_z.$$
If $f \in \mathcal{U}$ and $p \in [0,1]$, then $pf = pf + (1-p)z \in \mathcal{U}$.

To prove closure under transposition, we use induction as follows. First, in the base case, f is a rotation, and then $f^t = f^{-1} \in \mathcal{U}$, since f^{-1} is also a rotation, and $\mathcal{U}$ contains all rotations. If $f^t, g^t \in \mathcal{U}$, then $(f \circ g)^t = g^t \circ f^t \in \mathcal{U}$ and $(pf + (1-p)g)^t = pf^t + (1-p)g^t \in \mathcal{U}$. □

COROLLARY 3.6. *If $x_i \in [0,1]$ and $f_i \in \mathcal{U}$ for $1 \leq i \leq n$ with $\sum_{i=1}^n x_i \leq 1$, then $\sum_{i=1}^n x_i f_i \in \mathcal{U}$.*

Proof. The zero map z belongs to $\mathcal{U}$ so
$$\sum_{i=1}^n x_i f_i = \sum_{i=1}^n x_i f_i + \left(1 - \sum_{i=1}^n x_i\right) \cdot z \in \mathcal{U}$$
by Corollary 3.3. □

Let us take a first look at the structure of $\mathcal{U}$. In particular, we will see examples of linear operators that do *not* belong to the class of unital channels. To begin, we prove the valuable *trace lemma*:

LEMMA 3.7 (The trace lemma). *For any unital channel $f \in \mathcal{U}$, we have $\mathrm{tr}(f) \in [-1, 3]$.*

Proof. Each 3×3 rotation r is a normal matrix, so by Theorem 3.3 of [**8**], we can find an orthogonal matrix s such that srs^t is block diagonal, each block being either a 1×1 matrix consisting of a real eigenvalue of r or a 2×2 matrix of the form $\begin{pmatrix} a & b \\ -b & a \end{pmatrix}$.

Since we are in dimension three, only two cases are possible: either we have all eigenvalues, or we have a matrix of the form
$$srs^t = \begin{pmatrix} c & 0 & 0 \\ 0 & a & b \\ 0 & -b & a \end{pmatrix}$$

where $c = \pm 1$ is a real eigenvalue of r. Using $\det(s) = 1/\det(s^t)$ since $s^t = s^{-1}$,
$$\det(srs^t) = \det(s)\det(r)\det(s^t) = \det(r) = 1 = c(a^2 + b^2)$$
so we see that $c = 1$ and that (a, b) is a point on the unit circle. Then
$$\operatorname{tr}(r) = \operatorname{tr}(I \cdot r) = \operatorname{tr}(ss^t r) = \operatorname{tr}(srs^t) = 1 + 2a \in [-1, 3].$$
In the case of all eigenvalues, we get either a trace of -1 or 3. For an arbitrary unital f written as a convex sum of rotations (r_i),
$$\operatorname{tr}(f) = \operatorname{tr}\left(\sum x_i r_i\right) = \sum x_i \cdot \operatorname{tr}(r_i) \in [-1, 3]$$
by the linearity of the trace. $\square$

COROLLARY 3.8. *The antipodal map $a(x) = -x$ is not unital.*

Proof. By Lemma 3.7, the antipodal map is not unital because $\operatorname{tr}(a) = -3 \notin [-1, 3]$. $\square$

The antipodal map a takes any qubit x and "flips" it to $a(x) = -x$. That a is not a qubit channel says that there is no *single physical operation* capable of flipping an arbitrary qubit: 'universal' bit flipping is impossible. There is also something here of mathematical interest: a map that negates only one coordinate cannot be unital (if it were, then composing with a spin channel would imply that a is unital). Thus, we can negate any two of (r_x, r_y, r_z), but not only one and not all three.

PROPOSITION 3.9.
- If $f \in \mathcal{U}$ and $f^{-1} \in \mathcal{U}$ exists, then $-f \notin \mathcal{U}$.
- If an orthogonal matrix belongs to $\mathcal{U}$, it must be a rotation.
- For $f \in \mathcal{U}$, $f^{-1} \in \mathcal{U}$ iff f is a rotation.

Thus, no orthogonal matrix on $\mathbb{R}^3$ is unital unless it is a rotation. By contrast, on $\mathbb{R}^2$, the antipodal map is a rotation.

Proof. Let $f \in \mathcal{U}$ and suppose that it has an inverse $f^{-1} \in \mathcal{U}$. If $-f \in \mathcal{U}$, then $a = (-f) \circ f^{-1} \in \mathcal{U}$, which is impossible by Corollary 3.8. Thus, $-f \notin \mathcal{U}$.

Suppose now that some $f \in \mathcal{U}$ is defined by an orthogonal matrix. Then it is either a rotation or has a determinant of -1. In the latter case, $a \circ f$ is then a rotation, since it is orthogonal and has a determinant of $\det(a \circ f) = \det(a) \cdot \det(f) = -1 \cdot -1 = 1$. Thus, $a \circ f$ belongs to $\mathcal{U}$, and so does its inverse $f^{-1} \circ a$, since it is also a rotation (having $\det(f^{-1} \circ a) = +1$). By the previous result then, $-(a \circ f) \notin \mathcal{U}$ i.e. $f \notin \mathcal{U}$.

Finally, let $f \in \mathcal{U}$ with $f^{-1} \in \mathcal{U}$. By Corollary 3.4,
$$|x| = |f^{-1}(f(x))| \le |f(x)| \le |x|$$
which means that $|f(x)| = |x|$. Then f is a linear isometry and hence an orthogonal matrix. By the previous result, it must be a rotation. $\square$

We know that unital channels can be thought of as nondeterministically choosing between rotations – but can we understand them in relation to *classical* channels? We now establish a few senses in which unitality is the quantum analogue of a "binary symmetric channel." A classical channel is *binary symmetric* when its noise matrix has the form
$$\begin{pmatrix} 1-p & p \\ p & 1-p \end{pmatrix}$$

where p is the probability that a bit is flipped. Let $\Delta^2 = \{(x,y) : x, y \in [0,1]\ \&\ x + y = 1\}$.

PROPOSITION 3.10. *The class of binary symmetric channels $\mathcal{B}$ is the smallest class of functions of type $\Delta^2 \to \Delta^2$ such that*
- *The channel that flips all bits belongs to $\mathcal{B}$ i.e.*
$$\mathrm{flip} = \begin{pmatrix} 0 & 1 \\ 1 & 0 \end{pmatrix} \in \mathcal{B}$$
- *If $f, g \in \mathcal{B}$, then $f \circ g \in \mathcal{B}$, and*
- *If $f, g \in \mathcal{B}$ and $p \in [0,1]$, then $pf + (1-p)g \in \mathcal{B}$.*

Proof. Every binary symmetric channel belongs to $\mathcal{B}$ as follows. By closure under composition,
$$1 = \mathrm{flip}^2 = \mathrm{flip} \circ \mathrm{flip} \in \mathcal{B}$$
and this means every binary symmetric channel belongs to $\mathcal{B}$ since
$$\begin{pmatrix} 1-p & p \\ p & 1-p \end{pmatrix} = (1-p) \cdot 1 + p \cdot \mathrm{flip} \in \mathcal{B}.$$
For the converse, the flip channel is binary symmetric, and the binary symmetric channels are closed under composition and convex sum. $\square$

One connection between $\mathcal{U}$ and $\mathcal{B}$ is that each can be characterized as the set of functions which arise by taking convex sums of channels that preserve entropy:

PROPOSITION 3.11.
- *The entropy preserving classical channels on bits are the identity channel and the flip channel. Each channel in $\mathcal{B}$ is the convex sum of such channels.*
- *The entropy preserving quantum channels on qubits are the rotations. Each channel in $\mathcal{U}$ is the convex sum of such channels.*

Proof. First consider the classical case. If f preserves entropy, then in particular, $Hf(\bot) = \bot$, which means $f(\bot) = \bot$. Using parameters (a, b) for the noise matrix of f, we have
$$\begin{pmatrix} 1/2 \\ 1/2 \end{pmatrix} = (1/2, 1/2) \cdot \begin{pmatrix} a & 1-a \\ b & 1-b \end{pmatrix}$$
and so $a + b = 1$ i.e. f is a binary symmetric channel. Then the noise matrix is determined by a single parameter $p = b$. Because f preserves entropy, $H(x) = H((1-2p)x + p)$ for all $x \in [0,1]$. This is clearly true when $p = 0$ and $p = 1$. If p is another value that validates this equation for all x, then for $x = 1$, we get
$$H(1-p) = H(1) = 0$$
which implies that either $p = 0$ or $p = 1$. Thus, the only entropy preserving classical binary channels are the identity and the flip channel. Every binary symmetric channel is a convex sum of these two channels, as seen in the proof of Prop. 3.10.

In the quantum case, all we need to show is that the entropy preserving quantum channels are the rotations, since we know that each unital channel is a convex sum of rotations. Because every rotation preserves the Euclidean norm and the eigenvalues of a density operator with Bloch vector r are $(1 \pm |r|)/2$, we see that rotations

preserve entropy. Now suppose f is any operator that preserves entropy. Since f preserves entropy, we have
$$S(x) = H((1+|x|)/2) = H((1+|f(x)|)/2) = S(f(x))$$
On the interval $[1/2, 1]$, entropy is injective, so $|x| = |f(x)|$. This implies that $f(0) = 0$ and so f is unital and hence linear. Then f is a linear isometry in the Euclidean norm, which means f is an orthogonal matrix, and hence a rotation by Prop. 3.9. □

Another connection between $\mathcal{B}$ and $\mathcal{U}$ is that each is precisely the collection of *entropy increasing* channels: the set of channels whose output state never has entropy strictly less than that of its input state. This follows easily from the result above using the convexity of entropy and the fact that there is a unique state of maximal entropy. Yet another connection is that $\mathcal{B}$ and $\mathcal{U}$ arise as the Scott continuous channels with a Scott closed set of fixed points in the Bayesian and spectral orders on Δ^2 and Ω^2 respectively [4].

With so many similarities between unital channels and binary symmetric channels, a fair question then becomes: what is the significance of a binary symmetric channel? Suppose we are communicating in a noisy environment, sending bits with equal frequency and suffering an error rate of p. Then some channel with probability of error p models the environment – but which one? We should assume the worst, and take the channel that has *minimal capacity*. As explained in [5], such a channel is always binary symmetric. Thus, binary symmetric channels – and entropy increasing channels in general – can be thought of as providing *conservative* models of noise.

4. The definition of scope

Each basis of the state space provides a different way to represent information. For each representation, there is an associated classical channel whose capacity measures the amount of information that can be transmitted through a quantum channel when that particular representation is used. The *scope* of a quantum channel is the range of classical capacities achieved as the sender and receiver vary over all possible representations. Let us illustrate the idea for qubits:

EXAMPLE 4.1. *Scope*. Alice and Bob fix a basis $\{|\psi\rangle, |\phi\rangle\}$ of the state space:
- The state $|\psi\rangle = a|0\rangle + b|1\rangle$ represents '0'
- The state $|\phi\rangle = c|0\rangle + d|1\rangle$ represents '1'

This choice of basis defines a classical channel:
- Alice sends a qubit $|*\rangle$ representing '0' or '1' to Bob.
- As the qubit $|*\rangle$ travels, it interacts with the environment, changing to $\varepsilon(|*\rangle\langle*|)$
- Bob receives and measures the qubit in the $\{|\psi\rangle, |\phi\rangle\}$ basis, obtaining a '0' or '1'.

This classical channel in turn has a capacity given by
$$\mathrm{C}(x,y) = \log_2\left(2^{\frac{\bar{x}H(y)-\bar{y}H(x)}{x-y}} + 2^{\frac{yH(x)-xH(y)}{x-y}}\right)$$
where $x = P(0|0)$, $y = P(0|1)$, $H(x) = -x\log_2(x) - (1-x)\log_2(1-x)$ is the base two entropy and $\bar{x} = 1-x$. For instance, if the environment is modelled by the bit

flipping channel
$$\varepsilon = (1-p)I + p \cdot \varepsilon_x$$
of Example 2.13, then these probabilities are given by
$$x = P(0|0) = (1-p)|a|^4 + (2p+2)|a|^2|b|^2 + (1-p)|b|^4$$
$$y = P(0|1) = 1 - ((1-p)|c|^4 + (2p+2)|c|^2|d|^2 + (1-p)|d|^4)$$
The *scope* of ε is the range of capacities achieved as $\{|\psi\rangle, |\phi\rangle\}$ varies over *all* bases of the state space.

As is clear, calculating the scope of a quantum channel is not a trivial matter. However, by switching to the Bloch representation, not only can we see how to calculate scope for examples like bit flipping, we can develop a *systematic method* for calculating the scope of *any* unital qubit channel. To take our first step toward this, we need to understand how to calculate the classical channels associated to a qubit channel in its Bloch representation:

LEMMA 4.2. *If $\rho, \sigma \in \Omega^2$ are mixed states with respective Bloch vectors r and s, then*
$$\mathrm{tr}(\rho \cdot \sigma) = \frac{1 + (r,s)}{2}$$
where (r,s) is the Euclidean inner product on $\mathbb{R}^3$.

Proof. With $r = (r_x, r_y, r_z)$ and $s = (s_x, s_y, s_z)$, we can write
$$\rho = \frac{I}{2} + \frac{r_x\sigma_x + r_y\sigma_y + r_z\sigma_z}{2} \quad \& \quad \sigma = \frac{I}{2} + \frac{s_x\sigma_x + s_y\sigma_y + s_z\sigma_z}{2}$$
where σ_x, σ_y and σ_z are the Pauli spin operators. Multiplying ρ and σ and taking the trace gives
$$\mathrm{tr}(\rho \cdot \sigma) = \frac{1}{2} + \frac{1}{4} \cdot \mathrm{tr}[(r_x\sigma_x + r_y\sigma_y + r_z\sigma_z)(s_x\sigma_x + s_y\sigma_y + s_z\sigma_z)]$$
using linearity of the trace, $\mathrm{tr}(I/4) = 1/2$ and equation 2.1. Setting $\vec{\sigma} = (\sigma_x, \sigma_y, \sigma_z)$, we abbreviate the expression on the right as $\mathrm{tr}(\langle r, \vec{\sigma}\rangle \cdot \langle s, \vec{\sigma}\rangle)$ and find that
$$\begin{aligned}
\mathrm{tr}(\langle r,\vec{\sigma}\rangle \cdot \langle s,\vec{\sigma}\rangle) &= \mathrm{tr}(r_x s_x \cdot I + r_x s_y \sigma_x \sigma_y + r_x s_z \sigma_x \sigma_z) \\
&+ \mathrm{tr}(r_y s_x \sigma_y \sigma_x + r_y s_y \cdot I + r_y s_z \sigma_y \sigma_z) \\
&+ \mathrm{tr}(r_z s_x \sigma_z \sigma_x + r_z s_y \sigma_z \sigma_y + r_z s_z \cdot I) \\
&= r_x s_x \cdot \mathrm{tr}(I) + 0 + 0 \\
&+ 0 + r_y s_y \cdot \mathrm{tr}(I) + 0 \\
&+ 0 + 0 + r_z s_z \cdot \mathrm{tr}(I) \quad \text{(Using equation 2.2)} \\
&= 2 \cdot (r,s)
\end{aligned}$$
Substituting this into the original equation gives the desired result. $\square$

Suppose Alice attempts to send Bob a qubit represented by ρ. As the qubit travels, it suffers the effect of noise described by the quantum channel ε. Bob then receives $\varepsilon(\rho)$ and performs a measurement in *some* basis $\{|0\rangle, |1\rangle\}$. The measurement operators in this case are the projections $P_0 = |0\rangle\langle 0|$ and $P_1 = |1\rangle\langle 1|$ and form a complete set since $P_0 + P_1 = I$, so by standard quantum mechanics, the probability that Bob obtains the result 0 is
$$p_0 = \mathrm{tr}(P_0^\dagger P_0 \cdot \varepsilon(\rho))) = \mathrm{tr}(P_0 P_0 \cdot \varepsilon(\rho)) = \mathrm{tr}(P_0 \cdot \varepsilon(\rho))$$

while the probability that Bob obtains the result 1 is
$$p_1 = \text{tr}(P_1^\dagger P_1 \cdot \varepsilon(\rho))) = \text{tr}(P_1 P_1 \cdot \varepsilon(\rho)) = \text{tr}(P_1 \cdot \varepsilon(\rho))$$
where we recall that projections satisfy $P_i^2 = P_i$. Now both projections P_0 and P_1, being density operators, also have a Bloch vector associated with them, given by s and t, respectively. If r is the Bloch vector for ρ and f is the Bloch representation of ε, then the probabilities p_0 and p_1 can be succinctly written as
$$p_0 = \frac{1 + (s, f(r))}{2} \quad \& \quad p_1 = \frac{1 + (t, f(r))}{2}$$
Further, since $|0\rangle$ and $|1\rangle$ form a basis for the state space, $s + t = 0$, which helps us see that $p_0 + p_1 = 1$. Finally, if Alice and Bob use the same basis $\{r, -r\}$ to represent information and attempt to communicate in the presence of unital noise f, then
$$P(0|0) = \frac{1 + (r, f(r))}{2} = \frac{1 + (-r, f(-r))}{2} = P(1|1).$$
These probabilities define a binary channel (a, b) with $a = P(0|0)$ and $b = P(0|1)$ whose capacity is given by
$$C(a, b) = \log_2 \left(2^{\frac{\bar{a}H(b) - \bar{b}H(a)}{a - b}} + 2^{\frac{bH(a) - aH(b)}{a - b}} \right)$$
where $C(a, a) := 0$ and $H(x) = -x \log_2(x) - (1 - x) \log_2(1 - x)$ is the base two entropy.

DEFINITION 4.3. Let f be a unital channel. For each $r \in \partial \mathbb{B}^3$, the associated classical channel is
$$x_f(r) = \left(\frac{1 + (r, f(r))}{2}, \frac{1 - (r, f(r))}{2} \right)$$
where $\partial \mathbb{B}^3 = \{x \in \mathbb{B}^3 : |x| = 1\}$ is the set of pure states.

The question we want to answer is: what range of capacities is achieved as $\{r, -r\}$ varies over *all* bases of the state space?

THEOREM 4.4. *If f is a unital channel, then the set of achievable capacities is given by*
$$\{C(x_f(r)) : r \in \mathbb{B}^3, |r| = 1\} = \left[1 - H\left(\frac{1 + m^-}{2}\right), 1 - H\left(\frac{1 + m^+}{2}\right) \right]$$
where
$$m^+ = \sup_{|x| = 1} |(x, f(x))| \quad \& \quad m^- = \inf_{|x| = 1} |(x, f(x))|.$$

Proof. Let $\{r, -r\}$ be the Bloch vectors of a basis used to represent the classical bits '0' and '1' respectively. Then the associated classical channel is described by
$$a = P(0|0) = \frac{1 + (r, f(r))}{2} \quad \& \quad b = P(0|1) = \frac{1 + (r, f(-r))}{2} = \frac{1 - (r, f(r))}{2}$$
By the *nonexpansivity* of f in Corollary 3.4, we have $a, b \in [0, 1]$. The channel (a, b) is binary symmetric so its capacity is $1 - H(a)$. Let us begin by rewriting the expression for capacity. The function $\varphi : [-1, 1] \to [0, 1]$ defined by
$$\varphi(x) = 1 - H\left(\frac{1 + x}{2}\right)$$

restricts to a bijection on either of the halves $[-1, 0]$ or $[0, 1]$: it is strictly decreasing on the former and strictly increasing on the latter. Now define $p_f : \partial \mathbb{B}^3 \to [-1, 1]$ by $p_f(r) = (r, f(r))$. The capacity achieved when information is represented in the $\{r, -r\}$ basis is then given by $\varphi(p_f(r))$.

To find the largest and smallest values of the function $\varphi \circ p_f$, we proceed as follows. First, $|p_f| : \partial \mathbb{B}^3 \to [0, 1]$ is continuous on the compact set $\partial \mathbb{B}^3$, so there are points u and v such that $|p_f(u)| = m^+$ and $|p_f(v)| = m^-$. Since φ is strictly increasing on $[0, 1]$ and $|p_f|$ maps into $[0, 1]$, the smallest value of $\varphi \circ | \cdot | \circ p_f$ is

$$\varphi(|p_f(v)|) = \inf_{|x|=1} \varphi(|p_f(x)|) = 1 - H\left(\frac{1 + m^-}{2}\right)$$

while its largest value is

$$\varphi(|p_f(u)|) = \sup_{|x|=1} \varphi(|p_f(x)|) = 1 - H\left(\frac{1 + m^+}{2}\right)$$

However, φ is symmetric about zero: $\varphi(x) = \varphi(-x)$ for all $x \in [-1, 1]$. Thus, $\varphi(p_f(x)) = \varphi(|p_f(x)|)$ for all x. This proves that all achievable capacities lie in the indicated range, that $\{v, -v\}$ is a basis for achieving the smallest value of capacity and that $\{u, -u\}$ is a basis for achieving the largest capacity. Finally, because $\partial \mathbb{B}^3$ is a *connected* set, all capacities in between the maximum and minimum are also achievable. □

Notice that we were able to completely characterize the range of capacities achievable by a unital channel using *binary symmetric* channels. This is another reason they seem to provide a classical counterpart to unitality.

DEFINITION 4.5. For a unital channel $f \in \mathcal{U}$, we define

$$f^+ = 1 - H\left(\frac{1 + m^+}{2}\right) \quad \& \quad f^- = 1 - H\left(\frac{1 + m^-}{2}\right)$$

where $m^+ = \sup_{|x|=1} |(x, f(x))|$ and $m^- = \inf_{|x|=1} |(x, f(x))|$. We define

$$s(f) = [f^-, f^+]$$

and call this the *scope* of f.

The scope $s(f)$ of a channel f is the range of its achievable capacities. It measures how much the capacity of f is capable of varying as different bases are used to represent classical bits, and thus how representative a particular value of capacity, such as f^+, is of a channel's behavior. Let us pause to consider an interesting class of channels whose scope *always* has maximum length.

EXAMPLE 4.6. *Projective measurements.* Any *projective measurement*

$$\varepsilon(\rho) = P_0 \rho P_0 + P_1 \rho P_1$$

for projections $P_0 + P_1 = I$ defines a unital channel ε that is *idempotent*: $\varepsilon^2 = \varepsilon$. Thus, its Bloch representation f_ε satisfies

$$f_\varepsilon^2 = f_\varepsilon \circ f_\varepsilon = [\![\varepsilon]\!] \circ [\![\varepsilon]\!] = [\![\varepsilon \circ \varepsilon]\!] = [\![\varepsilon]\!] = f_\varepsilon$$

and so is also idempotent. Let us now calculate the scope of an idempotent unital channel.

Given a unital channel $f : \mathbb{B}^3 \to \mathbb{B}^3$ with $f^2 = f$, there are three possibilities: either

(a) $f = I$,
(b) $f = 0$ or
(c) $(\exists x, y \in \mathbb{B}^3)\ f(x) \neq 0\ \&\ f(y) \neq y$.

In the first two cases, we have $s(f) = [1,1]$ and $s(f) = [0,0]$ respectively. In the last, we set $s = y - f(y)/|y - f(y)|$ and $t = f(x)/|f(x)|$. Then $s, t \in \partial \mathbb{B}^3$ are both pure states. By the idempotence of f,

$$m^- \leq |(s, f(s))| = \frac{|(s, f(y) - f^2(y))|}{|y - f(y)|} = |(s, 0)| = 0$$

and

$$|(t, f(t))| = \frac{|(f(x), f^2(x))|}{|f(x)|^2} = \frac{|(f(x), f(x))|}{|f(x)|^2} = \frac{|(f(x), f(x))|}{|f(x)|^2} = 1 \leq m^+$$

so the scope of f is

$$s(f) = [0, 1]$$

by Theorem 4.4.

Of course, we were only able to calculate the scope in the preceding example because the values m^+ and m^- could be deduced from the fact that the channel is question was idempotent. Is there a way to systematically calculate the scope of *any* unital channel?

5. The calculation of scope

A real $n \times n$ matrix A is *symmetric* when $A = A^t$. The eigenvalues of a symmetric matrix are real and there are exactly n of them: $(\lambda_1, \ldots, \lambda_n)$, though they are not all necessarily distinct. A standard fact about symmetric matrices is that

(5.1) $$\sup_{|x|=1} (x, Ax) = \max_{1 \leq i \leq n} \lambda_i \quad \& \quad \inf_{|x|=1} (x, Ax) = \min_{1 \leq i \leq n} \lambda_i.$$

This is quite fortunate for us.

First, each $f \in \mathcal{U}$ can be written as $f(x) = Mx$ where M is a real 3×3 matrix. If the matrix M happens to be symmetric, then by Theorem 4.4 we can calculate the scope of f by simply finding the largest and smallest eigenvalues of M. This in turn requires solving the characteristic equation of M, which means finding the zeroes of a third degree polynomial with real coefficients, and even a formula exists for these[1].

DEFINITION 5.1. A unital channel is called *symmetric* when $f = f^t$. The class of symmetric unital channels is denoted $\mathcal{S}$.

PROPOSITION 5.2. *Let f be a symmetric unital channel with eigenvalues $\lambda_1 \leq \lambda_2 \leq \lambda_3$. Then*

$$s(f) = \left[\frac{(1 + \mathrm{sgn}(\lambda_1 \lambda_3))}{2} \left(1 - H\left(\frac{1 + \min |\lambda_i|}{2} \right) \right), 1 - H\left(\frac{1 + \max |\lambda_i|}{2} \right) \right]$$

where $\mathrm{sgn}(x) = x/|x|$ *for* $x \neq 0$ *and* $\mathrm{sgn}(0) = 0$.

---
[1] There is a formula for fourth degree polynomials too, but not five.

Proof. By Theorem 4.4, we know that
$$s(f) = \left[1 - H\left(\frac{1+m^-}{2}\right), 1 - H\left(\frac{1+m^+}{2}\right)\right]$$
It is a standard result of linear algebra that $m^+ = \max |\lambda_i|$ when f is symmetric. Thus, to finish the proof we only need to compare left endpoints of the intervals. Let x and y be eigenvectors for λ_1, λ_3 respectively.

If $\mathrm{sgn}(\lambda_1 \lambda_3) = 0$, then either $\lambda_1 = 0$ or $\lambda_3 = 0$, and we get
$$0 \leq m^- \leq \min\{|(x, f(x))|, |(y, f(y))|\} = \min\{|\lambda_1|, |\lambda_3|\} = 0$$
which means $m^- = 0 = \min |\lambda_i|$, so the formula in the statement of the theorem holds in this case. Then we can assume $\lambda_i \neq 0$ for all i.

If $\mathrm{sgn}(\lambda_1 \lambda_3) = -1$, then λ_1 and λ_3 have opposite signs and the formula has a left endpoint of zero. By the continuity of $x \mapsto (x, f(x))$ and the connectedness of the unit sphere,
$$(x, f(x)) = \lambda_1 \;\&\; (y, f(y)) = \lambda_3 \implies (\exists z)\, (z, f(z)) = 0 \implies m^- = 0$$
so we see that the formula gives the correct value of $s(f)$.

If $\mathrm{sgn}(\lambda_1 \lambda_3) = +1$, then λ_1 and λ_3 have the same signs, and λ_2 has the same sign since it is the middle eigenvalue. If the overall sign is positive, then the formula gives $s(f)$ by equation (5.1). If it is negative, then $g = -f$ is a symmetric matrix with positive eigenvalues $|\lambda_1|, |\lambda_2|, |\lambda_3|$ so
$$\inf_{|x|=1} (x, g(x)) = \min |\lambda_i|$$
by equation (5.1). We claim that this quantity is equal to m^-. To prove this, first note that
$$\min |\lambda_i| = \inf_{|x|=1} (x, g(x)) = \inf_{|x|=1} -(x, f(x)) = -\sup_{|x|=1} (x, f(x))$$
which implies that $(x, f(x)) < 0$ for all x. Then $(x, g(x)) = -(x, f(x)) = |(x, f(x))|$, which gives $m^- = \min |\lambda_i|$ as desired. $\square$

EXAMPLE 5.3. *The bit flipping channels.* Consider the bit flipping channel
$$f_x = (1-p)I + ps_x = \begin{pmatrix} 1 & 0 & 0 \\ 0 & 1-2p & 0 \\ 0 & 0 & 1-2p \end{pmatrix}$$
from Example 2.13. It is diagonal hence symmetric and its eigenvalues are simply the elements along the diagonal. Thus, by Prop. 5.2, its scope is
$$s(f_x) = \begin{cases} [1 - H(p), 1] & \text{if } p \leq 1/2 \\ [0, 1] & \text{if } p \geq 1/2 \end{cases}$$
where we use the equality $H(1-p) = H(p)$. The answer is the same for phase flipping and bit-phase flipping. For the depolarization channel
$$d = (1-p)I = \begin{pmatrix} 1-p & 0 & 0 \\ 0 & 1-p & 0 \\ 0 & 0 & 1-p \end{pmatrix}$$
of Example 2.13, the scope is
$$s(d) = [1 - H(p), 1 - H(p)]$$

a single value of capacity, regardless of which basis is used to represent information.

Classically, if the environment flips bits with equal probability, then it is not possible to transmit any information: the resulting channel

$$\frac{1}{2} \cdot I + \frac{1}{2} \cdot \text{flip} = \begin{pmatrix} 1/2 & 1/2 \\ 1/2 & 1/2 \end{pmatrix}$$

has capacity zero. What happens in the quantum case? If one of the spin channels is applied with probability $1/2$, is it still possible to transmit information?

EXAMPLE 5.4. *Random qubit flipping?* Returning to the bit flipping channel f_x of the previous example with $p = 1/2$, we have

$$f_{1/2} = \frac{I + s_x}{2} = \begin{pmatrix} 1 & 0 & 0 \\ 0 & 0 & 0 \\ 0 & 0 & 0 \end{pmatrix}$$

This channel is idempotent and is neither I nor 0, so from Example 4.6, we know it has scope $s(f_{1/2}) = [0,1]$. Thus, there is a way to represent information so that *any* classical capacity can be achieved through the channel $f_{1/2}$.

In particular, if bits are coded in the $\{|0\rangle, |1\rangle\}$ basis, which has Bloch vectors $\pm e_3 = (0, 0, \pm 1)$, then $f_{1/2}$ has the same effect that it has classically: no information can be transmitted. However, if we use the basis $\{|+\rangle, |-\rangle\}$, which has Bloch vectors $\pm e_1 = (\pm 1, 0, 0)$, then we can transmit information *perfectly*.

The problem of "randomly flipping a qubit," so that no information can be transmitted in any basis, can be solved by applying the channel $f_{1/2}$ *followed by* the channel $(I + s_y)/2$. Intuitively, it is not enough to 'flip bits' – one must also 'flip phases'. This is explained in more detail in [**5**].

EXAMPLE 5.5. *The intercept-resend attack in quantum cryptography.* In this attack, an eavesdropper randomly chooses between the $\{|0\rangle, |1\rangle\}$ and $\{|+\rangle, |-\rangle\}$ bases and then uses this choice to perform a projective measurement on a qubit being sent from Alice to Bob.

As shown in [**5**], $f = (I + s_x)/2$ is the Bloch representation of a measurement in the $\{|+\rangle, |-\rangle\}$ basis, while $g = (I + s_y)/2$ is the Bloch representation of a measurement in the $\{|0\rangle, |1\rangle\}$ basis. Thus, the action of the eavesdropper causes noise of the form

$$\frac{1}{2}f + \frac{1}{2}g = \frac{1}{2}I + \frac{1}{4}s_x + \frac{1}{4}s_y = \begin{pmatrix} 1/2 & 0 & 0 \\ 0 & 1/2 & 0 \\ 0 & 0 & 0 \end{pmatrix}$$

and since the resulting channel is symmetric, it has scope

$$[0, 1 - H(3/4)] = [0, 1 - H(1/4)]$$

In particular, while the act of eavesdropping causes noise which reduces the amount of information that can be transmitted between sender and receiver, it *does not* reduce it to zero.

The next channel has the interesting property that entangled states can be used to reduce the probability of error when transmitting classical information through it:

EXAMPLE 5.6. *The two-Pauli channel.* The two-Pauli channel [1] is
$$f_x = x \cdot I + \left(\frac{1-x}{2}\right) s_x + \left(\frac{1-x}{2}\right) s_y$$
where $x \in [0, 1]$, so we have
$$f_x = \begin{pmatrix} x & 0 & 0 \\ 0 & x & 0 \\ 0 & 0 & 2x-1 \end{pmatrix}$$
By Prop 5.2,
$$s(f_x) = \begin{cases} [0, 1 - H(x/2)] & \text{if } 0 \leq x \leq 1/3 \\ [0, 1 - H((1+x)/2)] & \text{if } 1/3 \leq x \leq 1/2 \\ [1 - H(x), 1 - H((1+x)/2)] & \text{if } 1/2 < x \leq 1 \end{cases}$$
doing a case-by-case analysis.

For symmetric channels, Prop. 5.2 gives us a systematic way to calculate scope. But not all unital channels are symmetric. How do we calculate the scope of an arbitrary unital channel? A natural idea is to try and show that every unital channel f can be represented by a symmetric channel $\varphi(f)$ that has the same scope as f. We would like this representation to be natural as follows:

- If f is symmetric, then there is no need to represent it differently, so we would like $\varphi(f) = f$.
- When a channel has been decomposed into a convex sum of simpler parts, we would like it to be easy to calculate φ, so we would like
$$\varphi(pf + (1-p)g) = p\,\varphi(f) + (1-p)\varphi(g).$$
That is, φ should preserve convex sums.
- Because $(x, f(x)) = (x, f^t(x))$, f and f^t define the same classical channel in any given basis, so we should have
$$\varphi(f^t) = \varphi(f)$$
and by the symmetry of $\varphi(f)$, we then have $\varphi(f^t) = \varphi(f) = \varphi(f)^t$. That is, φ should preserve the transpose operation.

Not only is the problem of obtaining such a φ solvable – it has a *unique* solution.

THEOREM 5.7. *Let X be a nonempty subset of a real vector space that is closed under convex sums, and let $* : X \to X$ be a convex linear involution with* $\mathrm{fix}(*) := \{x \in X : x^* = x\}$. *Then there is a convex linear retraction* $\varphi : X \to \mathrm{fix}(*)$ *which preserves $*$. Furthermore, there is only one function which has these properties, and it is given by*
$$\varphi(x) = \frac{1}{2}x + \frac{1}{2}x^*$$
for $x \in X$.

Proof. It is clear that φ as defined in the statement of the theorem has all the properties stated; what we need to do is establish its uniqueness. To this end, suppose φ is a convex linear retraction of X onto $\mathrm{fix}(*)$ that preserves $*$. Let $x \in X$. First notice that by the convex linearity of $*$,
$$\left(\frac{1}{2}x + \frac{1}{2}x^*\right)^* = \frac{1}{2}x^* + \frac{1}{2}(x^*)^* = \frac{1}{2}x^* + \frac{1}{2}x = \frac{1}{2}x + \frac{1}{2}x^*$$

is a fixed point of $*$, and therefore a fixed point of φ, since φ is a retraction onto fix($*$). Then

$$\begin{aligned}
\varphi(x) &= \varphi\left(\frac{1}{2}x + \frac{1}{2}x\right) \\
&= \frac{1}{2}\varphi(x) + \frac{1}{2}\varphi(x) &&\text{(Convex Linearity)} \\
&= \frac{1}{2}\varphi(x) + \frac{1}{2}\varphi(x)^* &&(\varphi(x) \in \text{fix}(*)) \\
&= \frac{1}{2}\varphi(x) + \frac{1}{2}\varphi(x^*) &&(\varphi \text{ preserves } *) \\
&= \varphi\left(\frac{1}{2}x + \frac{1}{2}x^*\right) &&\text{(Convex Linearity)} \\
&= \frac{1}{2}x + \frac{1}{2}x^* &&\text{(Fixed point of } \varphi)
\end{aligned}$$

which finishes the proof. □

The above result applies for instance to a subset of real matrices closed under convex sum and the transpose operation. In particular, we have

COROLLARY 5.8. *There is a unique function from $\mathcal{U}$ to $\mathcal{S}$ that preserves convex sum, transpose and retracts $\mathcal{U}$ onto $\mathcal{S}$. It is given by*

$$\varphi(f) = \frac{1}{2}f + \frac{1}{2}f^t$$

for $f \in \mathcal{U}$.

Proof. For $f \in \mathcal{U}$, $f^t \in \mathcal{U}$ by Prop. 3.5. We now apply Theorem 5.7 with X the subset of unital channels and $*$ the transpose operation. □

From an information theoretic viewpoint, f and $\varphi(f)$ are identical:

THEOREM 5.9. *Let $f \in \mathcal{U}$ be an arbitrary unital channel. Then*
- *For each $x \in \mathbb{B}^3$, $(x, f(x)) = (x, \varphi(f)x)$. Thus, in any basis, f and $\varphi(f)$ define the same classical channel.*
- *The channels f and $\varphi(f)$ have the same scope.*

The scope of a unital channel f is found by calculating the largest and smallest eigenvalues of $\varphi(f)$.

Proof. The entire result is obvious once we prove the first equality:

$$\begin{aligned}
(x, \varphi(f)x) &= \frac{1}{2}(x, f(x)) + \frac{1}{2}(x, f^t(x)) \\
&= \frac{1}{2}(x, f(x)) + \frac{1}{2}(f(x), x) &&\text{(Property of the transpose)} \\
&= \frac{1}{2}(x, f(x)) + \frac{1}{2}(x, f(x)) &&\text{(Real inner product)} \\
&= (x, f(x))
\end{aligned}$$

□

The function $\varphi(f)$ extracts all the classical information about f as a channel into a form where we can then systematically obtain it as a routine eigenvalue calculation. An example will make the value of this technique clear.

EXAMPLE 5.10. Consider the basic rotation $r_x(\theta)$, which is only symmetric if it is an involution.

$$\varphi(r_x(\theta)) = \frac{1}{2}\begin{pmatrix} 1 & 0 & 0 \\ 0 & \cos\theta & -\sin\theta \\ 0 & \sin\theta & \cos\theta \end{pmatrix} + \frac{1}{2}\begin{pmatrix} 1 & 0 & 0 \\ 0 & \cos\theta & \sin\theta \\ 0 & -\sin\theta & \cos\theta \end{pmatrix} = \begin{pmatrix} 1 & 0 & 0 \\ 0 & \cos\theta & 0 \\ 0 & 0 & \cos\theta \end{pmatrix}$$

so the largest eigenvalue of $\varphi(f)$ is 1 and the smallest is $\cos\theta$, meaning that the scope of $r_x(\theta)$ can be any interval of the form $[a,1]$ where $a \in [0,1]$ depends on θ.

It is remarkable that the scope of any unital channel f can be found by calculating the scope of a symmetric channel $\varphi(f)$. To illustrate why, let us point out that while any convex sum of rotations gives rise to a unital channel, only a convex sum of *involutive* rotations can give rise to a symmetric unital channel. The proof of this requires delving deeper into the structure of unital channels themselves: we start by using the trace lemma (Lemma 3.7) to give a new and elementary proof of the well-known "complete positivity" conditions [2]:

LEMMA 5.11. *A diagonal matrix*

$$\begin{pmatrix} \lambda_1 & 0 & 0 \\ 0 & \lambda_2 & 0 \\ 0 & 0 & \lambda_3 \end{pmatrix}$$

is unital if and only if $|\lambda_i| \leq 1$ *for each* $i \in \{1,2,3\}$ *and if the following four inequalities are satisfied:*

(i) $1 + \lambda_1 + \lambda_2 + \lambda_3 \geq 0$
(ii) $1 + \lambda_1 - \lambda_2 - \lambda_3 \geq 0$
(iii) $1 - \lambda_1 + \lambda_2 - \lambda_3 \geq 0$
(iv) $1 - \lambda_1 - \lambda_2 + \lambda_3 \geq 0$

Proof. ($\Rightarrow$): By induction, each unital channel f is nonexpansive ($|f(x)| \leq |x|$), so $|\lambda_i| \leq 1$. By the trace lemma, the channel

$$f = \begin{pmatrix} \lambda_1 & 0 & 0 \\ 0 & \lambda_2 & 0 \\ 0 & 0 & \lambda_3 \end{pmatrix}$$

must have

$$\text{tr}(f) = \lambda_1 + \lambda_2 + \lambda_3 \geq -1$$

which gives inequality (i). The last three inequalities follow from the fact that $s_x \cdot f$, $s_y \cdot f$, $s_z \cdot f$ are unital and also have a trace that exceeds -1.

($\Leftarrow$) Denoting the nonnegative expressions in inequalities (i)-(iv) by x_i, we then have

$$\begin{pmatrix} \lambda_1 & 0 & 0 \\ 0 & \lambda_2 & 0 \\ 0 & 0 & \lambda_3 \end{pmatrix} = \sum_{i=1}^{4} \frac{x_i}{4} \cdot r_i$$

where (r_1, r_2, r_3, r_4) are the spin channels (I, s_x, s_y, s_z). □

PROPOSITION 5.12.

(a) A channel $f \in \mathcal{U}$ is an involution iff there is a rotation r and an $s \in \{I, s_x, s_y, s_z\}$ such that $f = r \cdot s \cdot r^{-1} = r \cdot s \cdot r^t$. In particular, all unital involutions are symmetric.

(b) A unital channel $f \in \mathcal{U}$ is symmetric iff it is a convex sum of involutive rotations.

Proof. (a) Since $f \in \mathcal{U}$ is an involution, $f^{-1} = f \in \mathcal{U}$, so by Prop. 3.9, f is a rotation. But as a rotation, $f^{-1} = f^t$, while as an involution, $f = f^{-1}$, thus $f = f^t$ is symmetric. Since f is symmetric, there is an orthogonal matrix r

$$rfr^t = \begin{pmatrix} \lambda_1 & 0 & 0 \\ 0 & \lambda_2 & 0 \\ 0 & 0 & \lambda_3 \end{pmatrix}$$

Because r is an odd dimensional matrix, we can assume r is a rotation, by replacing r with $-r$ if necessary. Each λ_i must be either 1 or -1 since they are all real and each is an eigenvalue of a rotation. However, their product must be one, since it is the determinant of f. This proves the desired result.

(b) ($\Rightarrow$) If f is symmetric, then as in the proof of (a), we can find a rotation r such that

$$rfr^t = \begin{pmatrix} \lambda_1 & 0 & 0 \\ 0 & \lambda_2 & 0 \\ 0 & 0 & \lambda_3 \end{pmatrix}$$

Because rfr^t is unital, the diagonal map on its right is unital and so must be a convex sum of spin channels as seen in the proof of Lemma 5.11. Conjugating both sides of this equation by r now shows that f is a convex sum of four involutions, the four involutions forming a copy of the Klein four group.

($\Leftarrow$) If f is a convex sum of involutions

$$f = \sum_{i=1}^n x_i \cdot r_i$$

then by part (a), each r_i is symmetric and we see that

$$f^t = \sum_{i=1}^n x_i \cdot r_i^t = \sum_{i=1}^n x_i \cdot r_i = f$$

f is also symmetric. $\square$

Since every symmetric channel is a convex sum of four involutions which comprise a copy of the Klein four group, the same is true of $\varphi(f) = (f + f^t)/2$ for *any* unital channel f.

6. Scope and the Holevo capacity

A standard way of measuring the capacity of a quantum channel in quantum information theory is the Holevo capacity; it is sometimes called the product state capacity since input states are not allowed to be entangled across two or more uses of the channel.

DEFINITION 6.1. For a trace preserving quantum operation f, the *Holevo capacity* is given by
$$C(f) = \sup_{\{x_i, \rho_i\}} \left[S\left(f\left(\sum_i x_i \rho_i \right) \right) - \sum_i x_i \cdot S(f(\rho_i)) \right]$$
where the supremum is taken over all ensembles $\{x_i, \rho_i\}$ of possible input states ρ_i to the channel.

The possible input states ρ_i to the channel are in general mixed and the x_i are probabilities with $\sum_i x_i = 1$.

THEOREM 6.2. *If f is a unital channel with scope $s(f) = [f^-, f^+]$, then $f^+ \leq C(f)$. If f is symmetric, then $f^+ = C(f)$.*

Proof. Remembering that f is map on the Bloch sphere $\mathbb{B}^3$, and using the correspondence between density operators and the Bloch sphere, the Holevo capacity of f is given by
$$C(f) = \sup_{\{x_i, r_i\}} \left[H\left(\frac{1 + |f(\sum_i x_i r_i)|}{2} \right) - \sum_i x_i \cdot H\left(\frac{1 + |f(r_i)|}{2} \right) \right]$$
where r_i are Bloch vectors for density operators in an ensemble, and we recall that eigenvalues of a density operator with Bloch vector r are $(1 \pm |r|)/2$. By the continuity of $|f|$, there is a pure state $r \in \mathbb{B}^3$ for which
$$|f(r)| = \sup_{|x|=1} |f(x)|$$
In order to keep this proof self-contained, we first repeat the proof from [4] that
$$C(f) = 1 - H((1 + |f(r)|)/2)$$
Setting $r_1 = r$, $r_2 = -r$ and $x_1 = x_2 = 1/2$ defines an ensemble for which the expression maximized in the definition of $C(f)$ reduces to $1 - H((1 + |f(r)|)/2)$. Notice that in this step we explicitly make use of the fact that f is unital: $f(0) = 0$. This proves $1 - H((1 + |f(r)|)/2) \leq C(f)$.

For the reverse inequality, any term in the supremum is clearly bounded from above by
$$1 - \sum_i x_i \cdot H\left(\frac{1 + |f(r_i)|}{2} \right)$$
since $H(x) \leq 1$. Because
$$|f(r_i)| \leq \sup_{x \in \mathbb{B}^3} |f(x)| = \sup_{|x|=1} |f(x)| = |f(r)|$$
we have
$$H\left(\frac{1 + |f(r_i)|}{2} \right) \geq H\left(\frac{1 + |f(r)|}{2} \right)$$
which then gives $C(f) \leq 1 - H((1 + |f(r)|)/2)$ and thus that these two expressions are equal.

To prove $f^+ \leq C(f)$, note that for $|x| = 1$, we have
$$|(x, f(x))| = |x| \cdot |f(x)| \cdot |\cos \theta| \leq |f(x)|$$
so that
$$m^+ = \sup_{|x|=1} |(x, f(x))| \leq \sup_{|x|=1} |f(x)| = |f(r)|$$

and thus $f^+ = 1 - H((1+m^+)/2) \leq C(f)$. If f is symmetric, then $m^+ = |f(r)|$, so $f^+ = C(f)$. □

As the last proof makes clear, $f^+ = C(f)$ iff $m^+ = \sup |f(x)|$, so for instance, we also have equality for rotations r, since r on $\mathbb{R}^3$ has at least one real eigenvalue λ with $|\lambda| = 1$. It is worth pointing out that there are unital channels f for which $f^+ < C(f)$. Any nonzero skew-symmetric unital channel has positive capacity but scope $[0,0]$, as we now show:

PROPOSITION 6.3.
  (i) *A unital channel f has scope $s(f) = [0,0]$ if and only if $f^t = -f$.*
  (ii) *A unital channel f has scope $s(f) = [1,1]$ if and only if $f = I$.*

Proof. (i) First suppose that f is symmetric. If it achieves capacity zero in all bases, then its largest and smallest eigenvalues must be zero. Thus, all its eigenvalues are zero, which means that it is the zero matrix.

Now consider an arbitrary f that achieves zero capacity in every basis. Since $\varphi(f) = (f + f^t)/2$ and f have the same scope, $\varphi(f)$ is symmetric and achieves capacity zero in all bases, so $\varphi(f) = 0$, which finishes the proof.

(ii) Suppose f is symmetric. Then all of its eigenvalues have magnitude one. If two of its eigenvalues had opposite signs, f would achieve capacity zero in some basis by continuity. Thus, its eigenvalues are either all 1 or all -1. In the first case, $f = I$, while in the second $f = -I$, the antipodal map, which is impossible by Corollary 3.8.

For an arbitrary f with scope $[1,1]$, we know $\varphi(f)$ also has scope $[1,1]$, so by the previous remark $\varphi(f) = I$. For $|x| = 1$,

$$(x, f(x)) = (x, \varphi(f)(x)) = (x, x) = 1$$

so $1 = |(x, f(x))| \leq |f(x)| \leq 1$ gives $|f(x)| = 1$. Since f preserves the norm for unit vectors and is a linear mapping, it does so for all vectors in $\mathbb{R}^3$ and must be a linear unital isometry and hence a rotation. As in the proof of Lemma 3.7, take an orthogonal matrix r such that

$$rfr^t = \begin{pmatrix} 1 & 0 & 0 \\ 0 & a & b \\ 0 & -b & a \end{pmatrix}$$

with $a^2 + b^2 = 1$. The trace of f is then

$$\operatorname{tr}(f) = \operatorname{tr}(rfr^t) = 1 + 2a$$

On the other hand, since $\varphi(f) = I$,

$$\operatorname{tr}(f) = \frac{\operatorname{tr}(f) + \operatorname{tr}(f^t)}{2} = \operatorname{tr}(\varphi(f)) = \operatorname{tr}(I) = 3.$$

Equating both terms, $1 + 2a = 3$ and so $a = 1$, $b = 0$ and $f = I$. □

Of course, we have not yet seen that $\mathcal{U}$ contains any *nonzero* skew symmetric matrices, but they do actually exist:

PROPOSITION 6.4. *Let E_{ij} denote the 3×3 matrix with a one in the (i,j) position and a zero in every other location.*
  (i) *The matrices E_{ij} and $-E_{ij}$ are unital for each $i, j \in \{1, 2, 3\}$.*

(ii) *The skew-symmetric matrix*
$$\begin{pmatrix} 0 & -a & -b \\ a & 0 & -c \\ b & c & 0 \end{pmatrix}$$
is unital provided $a, b, c \geq 0$ and $a + b + c \leq 1/2$.

Proof. (i) The three matrices
$$f = r_x(\pi/2) = \begin{pmatrix} 1 & 0 & 0 \\ 0 & 0 & -1 \\ 0 & 1 & 0 \end{pmatrix}, \quad g = r_y(\pi/2) = \begin{pmatrix} 0 & 0 & 1 \\ 0 & 1 & 0 \\ -1 & 0 & 0 \end{pmatrix},$$
$$h = r_z(\pi/2) = \begin{pmatrix} 0 & -1 & 0 \\ 1 & 0 & 0 \\ 0 & 0 & 1 \end{pmatrix}$$
as well as their transposes are all unital, since each is a rotation. Each E_{ii} is unital since
$$E_{11} = (f + f^t)/2, \quad E_{22} = (g + g^t)/2, \quad E_{33} = (h + h^t)/2$$
and each $-E_{ii}$ is unital since
$$-E_{11} = E_{11} \cdot s_y, \quad -E_{22} = E_{22} \cdot s_x, \quad -E_{33} = E_{33} \cdot s_y$$
where s_x and s_y are the spin channels. Noting that $M \cdot E_{ii}$ gives us the matrix with the same i^{th} column as M and zeroes elsewhere, we see that the matrices $\pm E_{1j}$ are unital since
$$E_{12} = h^t \cdot E_{22}, \quad -E_{12} = h \cdot E_{22}, \quad E_{13} = g \cdot E_{33}, \quad -E_{13} = g^t \cdot E_{33},$$
the $\pm E_{2j}$ are unital since
$$E_{21} = h \cdot E_{11}, \quad -E_{21} = h^t \cdot E_{11}, \quad E_{23} = f^t \cdot E_{33}, \quad -E_{23} = f \cdot E_{33},$$
and the $\pm E_{3j}$ are unital since
$$E_{31} = g^t \cdot E_{11}, \quad -E_{31} = g \cdot E_{11}, \quad E_{32} = f \cdot E_{22}, \quad -E_{32} = f^t \cdot E_{22}.$$
(ii) We can write this skew-symmetric matrix as
$$\begin{pmatrix} 0 & -a & -b \\ a & 0 & -c \\ b & c & 0 \end{pmatrix} = a \cdot E_{21} + a(-E_{12}) + b \cdot E_{31} + b \cdot (-E_{13}) + c \cdot E_{32} + c \cdot (-E_{23})$$
which must be unital by Corollary 3.6. □

Thus, any nonzero skew-symmetric unital channel f has $f^+ = 0 < C(f)$ since $C(f) = 0$ only when $f \equiv 0$. We also have the following interesting corollary:

COROLLARY 6.5. *If $f \in \mathcal{U}$ and $p \in [-1/3, 1]$, then $pf \in \mathcal{U}$. Further, this is the largest range over which scalar multiplication is possible: for any $p \in \mathbb{R}$, if $pf \in \mathcal{U}$ for all $f \in \mathcal{U}$, then $p \in [-1/3, 1]$.*

Proof. Let $f \in \mathcal{U}$. First note that
$$-\frac{1}{3}f = \frac{1}{3}f(-E_{11}) + \frac{1}{3}f(-E_{22}) + \frac{1}{3}f(-E_{33})$$

is unital, as a convex sum of unital channels. Given $p \in [-1/3, 0]$, there is $q \in [0,1]$ with $p = -q/3$ so $pf = q(-f/3) \in \mathcal{U}$ by Prop. 3.5. To see that this is the largest range, we must have
$$3p = \text{tr}(p \cdot I) \in [-1, 3]$$
by Lemma 3.7, so $p \in [-1/3, 1]$, finishing the proof. □

7. Adaptive quantum communication

In quantum cryptography, the number of bits we can transmit requires a key of the same size, so the speed at which we transmit information depends on how fast we can generate keys. If the error rate within a session of QKD is too high, we have to start over: this slows the key generation rate. Key generation rates are important not only because it is desirable to communicate as fast as possible, but also because there are times when it is *the only* way for communication to be possible: for instance, in freespace, we have at best 5-6 minutes to transmit information to a satellite before it is out of reach. By *minimizing the error rate*, we can avoid restarting and speed up the rate at which information is transmitted. We now indicate how the theory of scope can be used to develop a method for minimizing the error rate in quantum cryptography – we call it *adaptive quantum cryptography*.

  (i) For each $i \in \{1, 2, 3\}$, Alice sends many 0's prepared in the e_i basis to Bob
 (ii) For each $i \in \{1, 2, 3\}$, Bob measures one third of them in the e_1 basis, one third in the e_2 basis and one third in the e_3 basis
(iii) Bob calculates the channel f which governs the noise in the environment. First, he uses the measurement results from (ii) to estimate the probability p_{ij} that e_i is received when e_j is sent. But since the element f_{ij} of f located at position (i, j) is related to p_{ij} via
$$p_{ij} = \frac{1 + (e_i, f(e_j))}{2} = \frac{1 + f_{ij}}{2}$$
these probabilities allow Bob to construct f!
 (iv) Bob calculates the scope $s(f)$ and an eigenvector r associated to f^+,
  (v) Alice and Bob agree to engage in QKD using r and a forty five degree rotation of r – this requires use of a private key so that Bob can transmit the information about r to Alice, but once done, all future communication will be at a faster rate assuming a stable environment.

EXAMPLE 7.1. Consider a typical case like bit flipping with σ_y
$$f = \begin{pmatrix} 1 - 2p & 0 & 0 \\ 0 & 1 & 0 \\ 0 & 0 & 1 - 2p \end{pmatrix}$$
In QKD, random bits are randomly coded in the bases e_1 and e_3, so the error rate is
$$\frac{1}{2}\left(\frac{1 + (\mp e_1, f(\pm e_1))}{2}\right) + \frac{1}{2}\left(\frac{1 + (\mp e_3, f(\pm e_3))}{2}\right) = \frac{1}{2} \cdot p + \frac{1}{2} \cdot p = p,$$
while with adaptive communication, the bases used are e_2 and without loss of generality e_1, so the error rate is
$$\frac{1}{2}\left(\frac{1 + (\mp e_1, f(\pm e_1))}{2}\right) + \frac{1}{2}\left(\frac{1 + (\mp e_2, f(\pm e_2))}{2}\right) = \frac{1}{2} \cdot p + \frac{1}{2} \cdot 0 = \frac{p}{2}.$$

Of course, depolarization $f(r) = p \cdot r$ for $p \in [0,1]$ is an example of an effect where the error rate cannot be improved upon, but in many cases, it will be. For such a scheme to be physically realized, numerous questions must be answered:

(a) We can test for unitality using the complete positivity conditions (Lemma 5.11), but what do we do if the noise in the environment is not unital? How necessary is a theory of scope for non-unital channels?

(b) To calculate scope, we need to be able to solve a cubic equation – there are well known formulas for doing so, is this the best way to solve it, or are there new techniques that can be applied in this particular case?

(c) How many 'test bits' does Alice have to send Bob in step (i) in order to make sure that Bob obtains a channel that accurately models the environment?

(d) Can we *quantify* the degree of improvement in the error rate for a given quantum channel when using adaptive quantum communication?

Other issues perhaps also of interest include: when the eavesdropper knows we are adaptively communicating, they will adapt too, now what? Or: if we are trying to perform QKD from the ground to a low earth orbit satellite via entanglement, it is possible that relativistic effects could adversely affect our ability to communicate – or that relativistic effects could be used to provide a new way to help prevent eavesdropping

8. Closing

In general, a qubit channel $f : \mathbb{B}^3 \to \mathbb{B}^3$ has the form $f(x) = Mx + b$ where M is a 3×3 real matrix and $b \in \mathbb{B}^3$. Such a channel is unital iff $b = 0$, so there are many qubit channels that are not unital, amplitude damping being a notable example. Developing a method for calculating the scope of a nonunital qubit channel is a difficult but high priority. As a first step toward this, the algebraic structure of nonunital qubit channels has to be uncovered.

Part of the value of the definition of scope is that when a certain capacity is achieved, we can point to a definite procedure the sender and receiver should follow in order to achieve that capacity: prepare in basis r, send and then measure in basis r. It is possible though to challenge the definition of scope. For instance, what if the sender prepares in basis r, sends and then the receiver measures in basis s? With a definition of scope that allows for *two* different bases, the least value of capacity ceases to have meaning: for a given unital f, the receiver can choose a pure state s that is Euclidean orthogonal to $f(r)/|f(r)|$ ensuring that $(1 + (s, f(r)))/2 = 1/2$, so the smallest capacity achievable is *always* zero. For symmetric channels, it can be shown that the upper bound in such a definition can be achieved with a *single* basis and is equal to f^+. For nonsymmetric channels, like non-zero skew-symmetric channels, one can achieve a higher capacity than f^+.

Aside from adaptive quantum communication, whose precise operational details will be the subject of future research, there are other uses of scope that may be possible. One is the role of scope in classifying physical effects according to the degree that they disturb the state of a system. For instance, a 'weak effect' would be a channel with scope close to $[1,1]$. One reason for this is given in Prop. 6.3: the only channel with scope $[1,1]$ is the identity. Other examples are *projective measurements*: they always have scope $[0,1]$, which is a maximum distance from $[1,1]$, indicating that the disturbance caused by such an effect is extreme. By contrast, the Holevo capacity does not distinguish between channels in as precise a manner as scope: for instance, it assigns the value 1 to any rotation, whereas the scope of a rotation always depends on the angles involved.

9. Acknowledgements

We thank Tanner Crowder for helping us find the complete positivity conditions in the literature, and for many valuable discussions while this paper was being written; Johnny Feng, who first asked if the antipodal map was a qubit channel; and Marco Lanzagorta, who upon hearing that it was not a qubit channel would not stop talking about it until we both realized something really cool: that 'universal' bit flipping is physically impossible.

References

[1] C. H. Bennett, C. A. Fuchs, and J. A. Smolin, *Entanglement-enhanced classical communication on a noisy quantum channel*, in Quantum Communication, Computing and Measurement, O. Hirota, A. S. Holevo, and C. M. Caves, Eds. New York: Plenum, 1997, pp. 79–88.

[2] P.S. Bourdon and H.T. Williams, *Unital quantum operations on the Bloch ball and Bloch region*, Physical Review A, Vol. 69, Article 022314, 2004.

[3] L. J. Landau and R. F. Streater, *On Birkhoff's theorem for doubly stochastic completely positive maps of matrix algebras*, Linear Algebra and its Applications **193**, p. 107–127, 1993.

[4] K. Martin. *A domain theoretic model of qubit channels*. ICALP 2008, Lecture Notes in Computer Science, Vol. 5126, p. 283–297, 2008.

[5] K. Martin. *How to randomly flip a quantum bit*. Electronic Notes in Theoretical Computer Science, Volume 270, Issue 1, p. 81–97, Elsevier Science, 2011.

[6] M. Nielsen and I. Chuang, *Quantum computation and quantum information*. Cambridge University Press, 2000.

[7] M. B. Ruskai, S. Szarek and E. Werner. *An analysis of completely-positive trace-preserving maps on $\mathcal{M}_2$*. Linear Algebra and its Applications, Volume 347, Issues 1-3, p. 159–187, 2002.

[8] D. Serre. *Matrices: Theory and applications*. Springer-Verlag, Graduate Texts in Mathematics, 2000.

[9] C. E. Shannon. *A mathematical theory of communication*. Bell Systems Technical Journal 27, 379–423 and 623–656, 1948.

Center for High Assurance Computer Systems, Naval Research Laboratory Washington, DC 20375.
Email address: `kmartin@itd.nrl.navy.mil`

Spacetime geometry from causal structure and a measurement

Keye Martin and Prakash Panangaden

ABSTRACT. The causal structure of spacetime defines a partial order on the events of spacetime. In an earlier paper, using techniques from domain theory, we showed that for globally hyperbolic spacetimes one could reconstruct the topology from the causal structure. However, the causal structure determines the metric only up to a local rescaling (a conformal transformation); in a four dimensional spacetime, the metric tensor has ten components, and thus effectively only nine are determined by the causal structure. After establishing the relationship between measurement in domain theory, the concept of global time function and the Lorentz distance, we are able to domain theoretically recover the final tenth component of the metric tensor, thereby obtaining causal reconstruction of not only the topology of spacetime, but also its geometry.

1. Introduction

The study of spacetime structure from an abstract viewpoint – i.e., not from the viewpoint of solving differential equations – was initiated by Penrose [18] in a dramatic paper in which he showed a fundamental inconsistency of gravity: all the spacetimes satisfying some general conditions develop singularities. Penrose's paper initiated a whole new way of studying general relativity: an abstract approach using ideas of differential topology and geometry rather than looking for solutions of Einstein's equations.

It was known since Chandrasekhar [3] that since gravity is universal and inherently attractive, a gravitating mass of sufficient size will eventually collapse. It was widely believed that the collapse phenomenon discovered by Chandrasekhar was an artifact of special symmetry assumptions and that in a realistic situation perturbations would prevent the appearance of singularities. Penrose dashed this hope by showing that singularities arise *generically*. What Penrose showed was that any such collapse eventually leads to a singularity where the mathematical description of spacetime as a continuum breaks down. This leads to the need to reformulate gravity. Part of the motivation for the search for a quantum theory of gravity is

---

1991 *Mathematics Subject Classification.* Primary 83C99; Secondary 06B30.

Key words and phrases. Causal structure, topology, geometry, spacetime, domain theory, measurement.

Research supported by a grant from the Office of Naval Research.

the hope that this elusive theory will resolve the problem of gravitational collapse. A good discussion of the history of these ideas is in a recent book by Hawking and Penrose [9].

Since the first singularity theorems [18, 8] causality has played a key role in understanding spacetime structure. The analysis of causal structure relies heavily on techniques of differential topology [19]. For the past decade Sorkin and others [21] have pursued a program for quantization of gravity based on causal structure. In this approach the causal relation is regarded as the fundamental ingredient and the topology and geometry are secondary.

In a paper that appeared in 2006 [15], we prove that the causality relation is much more than a relation – it turns a globally hyperbolic spacetime into what is known as a *bicontinuous poset*. The order on a bicontinuous poset allows one to define an intrinsic topology called *the interval topology*. On a globally hyperbolic spacetime, the interval topology is the manifold topology. Theorems that reconstruct the spacetime topology have been known [19] and Malament [12] has shown that the class of time-like curves determines the causal structure. We establish these results as well though in a purely order theoretic fashion: there is no need to know what "smooth curve" means.

Our more abstract stance also teaches us something *new*: a globally hyperbolic spacetime *itself* can be reconstructed in a purely order theoretic manner, beginning from only a countable dense set of events and the causality relation. The ultimate reason for this is that the category of globally hyperbolic posets, which contains the globally hyperbolic spacetimes, is *equivalent* to a very special category of posets called *interval domains*. This provides a profound connection between domain theory, first introduced for the purposes of assigning semantics to programming languages, and general relativity, a theory meant to explain gravity. Even from a purely mathematical perspective this equivalence is surprising, since globally hyperbolic spacetimes are usually not order theoretically complete, but interval domains always are.

While our previous work has focused on the role of domain theory in investigating qualitative aspects in relativity [16] – like the topology – in this paper, we investigate reconstructing quantitative aspects of spacetime structure: the metric. The theory of measurement was introduced by Martin in [13] as a way of incorporating quantitative information into domain theory. In this paper we will show how not only the topology, but the *geometry* of spacetime can be reconstructed order theoretically from the causal structure together with an appropriate measurement. The reason is that the Lorentz distance defines a Scott continuous function on the domain of spacetime intervals. What is even more interesting, though, is that our setting provides a way to *topologically* distinguish between Newtonian and relativistic notions of time. Every global time function defines a measurement on the domain of spacetime intervals, in particular, it is Scott continuous. The Lorentz distance is not only Scott continuous, but satisfies a stronger property, that it is interval continuous. An interval continuous function must assign zero to any element which approximates nothing. In all spacetimes there are non-empty intervals that correspond to a null line segment; these do not approximate anything (but they are not maximal either since they will contain other null sub-intervals) and indeed their "length" in the Lorentz metric is zero. Thus, no interval continuous function on the domain of spacetime intervals can ever be a measurement and the reason for this has entirely to do with relativity: a clock moving at the speed of light records no time as having elapsed, so an interval continuous function is incapable of distinguishing between a single event and a null interval. In Section 7 we discuss this point at length.

2. Domains, continuous posets and topology

We review some basic concepts which can be found, for example in the comprehensive book "Continuous Lattices and Domains" [7]. Occasionally our terminology differs; we will point out such occasions.

A *poset* is a partially ordered set, i.e., a set together with a reflexive, antisymmetric and transitive relation.

DEFINITION 2.1. *Let $(P, \sqsubseteq)$ be a partially ordered set. A nonempty subset $S \subseteq P$ is* directed *if $(\forall x, y \in S)(\exists z \in S)\, x, y \sqsubseteq z$. The* supremum *of $S \subseteq P$ is the least of all its upper bounds provided it exists. This is written $\bigsqcup S$.*

These ideas have duals that will be important to us: a nonempty $S \subseteq P$ is *filtered* if $(\forall x, y \in S)(\exists z \in S)\, z \sqsubseteq x, y$. The *infimum* $\bigwedge S$ of $S \subseteq P$ is the greatest of all its lower bounds provided it exists.

DEFINITION 2.2. *For a subset X of a poset P, set*
$$\uparrow\! X := \{y \in P : (\exists x \in X)\, x \sqsubseteq y\} \ \& \ \downarrow\! X := \{y \in P : (\exists x \in X)\, y \sqsubseteq x\}.$$
We write $\uparrow\! x = \uparrow\!\{x\}$ and $\downarrow\! x = \downarrow\!\{x\}$ for elements $x \in X$.

A partial order allows for the derivation of several intrinsically defined topologies. Here is our first example.

DEFINITION 2.3. *A subset U of a poset P is* Scott open *if*
  (i) *U is an upper set: $x \in U$ & $x \sqsubseteq y \Rightarrow y \in U$, and*
  (ii) *U is inaccessible by directed suprema: For every directed $S \subseteq P$ with a supremum,*
$$\bigsqcup S \in U \Rightarrow S \cap U \neq \emptyset.$$
The collection of all Scott open sets on P is called the Scott topology.

Closely related to directed sets are ideals.

DEFINITION 2.4. *An* ideal *I in a poset is a directed set such that if $x \in I$ and $y \leq x$, then $y \in I$. A set with the latter property is called a* lower set.

Posets can have a variety of completeness properties. The following completeness condition has turned out to be particularly useful in applications.

DEFINITION 2.5. *A* dcpo *is a poset in which every directed subset has a supremum. The* least element *in a poset, when it exists, is the unique element $\bot$ with $\bot \sqsubseteq x$ for all x.*

If one takes any poset the collection of ideals ordered by inclusion forms a dcpo. This means that the union of any directed family of ideals is an ideal. It is easy to check this explicitly from the definition.

The set of *maximal elements* in a dcpo D is
$$\max(D) := \{x \in D : \uparrow\! x = \{x\}\}.$$
Each element in a dcpo has a maximal element above it; this follows at once from Zorn's Lemma and indeed is equivalent to it and hence to the Axiom of Choice.

DEFINITION 2.6. *For elements x, y of a poset, write $x \ll y$ iff for all directed sets S with a supremum,*
$$y \sqsubseteq \bigsqcup S \Rightarrow (\exists s \in S)\, x \sqsubseteq s.$$
We set $\downarrow\!x = \{a \in D : a \ll x\}$ and $\uparrow\!x = \{a \in D : x \ll a\}$.

For the symbol "$\ll$," read "approximates." A number of basic properties are immediate from the definition. For example, the fact that the relation is transitive and the following:
$$x \leq y \ll z \leq w \Rightarrow x \ll w.$$

DEFINITION 2.7. *A* basis *for a poset D is a subset B such that $B \cap \downarrow\!x$ contains a directed set with supremum x for all $x \in D$. A poset is* continuous *if it has a basis. A poset is ω-continuous if it has a countable basis.*

Continuous posets have an important property, they are *interpolative*.

PROPOSITION 2.8. *If $x \ll y$ in a continuous poset P, then there is $z \in P$ with $x \ll z \ll y$.*

Proof: Consider the set $K = \{u \mid \exists v.\, u \ll v \ll y\}$. This set is clearly not empty since x is the supremum of elements that approximate it so for some w we have $w \ll x \ll y$. Let $u_1, u_2 \in K$ then we have $u_1 \ll v_1 \ll y$ and $u_2 \ll v_2 \ll y$ for some v_1, v_2. Since $\downarrow\!y$ is directed there is some $v \ll y$ with $v_1, v_2 \leq v$. Thus $u_1, u_2 \ll v$ and since $\downarrow\!v$ is directed we have an element $u \ll v$ with $u_1, u_2 \leq u$. Now since $u \ll v \ll y$, $u \in K$, hence K is a directed set. Now clearly $\bigsqcup K = z$ so by the definition of $x \ll z$ we have that there is some $w \in K$ with $x \leq w$ which means that there is some z such that $x \leq w \ll z \ll y$. □

This proof is taken from [10]. A very short proof using ideals can be found in [7].

This enables a clear description of the Scott topology.

THEOREM 2.9. *The collection $\{\uparrow\!x : x \in D\}$ is a basis for the Scott topology on a continuous poset.*

Proof: From the interpolation property it easily follows that sets of the form $\uparrow\!x$ are Scott open. If U is any Scott open set and $x \in U$ then the directed set $\downarrow\!x$ must intersect U, since $\bigsqcup \downarrow\!x = x \in U$. Let $y \in U \cap \downarrow\!x$, then $\uparrow\!y \subset U$, thus for any point x in U we can find a set of the form $\uparrow\!y$ containing x and contained in U so these sets form a basis for the Scott topology. □

DEFINITION 2.10. *A* continuous dcpo *is a continuous poset which is also a dcpo. A* domain *is a continuous dcpo.*

The next example is due to Scott[20] and worth keeping in mind when we consider the analogous construction for globally hyperbolic spacetimes.

EXAMPLE 2.11. *The collection of compact intervals of the real line*
$$\mathbb{IR} = \{[a, b] : a, b \in \mathbb{R}\ \&\ a \leq b\}$$
ordered under reverse inclusion
$$[a, b] \sqsubseteq [c, d] \Leftrightarrow [c, d] \subseteq [a, b]$$
is an ω-continuous dcpo:

- For directed $S \subseteq \mathbf{IR}$, $\bigsqcup S = \bigcap S$,
- $I \ll J \Leftrightarrow J \subseteq int(I)$, and
- $\{[p,q] : p, q \in \mathbb{Q}\ \&\ p \leq q\}$ is a countable basis for $\mathbf{IR}$.

In the above $\mathbb{Q}$ stands for the rationals. The domain $\mathbf{IR}$ is called the interval domain.

We also have $\max(\mathbf{IR}) \simeq \mathbb{R}$ in the Scott topology. More precisely the subspace topology that $\max(\mathbf{IR})$ inherits from the domain equipped with the Scott topology is homeomorphic to the reals with its usual topology. Approximation can help explain why:

EXAMPLE 2.12. *A basic Scott open set in $\mathbf{IR}$ is*
$$\mathord{\uparrow}[a,b] = \{x \in \mathbf{IR} : x \subseteq (a,b)\}.$$

One of the interesting things about $\mathbf{IR}$ is that it is a domain that is derived from an underlying poset with an abundance of order theoretic structure. Part of this structure is that the real line is *bicontinuous*, a fundamental notion in the present work:

DEFINITION 2.13. *A continuous poset P is* bicontinuous *if*
- *For all $x, y \in P$, $x \ll y$ iff for all filtered $S \subseteq P$ with an infimum,*
$$\bigwedge S \sqsubseteq x \Rightarrow (\exists s \in S)\, s \sqsubseteq y,$$
and
- *For each $x \in P$, the set $\mathord{\Uparrow} x$ is filtered with infimum x.*

In order to clarify the above definition we deconstruct it as follows. Given a continuous poset with its approximation relation $\ll$ we define the dual relation $\ll^{op}$ by
$$x \ll^{op} y \text{ iff } \inf S \leq x \text{ implies } S \cap \mathord{\downarrow} y \neq \emptyset$$
for any filtered set S with an infimum. Of course, there is no prima facie reason why $\ll$ and $\ll^{op}$ should be related. We can say that a poset is "dually continuous" if for every x the set $\{y | x \ll^{op} y\}$ is filtered and has x as its infimum. Our definition then amounts to saying that the poset is continuous, dually continuous and the two relations $\ll$ and $\ll^{op}$ coincide. In other work [**7**] the term "bicontinuous" is used for the situation where the two approximation relations do not coincide; such authors use the term "strongly bicontinuous" for what we have called bicontinuous. For us the present terminology seems more natural and leads to the pleasing theory of the interval topology described below.

EXAMPLE 2.14. *$\mathbb{R}$, $\mathbb{Q}$ are bicontinuous.*

DEFINITION 2.15. *On a bicontinuous poset P, sets of the form*
$$(a,b) := \{x \in P : a \ll x \ll b\}$$
form a basis for a topology called the interval topology.[1]

The proof that such sets form the base for a topology uses interpolation and bicontinuity and is given in our previous paper [**15**]. In contrast to a domain, a bicontinuous poset P has $\mathord{\Uparrow} x \neq \emptyset$ for each x, so it is rarely a dcpo.

---
[1]The term "interval topology" means something different in lattice theory.

3. The mathematical structure of spacetime

The mathematical structure used to define spacetime in general relativity is very rich and can be described in a sequence of layers. We give a quick overview of this structure emphasizing particularly the causal structure. This is standard material and is explained well in a number of text books. Ones that we recommend particularly are: *The Large-Scale Structure of Spacetime* by Hawking and Ellis [8], *Techniques of differential topology in relativity* by Penrose [19], *General Relativity* by Wald [23] and *Global Lorentzian Geometry* by Beem, Ehrlich and Easley [2]. A beautiful account of how some of these structures are related to the physics of particles and light rays is given in an article appropriately named, "The geometry of free fall and light propagation" by Ehlers, Pirani and Schild [4] which we highly recommend for a reader interested in the physical significance of the mathematics.

The basic ingredient of general relativity is an *event* which we take to be an undefined primitive concept in the same way that a point is taken as a primitive concept in the geometry of space. Note that an event is not to be understood, as in ordinary language, as the occurrence of some action but rather as a *potential* occurrence. This is just as a point in space is not necessarily the location of a physical entity but the place where a material particle *could* be. The collection of events is a set called spacetime. A set is, of course, no structure at all. It is the canvas on which we paint the rest of the structure.

The next level of mathematical structure is to make the spacetime into a topological space. It is at this point that one incorporates the fact that it is a 4-dimensional topological manifold. Here is the precise definition

DEFINITION 3.1. *A topological n-dimensional manifold $\mathcal{M}$ is a topological space equipped with a family of open sets $\{O_i\}_{i \in I}$ together with a family of continuous functions $\phi_i : O_i \to \mathbb{R}^n$ such that each ϕ_i is a homeomorphism of O_i onto its image. We assume that as a topological space M is connected, Hausdorff and has a countable basis.*

It is often assumed that a manifold is paracompact: this means that every open cover has a locally-finite refinement. It is a very useful technical condition that lies at the heart of partition-of-unity arguments and is crucially used to prove the existence of metrics. We will not be discussing anything at that level of detail so we will never mention paracompactness again except to note that a connected Hausdorff manifold is paracompact iff it has a countable basis. Given a manifold as we have defined it above, a pair $(O_i, \phi_i)_{i \in I}$ is called a *chart*; we will use the word "chart" ambiguously for the pair, for the set O_i and for the function ϕ_i. The collection of charts is called an *atlas*.

The next structure that one defines on spacetime is the differential structure. This allows one to "do calculus" or at least to define the notion of derivative. A manifold could be something like the surface of a sphere: there would be no sense in "adding" the points of a sphere. The formula for the derivative of a function f has in it the expression $f(x + \epsilon)$: what does $x + \epsilon$ mean if one is not working in a vector space? The notion of manifold is precisely designed to allow one to think of a manifold as locally like a vector space; that is what the charts are for. However, we have to be sure that the charts agree on the notion of derivative. This brings us to the concept of a *differential manifold*.

Consider what happens when two charts intersect: $V \stackrel{\text{def}}{=} O_i \cap O_j \neq \emptyset$. Define $U_i = \phi_i(O_i)$, $U_j = \phi_j(O_j)$, $U = \phi_i(V)$ and $W = \phi_j(V)$. Now the function $\phi_j \circ \phi_i^{-1}$ is a well-defined continuous function, in fact a homeomorphism, from U to W. Since U and W are open subsets of $\mathbb{R}^n$ it is clear what one means by saying that they are differentiable. Such functions are conveniently called *transition functions* as they allow one to translate between charts.

DEFINITION 3.2. *A manifold is said to be* smooth *if all the transition functions are infinitely differentiable.*

The charts allow one to endow patches of the topological space M with the structure of a vector space: exactly what one needs to define the notion of a derivative. The condition on the transition functions ensures that the notion of what is a differentiable function will not be chart dependent.

We will not review the entire apparatus of differential geometry here. However, the reader should be convinced that there is a clear strategy for developing the notions of the differential calculus on manifolds now. One uses the charts to move to $\mathbb{R}^n$ and uses the usual undergraduate calculus notions there. Thus, for example, it should be clear how one can define a smooth real-valued function on a manifold or a smooth function between two manifolds.

Once one has the notion of a smooth structure – another more snappy name for differential structure – one can define curves and tangent vectors. A *smooth curve* on M is a smooth function γ from some interval[2] $[a, b]$ to M. Note that the curve is not just the image of γ but γ itself: this is what one normally thinks of as a parametrized curve. Two different functions that happen to have the same image are different curves. One can now define the tangent vector to a curve in the usual way using the charts to move back and forth between M and $\mathbb{R}^n$. At each point there is a vector space now attached to the point: the collection of all tangent vectors at that point, it is called the *tangent space* at a point p. The whole apparatus of multi-linear algebra can now be brought to bear and one can define dual vectors at a point and indeed arbitrary tensors at every point. If V is the tangent space and V^* is the dual space one says that a tensor has type (p, q) if it belongs to $V \otimes V \otimes \ldots (p \text{ terms}) \ldots \otimes V \otimes V^* \otimes \ldots (q \text{ terms}) \ldots \otimes V^*$.

Now we come to the absolutely crucial part of the structure. Given a tangent space one can define a cone[3] which we call the forward or future light cone. Mathematically, any cone will do but, of course, the physics determines the forward light cone through the propagation of light rays emanating from a point. We similarly define another cone by taking the negatives of the vectors in the forward light cone: this is the backward or past light cone. We come to the first important restriction on the spacetimes that we consider.

DEFINITION 3.3. *A manifold is* time orientable *if it is possible to choose globally a consistent definition of future and past light cones.*

It might seem *prima facie* that every manifold will be time orientable but [8] gives examples showing that this is not the case. Essentially the same type of construction that one uses to produce the Mobius strip can produce a non time

---

[2]It could also be an open or half-open interval.
[3]Technically a cone C is a subset of a real vector space V which is closed under addition and multiplication by *positive* scalars and such that if both x and $-x$ are in C then x is zero.

orientable manifold. Henceforth, we assume that all manifolds we consider are time orientable.

A choice of future and past light cones defines the causal structure in the following way. Given a smooth curve we can determine whether its tangent vector at a point p through which the curve passes lies inside the future light cone, or on its boundary, or in the past light cone, or on its boundary or outside both cones. The tangent vector is said to be future timelike, future null, past timelike, past null or spacelike, respectively. Of course the tangent to a curve may be at different places all of the above. However, we are interested in curves that have a timelike or null tangent vector as these are the curves along which causal effects propagate.

DEFINITION 3.4. *A curve is said to be a* future-directed causal *curve if its tangent vector everywhere lies inside or on the boundary of the future light cone. A curve is said to be a* future-directed timelike *curve if its tangent vector is everywhere strictly inside the future light cone. A curve is said to be a* future-directed null *curve if its tangent vector is everywhere on the boundary of the future light cone.*

We usually work with future directed curves; there are analogous definitions for past directed curves.

The next structure that one usually defines is the *affine* structure. This defines what it means to "move a vector parallel to itself" along a curve: this is called parallel transport and the mathematical gadget that describes this is called the affine connection. We will not discuss the affine connection here. We remark in passing that it is used to define what it means to be a "straight line" or a geodesic on a manifold.

Finally we get to measure the length of a curve. This is done by a symmetric tensor. A *Lorentz metric* on a manifold is a symmetric, nondegenerate tensor field of type $(0,2)$ whose signature is $(-+++)$; it is traditionally denoted by g. The fact that it is of type $(0,2)$ means that given a vector v it assigns a number $g(v,v)$ which is quadratic in v: just what we expect for length squared. What is unusual is that some vectors have positive length and some non-zero vectors have zero length. Vectors that are timelike have negative length and null vectors have zero length. While those brought up on metric spaces may be disturbed by the indefiniteness of this kind of metric it is worth getting used to and accepting it as a reasonable definition. The physical fact that forces this is the *experimental* observation that the real world is covariant with respect to a particular group, the Lorentz group, and that the invariant for this group is indeed of the given signature.

Our presentation of the layers of spacetime structure is not how most textbooks present it. They tend to take the metric as fundamental and present all aspects of the structure in one shot, but this is conceptually confusing. As we have presented it each layer requires the previous layer for a proper definition. Here is how a spacetime is usually defined.

DEFINITION 3.5. *A spacetime is a real four-dimensional*[4] *smooth manifold $\mathcal{M}$ with a Lorentz metric g_{ab}.*

Of course, once one has a metric it encodes the light cones by telling you which vectors are timelike, null and spacelike through their "length" squared. But in order to even define smooth tensor fields, the differential structure must be in place and before that the topology.

---

[4]The results in the present paper work for any dimension $n \geq 2$ [**11**].

Let $(\mathcal{M}, g_{ab})$ be a time-orientable spacetime. Let $\Pi^+_\leq$ denote the future directed causal curves, and $\Pi^+_<$ denote the future directed time-like curves. These curves can be used to define order relations on spacetime. The following definitions are standard in the relativity literature.

DEFINITION 3.6. *For $p \in \mathcal{M}$,*
$$I^+(p) := \{q \in \mathcal{M} : (\exists \pi \in \Pi^+_<)\, \pi(0) = p, \pi(1) = q\}$$
and
$$J^+(p) := \{q \in \mathcal{M} : (\exists \pi \in \Pi^+_\leq)\, \pi(0) = p, \pi(1) = q\}$$
Similarly, we define $I^-(p)$ and $J^-(p)$.

We write the relation J^+ as
$$p \sqsubseteq q \equiv q \in J^+(p).$$
The following properties from [8] are very useful:

PROPOSITION 3.7. *Let $p, q, r \in \mathcal{M}$. Then*
  (i) *The sets $I^+(p)$ and $I^-(p)$ are open.*
  (ii) *$p \sqsubseteq q$ and $r \in I^+(q) \Rightarrow r \in I^+(p)$*
  (iii) *$q \in I^+(p)$ and $q \sqsubseteq r \Rightarrow r \in I^+(p)$*
  (iv) *$\mathrm{Cl}(I^+(p)) = \mathrm{Cl}(J^+(p))$ and $\mathrm{Cl}(I^-(p)) = \mathrm{Cl}(J^-(p))$, where Cl stands for topological closure.*

From the physical point of view there are a number of causality conditions that one can imagine imposing on a spacetime. Time orientability is a precondition for even discussing causality in any global sense. The basic causality condition is that there are no closed causal curves. In other words, we always assume the chronology conditions that ensure $(\mathcal{M}, \sqsubseteq)$ is a partially ordered set. There are a number of other, stronger conditions that one can impose; they are discussed at length in [8]. We will not mention all of them here. Two that we will mention are strong causality and global hyperbolicity. Intuitively, strong causality says that one cannot even come close to violating causality in the sense that for every point there is an open neighbourhood such that causal curves that leave it cannot reenter it. Thus not only do causal curves not come back to their starting point they do not come arbitrarily close to it.

We will not give the formal version of the above definition referring instead to [8] or [19]. There is a convenient topological characterization of strong causality. First we define the Alexandroff topology on a spacetime. It is the topology which has $\{I^+(p) \cap I^-(q) : p, q \in \mathcal{M}\}$ as a basis [19][5]. Penrose [19] proved the following important theorem.

THEOREM 3.8. *A spacetime $\mathcal{M}$ is strongly causal iff its Alexandroff topology is Hausdorff iff its Alexandroff topology is the manifold topology.*

This shows the topological significance of strong causality. The number of topologies proliferate when the spacetime lacks strong causality.

---

[5]This terminology is common among relativists but order theorists use the phrase "Alexandrov topology" to mean something else: the topology generated by the upper sets.

4. Global hyperbolicity

Beyond strong causality the most discussed conditions are *stable causality*, *causal simplicity* and *global hyperbolicity*. Stable causality is intuitive but requires some technical machinery to formalize properly. Roughly speaking, it says that the spacetime is still causal even if the light cones are opened out a little. Causal simplicity says that the sets $J^{\pm}(p)$ are always closed; effectively it means that spacetime does not have holes cut out of it. Global hyperbolicity is the strongest and the name does not give many clues to the uninitiated.

One of the most important things that a mathematical physicist wants to do is to solve the following problem. One is told the complete configuration of a system at some time and one wants to determine the complete future evolution. In relativity this means one is given a spacelike surface which is complete in some sense and one wants to be able to predict the evolution for all future times. If there is such a surface the spacetime is called globally hyperbolic. The use of the term "global" should now be clear but why "hyperbolic"? The equations one is most interested in are wave equations which are hyperbolic partial differential equations. What does one mean by a "complete" spacelike surface? A reasonable way of thinking of this is a surface such that every timelike curve extended indefinitely in both directions must hit this this surface exactly once. If there is such a surface it is called a Cauchy surface. The problem of determining the evolution from data on such a surface (also called an "initial value surface") is called the Cauchy problem (also called the "initial value problem").

Penrose has called *globally hyperbolic* spacetimes "the physically reasonable spacetimes" [**23**].

DEFINITION 4.1. *A spacetime $\mathcal{M}$ is* globally hyperbolic *if it is strongly causal and if $\uparrow a \cap \downarrow b$ is compact in the manifold topology, for all $a, b \in \mathcal{M}$.*

This is the most useful version of the definition but it gives none of the intuitions about solving the Cauchy problem.

Given the sets $I^{\pm}(p)$ we can define an irreflexive transitive relation $\prec$ by $p \prec q$ if $q \in I^{+}(p)$ (or $p \in I^{-}(q)$); in other words, there is a future directed *timelike* curve from p to q. In our previous paper we proved the following.

THEOREM 4.2 ([**16**]). *If $\mathcal{M}$ is globally hyperbolic, then $(\mathcal{M}, \sqsubseteq)$ is a bicontinuous poset with $\ll \; = \prec$ whose interval topology is the manifold topology.*

This gives a striking connection between the approximation relation in domain theory and the notion of timelike order.

This result motivates the following definition:

DEFINITION 4.3. *A poset $(X, \leq)$ is* globally hyperbolic *if it is bicontinuous and each interval $[a, b] = \{x : a \leq x \leq b\}$ is compact in the interval topology.*

This abstracts the spacetime concept of global hyperbolicity to posets.

Globally hyperbolic posets are very much like the real line. In fact, a well-known domain theoretic construction pertaining to the real line extends in perfect form to the globally hyperbolic posets:

THEOREM 4.4 ([**16**]). *The closed intervals of a globally hyperbolic poset X*

$$\mathbf{I}X := \{[a, b] : a \leq b \; \& \; a, b \in X\}$$

ordered by reverse inclusion

$$[a,b] \sqsubseteq [c,d] \equiv [c,d] \subseteq [a,b]$$

form a continuous domain with

$$[a,b] \ll [c,d] \equiv a \ll c \text{ \& } d \ll b.$$

The poset X has a countable basis iff $\mathbf{I}X$ is ω-continuous. Finally,

$$\max(\mathbf{I}X) \simeq X$$

where the set of maximal elements has the relative Scott topology from $\mathbf{I}X$ and X has the interval topology.

Globally hyperbolic posets also have rich enough structure that we can deduce many properties of spacetime from them *without* appealing to differentiable structure or geometry. Here is one such example:

DEFINITION 4.5. *Let $(X, \leq)$ be a globally hyperbolic poset. A subset $\pi \subseteq X$ is a* causal curve *if it is compact, connected and linearly ordered. We define*

$$\pi(0) := \bot \quad and \quad \pi(1) := \top$$

where $\bot$ and $\top$ are the least and greatest elements of π. For $P, Q \subseteq X$,

$$C(P,Q) := \{\pi : \pi \text{ causal curve}, \pi(0) \in P, \pi(1) \in Q\}$$

and call this the space of causal curves *between P and Q.*

Here we are adapting the definitions from spacetime geometry to arbitrary posets. A causal curve as just defined is not a function from $[0,1]$ to X as in the case of curves on manifolds but we are mimicking that definition. The $\top$ and $\bot$ refer to the top and bottom elements of the subset π viewed as a poset. When π is embedded into X these are points of X (not necessarily the top or bottom of X, of course); we are writing $\pi(0)$ and $\pi(1)$ to be suggestive of a curve in the geometric setting.

This definition is motivated by the fact that a subset of a globally hyperbolic spacetime $\mathcal{M}$ is the image of a causal curve iff it is the image of a continuous monotone increasing $\pi : [0, 1] \to \mathcal{M}$ iff it is a compact connected linearly ordered subset of $(\mathcal{M}, \sqsubseteq)$.

THEOREM 4.6 ([15]). *If $(X, \leq)$ is a separable globally hyperbolic poset, then the space of causal curves $C(P, Q)$ is compact in the Vietoris topology and hence in the upper topology.*

In addition, while events in spacetime are maximal elements of $\mathbf{I}\mathcal{M}$, causal curves are maximal elements in a higher order domain $C(\mathbf{I}\mathcal{M})$, called the *convex powerdomain* of $\mathbf{I}\mathcal{M}$ [15]. In addition, the fact that spacetime has a canonical domain theoretic model teaches us something new: from only a countable set of events and the causality relation, one can reconstruct spacetime in a purely order theoretic manner. Explaining how requires domain theory.

5. Spacetime from a discrete causal set

Given a set with an order relation $(X, \leq)$ we will use the notation $F \leq x$ where F is a finite subset of X and $x \in X$ to mean that every $y \in F$ satisfies $y \leq x$. We use this to define an abstract basis.

An *abstract basis* is a set $(C, \ll)$ with a *transitive* relation that is *interpolative* from the $-$ *direction*:

$$F \ll x \Rightarrow (\exists y \in C)\, F \ll y \ll x,$$

for all finite subsets $F \subseteq C$ and all $x \in F$. Suppose, though, that it is also interpolative from the $+$ *direction*:

$$x \ll F \Rightarrow (\exists y \in C)\, x \ll y \ll F.$$

Then we can define a new abstract basis of *intervals*

$$\mathrm{int}(C) = \{(a,b) : a \ll b\} = \, \ll \, \subseteq C^2$$

whose relation is

$$(a,b) \ll (c,d) \equiv a \ll c \,\&\, d \ll b.$$

We recall some basic facts about ideal completions. Given an abstract basis $(C, \ll)$ as above, recall that an *ideal* is a subset of C that is a lower set and also a directed set. The collection of ideals ordered by inclusion is a directed complete poset called the ideal completion of $(C, \ll)$.

Let $\mathbf{I}C$ denote the ideal completion of the abstract basis $\mathrm{int}(C)$.

THEOREM 5.1 ([**16**]). *Let C be a countable dense subset of a globally hyperbolic spacetime $\mathcal{M}$ and $\ll\, =\, I^+$ be timelike causality. Then*

$$\max(\mathbf{I}C) \simeq \mathcal{M}$$

where the set of maximal elements has the relative Scott topology and $\mathcal{M}$ has the manifold topology.

In "ordering the order" I^+, taking its completion, and then the set of maximal elements, we recover spacetime by reasoning only about the causal relationships between a countable dense set of events. One objection to this might be that we begin from a *dense* set C, and then order theoretically recover the space $\mathcal{M}$ – but *dense* is a topological idea so we need to know the topology of $\mathcal{M}$ before we can recover it! But the denseness of C can be expressed in purely causal terms:

$$C \text{ dense} \equiv (\forall x, y \in \mathcal{M})(\exists z \in C)\, x \ll z \ll y.$$

Now the objection might be that we still have to reference $\mathcal{M}$. We too would like to not reference $\mathcal{M}$ at all. However, some global property needs to be assumed, either directly or indirectly, in order to reconstruct $\mathcal{M}$.

Theorem 5.1 is very different from results like "Let $\mathcal{M}$ be a certain spacetime with relation $\leq$. Then the interval topology is the manifold topology." Here we identify, in abstract terms, a process by which a countable set with a causality relation determines a space. The process is entirely order theoretic in nature, spacetime is not required to understand or execute it (i.e., if we put $C = \mathbb{Q}$ and $\ll\, =\, <$, then $\max(\mathbf{I}C) \simeq \mathbb{R}$). In this sense, our understanding of the relation between causality and the topology of spacetime is now explainable independently of geometry. Ideally, one would now like to know what constraints on C in general imply that max(**I**C) is a manifold. However, that is only to clarify the relationship with standard relativity; the process above may offer a more flexible definition of spacetime in general that is applicable to different physical situations – we have not ruled that possibility out.

Finally, let us mention that the category of globally hyperbolic posets[6] is in fact *naturally isomorphic* to a special category of domains called interval domains [**16**]. Thus, questions about spacetime can be converted to domain theoretic form, where we can use domain theory to answer them, and then translate the answers back to the language of physics (and vice-versa). It also implies that causality between events is equivalent to an order on *regions* of spacetime. Most importantly, it means that a globally hyperbolic spacetime with causality is equivalent to a structure **I**X whose origins are "discrete." This can be taken as the formal explanation for why spacetime can be reconstructed from a countable dense set of events in a purely order theoretic manner.

6. Time and measurement

A *domain* is a partially ordered set with intrinsic notions of completeness and approximation defined by the order. A *measurement* is a function μ that to each informative object x assigns a "measure" μx of the information content in x. In many cases, μx will be a number, but it need not be. Let us now define measurement precisely before discussing it further. We use the notation σ_D and σ_E to mean the Scott topology of D and E viewed as a collection of open sets.

DEFINITION 6.1. *A function* $f : D \to E$ *between posets is* Scott continuous *if the inverse image of a Scott open set in E is Scott open in D.*

Scott continuity can be characterized order theoretically: a function $f : D \to E$ between posets is Scott continuous iff f is monotone,

$$(\forall x, y \in D)\, x \sqsubseteq y \Rightarrow f(x) \sqsubseteq f(y),$$

and preserves directed suprema:

$$f(\bigsqcup S) = \bigsqcup f(S),$$

for all directed $S \subseteq D$ with a supremum. In particular, for the domain $[0, \infty)^*$ of non-negative reals in their opposite order, a Scott continuous function $\mu : D \to [0, \infty)^*$ will satisfy

(1) For all $x, y \in D$, $x \sqsubseteq y \Rightarrow \mu x \geq \mu y$, and
(2) If (x_n) is an increasing sequence in D, then

$$\mu\left(\bigsqcup_{n \geq 1} x_n\right) = \lim_{n \to \infty} \mu x_n$$

provided (x_n) has a supremum.

DEFINITION 6.2. *A Scott continuous map* $\mu : D \to E$ *between posets is said to* measure the content *of* $x \in D$ *if*

$$x \in U \Rightarrow (\exists \varepsilon \in \sigma_E)\, x \in \mu_\varepsilon(x) \subseteq U,$$

---

[6]The morphisms are monotone functions that are continuous in the interval topology.

whenever $U \in \sigma_D$ is Scott open and
$$\mu_\varepsilon(x) := \mu^{-1}(\varepsilon) \cap \downarrow x$$
are the elements ε close to x in content. The map μ measures X if it measures the content of each $x \in X$.

DEFINITION 6.3. *A measurement is a Scott continuous map* $\mu : D \to E$ *between posets that measures* $\ker \mu := \{x \in D : \mu x \in \max(E)\}$.

We often refer to μ as simply "measuring" $x \in D$ or as measuring $X \subseteq D$ when it measures each element of X. The case $E = [0, \infty)^*$, the set of non-negative reals in their dual order, is of particular interest in this paper: in this case, for $\varepsilon > 0$ and assuming $\mu x = 0$, we define
$$\mu_\varepsilon(x) := \mu_{[0,\varepsilon)}(x) = \{y \in D : y \sqsubseteq x \ \& \ \mu y < \varepsilon\}$$
and see that a Scott continuous $\mu : D \to [0, \infty)^*$ measures the content of $x \in D$ when
$$x \in U \Rightarrow (\exists \varepsilon > 0) \ x \in \mu_\varepsilon(x) \subseteq U$$
for all Scott open $U \subseteq D$. The map μ is then a measurement when it measures the content of its kernel $\ker(\mu) = \{x \in D : \mu x = 0\}$, the elements with no uncertainty. All such elements are maximal in the information order $\sqsubseteq$ on D. Let us now explain the intuition behind this idea on a continuous poset D.

The order on D defines a clear sense in which one object has 'more information' than another: a *qualitative* view of information content. The definition of measurement attempts to identify those monotone mappings μ which offer a *quantitative* measure of information content in the sense specified by the order. The essential point in the definition of measurement is that μ measure content in a manner that is consistent with the particular view offered by the order. There are plenty of monotone mappings that are not measurements – and while some of them may measure information content in *some other sense*, each sense must first be specified by a different information order. The definition of measurement is then a minimal test that a function μ must pass if we are to regard it as providing a measure of information content.

We now consider a few properties that measures of information content have which arbitrary monotone mappings in general need not have: qualities that make them 'different' from maps that are simply monotone. Other such properties may be found in [**13**].

THEOREM 6.4 ([**13**]). *Let* $\mu : D \to [0, \infty)^*$ *be a measurement on a continuous poset.*

(i) *If* $x \in \ker(\mu)$, *then* $x \in \max(D) = \{x \in D : \uparrow x = \{x\}\}$.
(ii) *If* μ *measures the content of* $y \in D$, *then*
$$(\forall x \in D) \ x \sqsubseteq y \ \& \ \mu x = \mu y \Rightarrow x = y.$$
(iii) *If* μ *measures* $X \subseteq D$, *then*
$$\{\uparrow \mu_\varepsilon(x) \cap X : x \in X, \varepsilon > 0\}$$
is a basis for the Scott topology on X.

Unfortunately, within the realm of physics, it is normally far from trivial to prove that a function is actually a measurement. Let us remedy this now by considering an easy lemma but one that has striking applications to measurements.

LEMMA 6.5. *For a sequence (x_n) in a compact Hausdorff space X, the following are equivalent:*

(i) *The sequence (x_n) converges to x,*
(ii) *For any convergent subsequence (x_{n_k}) of (x_n), we have $x_{n_k} \to x$.*

Proof. (ii) $\Rightarrow$ (i): if (x_n) does not converge to x, then there is an open set $U \subseteq X$ with $x \in U$ such that for all k there is $n_k \geq k$ with $x_{n_k} \notin U$. By compactness of X, (x_{n_k}) has a convergent subsequence (y_n). Because (y_n) is a subsequence of (x_n), we have $y_n \to x$ by (ii), so eventually $y_n \in U$, in contrast to $x_{n_k} \notin U$. $\square$

It is difficult to believe that such an easy lemma could be useful. But in fact:

THEOREM 6.6. *Let $\mu : D \to [0, \infty)^*$ be a strictly monotone, Scott continuous function defined on a poset D. If τ is a Hausdorff topology on D such that*

(i) *every Scott open set is τ open,*
(ii) *every sequence (x_n) in $\downarrow x$ with $\mu x_n \to \mu x$ is contained in some τ-compact $K \subseteq \downarrow x$,*
(iii) *the function μ is continuous from (D, τ) to $[0, \infty)$ with the Euclidean topology,*

then μ measures all of D.

Proof. Let $x_n \sqsubseteq x$ with $\mu x_n \to \mu x$. Take a compact set K with $x_n \in K \subseteq \downarrow x$. Let (x_{n_k}) be any convergent subsequence of (x_n). Let us write $x_{n_k} \to y$. Then since K is closed, $y \in K$ and hence $y \sqsubseteq x$. However, since the sequence $\mu x_n \to \mu x$, we know that $\mu x_{n_k} \to \mu x$. Since μ is continuous with respect to τ, we get

$$\mu y = \mu \left(\lim_{k \to \infty} x_{n_k} \right) = \lim_{k \to \infty} \mu x_{n_k} = \mu x$$

and thus by strict monotonicity, $x = y$. Then every convergent subsequence of (x_n) converges to x and all of this happens in the compact Hausdorff space K. Thus, $x_n \to x$ in (D, τ).

If μ does not measure the content of $x \in D$, then there is a Scott open set $U \subseteq D$ and a sequence $x_n \sqsubseteq x$ with $\mu x_n \to \mu x$ and $x_n \notin U$. By our above remarks, $x_n \to x$ in (D, τ), and since U is τ open, we have $x_n \in U$ for all but a finite number of n, which is a contradiction. $\square$

Notice that the proof above also shows that the previous result holds for maps of the form $\mu : D \to E$, where $E = \mathbb{R}$ or $E = \mathbb{R}^*$.

DEFINITION 6.7. *A global time function $t : \mathcal{M} \to \mathbb{R}$ on a globally hyperbolic spacetime $\mathcal{M}$ is a continuous function such that $x < y \Rightarrow t(x) < t(y)$ and $t^{-1}(r) = \Sigma$ is a Cauchy surface for $\mathcal{M}$, for each $r \in \mathbb{R}$.*

Because global time functions always exist on a globally hyperbolic spacetime, each such spacetime admits a natural measurement on the domain of spacetime intervals:

THEOREM 6.8. *For any global time function $t : \mathcal{M} \to \mathbb{R}$ on a globally hyperbolic spacetime, the function $\Delta t : \mathbf{I}(\mathcal{M}) \to [0, \infty)^*$ given by $\Delta t[a, b] = t(b) - t(a)$ measures all of $\mathbf{I}(\mathcal{M})$. It is a measurement with $\ker(\Delta t) = \max(\mathbf{I}(\mathcal{M}))$.*

Proof. The function Δt inherits its monotonicity from that of t; it is Scott continuous because t is continuous with respect to the manifold topology and directed suprema in $\mathbf{I}(\mathcal{M})$ are calculated using limits in the manifold topology. To prove that Δt measures $\mathbf{I}(\mathcal{M})$, we will show that t measures the continuous poset $(\mathcal{M}, \leq)$ and that it also measures $(\mathcal{M}, \leq^*)$, whose order $\leq^*$ is given by $x \leq^* y \equiv y \leq x$.

We apply the remark following Theorem 6.6 to $t : \mathcal{M} \to \mathbb{R}$ as follows. (i) The Scott topology is contained in the manifold topology. (ii) Given any sequence $x_n \leq x$ with $t(x_n) \to t(x)$, we have $x_n \in J^+(\Sigma) \cap J^-(x) \subseteq \downarrow\! x$ for some $\Sigma = t^{-1}(r)$, where r exists because $t(x_n)$ has a limit and the set $J^+(\Sigma) \cap J^-(x)$ is compact [**23**]. By the remark after Theorem 6.6, t measures $(\mathcal{M}, \leq)$. Because $(\mathcal{M}, \leq)$ is *bicontinuous*, $t : (\mathcal{M}, \leq^*) \to \mathbb{R}^*$ measures the continuous poset $(\mathcal{M}, \leq^*)$, again by the remark after Theorem 6.6. □

What is so interesting about this proof is that in order to apply Theorem 6.6, we not only need continuity, strict monotonicity and the connection between causal structure and topology, we also make use of the Cauchy surface Σ, the latter of which implies that spacetime has an initial value formulation. Another point of interest is that the same technique used here to prove Δt is a measurement, Theorem 6.6, has also been used to show the same about capacity on the domain of binary channels and entropy on the domains of classical and quantum states [**17**].

7. The Lorentz distance

The *Lorentz distance* on a globally hyperbolic spacetime $\mathcal{M}$ is the function $d : \mathcal{M} \times \mathcal{M} \to [0, \infty)$ given by

$$d(a,b) = \sup_{\pi_{ab}} \mathrm{len}(\pi_{ab})$$

where the sup is taken over all causal curves π_{ab} that join a to b; when $a \not\leq b$, $d(a,b) := 0$. By global hyperbolicity [**2**], the supremum in the definition of d is finite, yields the length of the maximum geodesic joining causally related events and the function d is continuous as a map from the manifold topology to the usual topology on $[0, \infty)$. Physically, $d(a, b)$ measures the amount of time recorded by a clock that travels from a to b when $a \leq b$. Thus, the Lorentz distance is determined by a map between domains of the form

$$d : \mathbf{I}(\mathcal{M}) \to [0, \infty)^* :: [a, b] \mapsto d(a, b)$$

and for the remainder of this paper we shall regard it as such. Crucially, $d[a, b] > 0$ iff $a \ll b$.

LEMMA 7.1. *The function* $d : \mathbf{I}(\mathcal{M}) \to [0, \infty)^*$ *is Scott continuous.*

Proof. Monotonicity of d: First, the "reverse triangle inequality"

$$a \leq b \leq c \Rightarrow d[a, c] \geq d[a, b] + d[b, c]$$

holds since d is defined as a sup and $\mathrm{len}(\pi_{ab} + \pi_{bc}) = \mathrm{len}(\pi_{ab}) + \mathrm{len}(\pi_{bc})$; here the notation $\pi_{ab} + \pi_{bc}$ means the causal curve obtained by joining π_{ab} and π_{bc} at b. Of course, such a joined curve may not be smooth but Penrose has shown [**19**] that one can always "smooth out this curve" with an arbitrarily small change in length.

We now apply it twice as follows: given $[a, b] \sqsubseteq [c, d]$, we have $a \leq c \leq d \leq b$, so

$$d[a, b] \geq d[a, c] + d[c, b] \geq d[a, c] + (d[c, d] + d[d, b]) \geq d[c, d]$$

which proves monotonicity.

Scott continuity of d: By the separability of $\mathcal{M}$, $\mathbf{I}(\mathcal{M})$ is ω-continuous, so it is enough to consider an increasing sequence (x_n) of intervals $x_n = [a_n, b_n]$. Then we have
$$d\left(\bigsqcup x_n\right) = d[\lim a_n, \lim b_n] = \lim d[a_n, b_n] = \lim d(x_n)$$
where the second to last equality uses the continuity of d as a map $\mathcal{M} \times \mathcal{M} \to [0, \infty)$ given in [**2**]. □

We now give a "new" definition of the interval topology which applies to any continuous poset. For bicontinuous posets it is equivalent to the old definition but this new version applies more generally.

DEFINITION 7.2. *The* interval topology *on a continuous poset P exists when sets of the form*
$$(a,b) = \{x \in P : a \ll x \ll b\} \quad \& \quad {\uparrow}x = \{y \in P : x \ll y\}$$
form a basis for a topology on P.

A function between continuous posets is *interval continuous* when each poset has an interval topology and the inverse image of an interval open set is interval open. By the bicontinuity of $\mathcal{M}$, the interval topology on $\mathbf{I}(\mathcal{M})$ exists, so we can consider interval continuity for functions $\mathbf{I}(\mathcal{M}) \to [0, \infty)^*$.

LEMMA 7.3.
  (i) *For all $x, y \in \mathbf{I}(\mathcal{M})$, if $x \ll y$, then $dx > dy$, and*
  (ii) *For all $x \in \mathbf{I}(\mathcal{M})$, if $dx > r \geq 0$, then there is $y \in \mathbf{I}(\mathcal{M})$ with $x \ll y$ and $dy = r$.*

Proof. (i) Given $x = [a,b] \ll [c,d] = y$, we know $a \ll b \ll c \ll d$, so as in the proof of Lemma 7.1, we again apply the reverse triangle inequality twice, this time noting that all distances involved are positive since $d[s,t] > 0$ iff $s \ll t$.

(ii) First, each interval $x = [a, b]$ is a path connected subset of $\mathcal{M}$, since any pair of points $s, t \in [a, b]$ can be joined by a continuous curve that first moves forward in time from s to a and then backward in time from a to t. In particular, $x = [a, b]$ is connected.

Because $dx > 0$, $a \ll b$, and interpolation gives $p \in x$ with $a \ll p \ll b$. The set ${\uparrow}p \cap {\downarrow}b$ is directed with sup b, while ${\downarrow}p \cap {\uparrow}a$ is filtered with inf a, so $x = [a, b]$ contains increasing and decreasing sequences of approximations with limits b and a respectively. By the continuity of $d : \mathcal{M} \times \mathcal{M} \to [0, \infty)$, there is an interval w with $p \in w$ such that
$$x \ll w \quad \& \quad dx > dw > r$$
where we make use of (i) to ensure $dx > dw$.

If $r = 0$, then set $y = [p, p]$ and the proof is finished. If $r > 0$, then the restriction of $d : \mathcal{M} \times \mathcal{M} \to [0, \infty)$ to the *connected set* $w \times w$ yields a continuous function that assumes the value dw and the value $0 = d[p, p]$. Thus it must also assume $r \in [0, dw]$, which gives $(c, d) \in w \times w$ with $d(c, d) = r$. Since $r > 0$, $c \ll d$, so let $y = [c, d] \in \mathbf{I}(\mathcal{M})$. Then y is the desired interval: we have $dy = r$, while $x \ll w \sqsubseteq y$ gives $x \ll y$. □

Notice that (i) above says that d preserves the way-below relation between domains, while (ii) is a kind of converse to (i). Together with Scott continuity they yield the following:

THEOREM 7.4. *The Lorentz distance* $d : \mathbf{I}(\mathcal{M}) \to [0,\infty)^*$ *is not only Scott continuous, it is also interval continuous. Thus, it does not measure* $\mathbf{I}(\mathcal{M})$ *at any point of* $\ker(d)$.

Proof. Let U be a basic interval open set in $[0,\infty)^*$. If U is Scott open, then by Lemma 7.1, $d^{-1}(U)$ is Scott open and the proof is finished. If $U = (s,t)$, then consider an interval $x \in d^{-1}(U)$, for which we have $s < dx < t$. By Scott continuity of d, there is $a \ll x$ with $da < t$. By Lemma 7.3(ii), there is b with $x \ll b$ and $db = s$. We claim $x \in (a,b) \subseteq d^{-1}(s,t)$.

If $y \in (a,b)$, then $a \ll y \ll b$, so by Lemma 7.3(i), $da > dy > db = s$, which gives $t > da > dy > s$, and finally $y \in d^{-1}(s,t)$, finishing the proof. □

That the Lorentz distance is not a measurement is a direct consequence of the fact that a clock travelling at the speed of light records no time as having elapsed i.e. the set of null intervals is

$$\ker(d) \setminus \max(\mathbf{I}(\mathcal{M})) \neq \emptyset$$

but measurements always have the property that $\mu x = 0$ implies $x \in \max(D)$ (Theorem 6.4).

In fact, no interval continuous function $\mu : \mathbf{I}(\mathcal{M}) \to [0,\infty)^*$ can be a measurement: by interval continuity, $\mu x = 0$ for any x with $\uparrow x = \emptyset$. Just like the Lorentz distance, an interval continuous μ will also assign 0 to "null intervals." In this way, we see that interval continuity captures an essential aspect of the Lorentz distance. In addition, since Δt is a measurement, it cannot be interval continuous. This provides a surprising *topological* distinction between the Newtonian and relativistic concepts of time: d is interval continuous, Δt is not. Put another way, Δt can be used to reconstruct the *topology* of spacetime (Theorem 6.4(iii)), while d is used to reconstruct its *geometry*.

8. Spacetime geometry from a countable causal set

Let us return now to the reconstruction of spacetime (Section 5) from a countable dense set $(C, \ll)$. Specifically, we take the rounded ideal completion[7] $\mathbf{I}(C)$ of the abstract basis of *intervals*

$$\mathrm{int}(C) = \{(a,b) : a \ll b\} = \ll \,\subseteq C^2$$

whose relation is

$$(a,b) \ll (c,d) \equiv a \ll c \ \& \ d \ll b.$$

We are then able to recover spacetime as

$$\max(\mathbf{I}C) \simeq \mathcal{M}$$

where the set of maximal elements have the Scott topology. Let us now suppose that in addition to $\mathrm{int}(C)$ that we also begin with a countable collection of numbers l_{ab} chosen one for each $(a,b) \in \mathrm{int}(C)$ in such a way that the map

$$\mathrm{int}(C) \to [0,\infty)^* :: (a,b) \mapsto l_{ab}$$

---

[7] An ideal I is a directed downward-closed set, it is *rounded* if for any $x \in I$ there is a $y \in I$ with $x \ll y$.

is monotone. Then in the process of reconstructing spacetime, we can also construct the Scott continuous function $d : \mathbf{IC} \to [0, \infty)^*$ given by

$$d(x) = \inf\{l_{ab} : (a, b) \ll x\}.$$

In the event that the countable number of l_{ab} chosen are the Lorentz distances $l_{ab} = d[a, b]$, then the function d constructed above yields the Lorentz distance for any spacetime interval, the reason being that both are Scott continuous and are equal on a basis of the domain.

Thus, from a countable dense set of events and a countable set of distances, we can reconstruct the spacetime manifold together with its geometry in a purely order theoretic manner.

9. Conclusions

We have seen the following ideas in this paper:

(1) how to reconstruct the spacetime topology from the causal structure using purely order-theoretic ideas,
(2) an abstract order-theoretic definition of global hyperbolicity,
(3) that one can reconstruct spacetime, meaning its topology and *geometry*, from a countable dense subset,
(4) an equivalence of categories between a new category of interval domains and the category of globally hyperbolic posets.
(5) a topological distinction between Newtonian and relativistic notions of time.

Acknowledgments

The second author would like to thank the Office of Naval Research for supporting this research.

References

[1] S. Abramsky and A. Jung. Domain theory. In T. S. E. Maibaum S. Abramsky, D. M. Gabbay, editor, *Handbook of Logic in Computer Science, vol. III*. Oxford University Press, 1994.
[2] J. Beem, P. Ehrlich and K. Easley. *Global Lorentzian Geometry*. CRC Press, Second Edition, 1996.
[3] S. Chandrasekhar. The maximum mass of ideal white dwarfs. *Astrophysical Journal*, 74:81–82, 1931.
[4] Jürgen Ehlers and Felix Pirani and Alfred Schild. The geometry of free fall and light propagation. In Lochlainn O'Raiferteagh, editor, *General Relativity: Papers in Honour of J.L. Synge*. Pages 64-84, Clarendon Press 1972.
[5] Lisbeth Fajstrup. Loops, ditopology and deadlocks: Geometry and concurrency. *Math. Structures Comput. Sci.*, 10(4):459–480, 2000.
[6] Eric Goubault, Lisbeth Fajstrup, and Martin Raussen. Algebraic topology and concurrency. Department of Mathematical Sciences RR-98-2008, Aalborg University, 1998. Presented at MFCS 1998 London.
[7] G.Gierz, K.H.Hoffmann, K.Keimel, J.D.Lawson, M.Mislove, and D.S.Scott. *Continuous lattices and domains*. Number 93 in Encyclopedia of Mathematics and its Applications. Cambridge University Press, 2003.
[8] S.W. Hawking and G.F.R. Ellis. *The Large Scale Structure of Space-time*. Cambridge Monographs on Mathematical Physics. Cambridge University Press, 1973.
[9] S. W. Hawking and R. Penrose. *The Nature of Space and Time*. Princeton University Press, 1996.
[10] Peter Johnstone, *Stone Spaces*, Cambridge Studies in Advanced Mathematics, Volume 3, Cambridge University Press, 1982.

[11] P. S. Joshi. *Global aspects in gravitation and cosmology*. International Series of Monographs on Physics 87. Oxford Science Publications, 1993.

[12] David Malement. The class of continuous timelike curves determines the topology of spacetime. *J. Math. Phys.*, 18(7):1399–1404, 1977.

[13] Keye Martin. *A foundation for computation*. PhD thesis, Department of Mathematics, Tulane University, 2000.

[14] Keye Martin. The measurement process in domain theory. In *27th International Colloquium on Automata, Languages and Programming (ICALP'00)*, number 1853 in Lecture Notes In Computer Science, pages 116–126. Springer-Verlag, 2000.

[15] Keye Martin. *Compactness of the space of causal curves*. Journal of Classical and Quantum Gravity, volume 23, page 1241, 2006.

[16] Keye Martin and Prakash Panangaden. *A domain of spacetime intervals in general relativity*. Communications in Mathematical Physics, 267(3):563–586, November 2006.

[17] K. Martin and P. Panangaden. *A technique for verifying measurements*. Electronic Notes in Theoretical Computer Science, Vol. 218, p. 261–273, Elsevier Science, 2008.

[18] Roger Penrose. Gravitational collapse and space-time singularities. *Phys. Rev. Lett.*, 14:57–59, 1965.

[19] Roger Penrose. *Techniques of differential topology in relativity*. Society for Industrial and Applied Mathematics, 1972.

[20] Dana Scott. Outline of a mathematical theory of computation. Technical Monograph PRG-2, Oxford University Computing Laboratory, 1970.

[21] R. Sorkin. Spacetime and causal sets. In J. D'Olivo et. al., editor, *Relativity and Gravitation: Classical and Quantum*. World Scientific, 1991.

[22] R.D. Sorkin and E. Woolgar. A causal order for spacetimes with c^0 Lorentzian metrics: Proof of compactness of the space of causal curves. *Classical and Quantum Gravity*, 13:1971–1994, 1996.

[23] R.M. Wald. *General relativity*. The University of Chicago Press, 1984.

CENTER FOR HIGH ASSURANCE COMPUTER SYSTEMS (CODE 5540), NAVAL RESEARCH LABORATORY, WASHINGTON D.C. 20375, USA
E-mail address: `kmartin@itd.nrl.navy.mil`

SCHOOL OF COMPUTER SCIENCE, MCGILL UNIVERSITY, MONTREAL, QUEBEC H3A 2A7, CANADA
E-mail address: `prakash@cs.mcgill.ca`

Geometry of abstraction in quantum computation

Dusko Pavlovic

ABSTRACT. Quantum algorithms are sequences of abstract operations, performed on non-existent computers. They are in obvious need of categorical semantics. We present some steps in this direction, following earlier contributions of Abramsky, Coecke and Selinger. In particular, we analyze function abstraction in quantum computation, which turns out to characterize its classical interfaces. Intuitively, *classical data can be recognized as just those data that can be manipulated using variables*, i.e. copied, deleted, and abstracted over. A categorical framework of polynomial extensions provides a convenient language for specifying quantum algorithms, with a clearly distinguished classical fragment, familiar from functional programming.

As a running example, we reconstruct Simon's algorithm, which is a simple predecessor of Shor's quantum algorithms for factoring and discrete logarithms. The abstract specification in the framework of polynomial categories displays some of the fundamental program transformations involved in developing quantum algorithms, and points to the computational resources, whether quantum or classical, which are necessary for the various parts of the execution. The relevant resources are characterized as categorical structures. They are normally supported by the standard Hilbert space model of quantum mechanics, but in some cases they can also be found in other, nonstandard models. We conclude the paper by sketching an implementation of Simon's algorithm using just abelian groups and relations.

1. Introduction

1.1. What do quantum programmers do? They do a variety of things, but there is a "design pattern" that they often follow, based on the *Hidden Subgroup Problem (HSP)* [**36, 38**, sec. 5.4]. Shor's factoring and discrete log algorithms [**51**] are examples of this pattern, as well as Hallgren's algorithm for the Pell equation [**24**]. They all provide an exponential speedup with respect to the best known classical algorithms. The simplest member of the family is Simon's algorithm for period finding [**52**], which we use as the running example. The other HSP algorithms only differ in "domain specific" details, but yield to the same semantics.

The input for Simon's algorithm is an arbitrary function $f: \mathbb{Z}_2^m \to \mathbb{Z}_2^n$, where $(\mathbb{Z}_2, \oplus, 0)$ is the group with two elements, and $\oplus$ is the "exclusive or" operation. The task is to find the period of f, if it exists, i.e. a bitstring $c \in \mathbb{Z}_2^m$ such that

---

2010 *Mathematics Subject Classification.* Primary 81P68, 18D10, 18D35; Secondary 18C50, 68Q55, 68Q12.

Supported by ONR.

$f(x \oplus c) = f(x)$ for all $x \in \mathbb{Z}_2^m$. For simplicity, let us assume that there is exactly one such c, as the discussion of the other cases does not bring in anything essential.

Since f is arbitrary, one cannot ascertain that a bitstring c is a solution without computing the value of $f(x)$ for every $x \in \mathbb{Z}_2^m$. But a quantum computer can compute all such values at once! This is called *quantum parallelism*, and is one of the first things explained to quantum programmers' apprentices [**38**, Sec. 1.4.2][1].

Mathematically speaking, the main capability of a quantum computer is that it can evaluate unitary operators. In fact, the quantum programmers themselves view their computers as *quantum oracles*, i.e. black boxes that compute unitaries. If the inputs of a function are represented as the basis vectors of a Hilbert space, and the function itself is captured as a unitary operator over it, then the quantum computer can compute all values of the function at once, by evaluating this unitary over a suitably generated combination of the basis vectors. Simon's algorithm shows how to extract the information about the period of the function from the projections of the resulting mixture.

But how do we represent a function $f : \mathbb{Z}_2^m \to \mathbb{Z}_2^n$ by a unitary operator? For an involutive function $g : B \to B$, the answer is easy: define $U_g : \mathbb{C}^B \longrightarrow \mathbb{C}^B$ by setting $U_g|b\rangle = |g(b)\rangle$, where $|b\rangle \in \mathbb{C}^B$ are the basis vectors indexed by $b \in B$. The fact that U_g is unitary follows from $g \circ g = \mathrm{id}_B$. For a general $f : \mathbb{Z}_2^m \to \mathbb{Z}_2^n$, first define a corresponding involution f', and then extract the unitary U_f:

(1.1)
$$\frac{f : \mathbb{Z}_2^m \longrightarrow \mathbb{Z}_2^n : x \mapsto f(x)}{\dfrac{f' : \mathbb{Z}_2^{m+n} \longrightarrow \mathbb{Z}_2^{m+n} : x, y \mapsto x, y \oplus f(x)}{U_f : \mathbb{C}^{\mathbb{Z}_2^{m+n}} \longrightarrow \mathbb{C}^{\mathbb{Z}_2^{m+n}} : |x, y\rangle \mapsto |x, y \oplus f(x)\rangle}}$$

where the basis vectors $|x, y\rangle$ of $\mathbb{C}^{\mathbb{Z}_2^{m+n}}$ are indexed by the bitstrings x of length m concatenated with the bitstrings y of length n. The values of the function f are recovered from $U_f|x, 0\rangle = |x, f(x)\rangle$.

The other conceptual component of Simon's algorithm, and of all HSP-algorithms, is a standard application of transform theory [**55**]: transform the inputs into another domain, where the computation is easier, compute the outputs there, and then transform them back[2]. In our special case, U_f is thus precomposed and postcomposed with a suitable version of the Fourrier transform, which for $\mathbb{Z}_2$ boils down to the Hadamard-Walsh tranform $H^{\otimes m}|z\rangle = \sum_{x \in \mathbb{Z}_2^m}(-1)^{x \cdot z}|x\rangle$. Here $x \cdot z = \sum_{i=1}^m x_i z_i$ denotes the inner product in $\mathbb{Z}_2^m$, and we ignore the renormalizing factor $2^{-\frac{m}{2}}$. This transform is applied to the first m arguments of U_f, to generate the desired superposition of all inputs of f. The quantum computer thus computes the following

---

[1]The gentle reader will hopefully appreciate that the explanations given to the apprentices tend to be oversimplified. A more precise explanation of quantum parallelism would have to mention that only a limited amount of information about the superposition of the values $f(x)$ can be extracted [**54**, Sec. 6.2]

[2]E.g., Laplace's transform maps a differential equation into a polynomial equation over the field, generated by the convolution ring in which the original equation was stated [**44**]. The solutions of the polynomial equation are then mapped back by the inverse Laplace transform.

vector[3]:

$$\begin{aligned} Simon &= (H^{\otimes m} \otimes \mathrm{id})U_f(H^{\otimes m} \otimes \mathrm{id})\,|0,0\rangle \\ &= \sum_{z,x\in\mathbb{Z}_2^m} (-1)^{x\cdot z}|z,f(x)\rangle \end{aligned}$$

When we measure the first component of this vector, it collapses to a single $|z\rangle$, i.e. we get $\gamma_z = |z\rangle \otimes \sum_{x\in\mathbb{Z}_2^m}(-1)^{x\cdot z}|f(x)\rangle$. By assumption, there is exactly one $c \in \mathbb{Z}_2^m$ such that $f(x \oplus c) = f(x)$ holds for all x. The coefficient of each of the basis vectors $|z,f(x)\rangle = |z,f(x\oplus c)\rangle$ is thus $\gamma_z^x = (-1)^{x\cdot z} + (-1)^{(x\oplus c)\cdot z} = (-1)^{x\cdot z}(1 + (-1)^{c\cdot z})$. It follows that $(\forall x \in \mathbb{Z}_2^m.\ \gamma_z^x \neq 0) \iff c \cdot z = 0$. Each time that we run the algorithm, we can thus extract a linear equation in c. After m runs, we can thus compute c. (The probability that at some step $k \leq m$ we may get an equation dependent on the previous ones is 0, because z are chosen randomly, and the measure of every proper linear subspace of $\mathbb{Z}_2^m$ is 0.) On the other hand, in order to convince ourselves classically that c is the period of f, we should to compute all values f, which requires 2^m steps, since f is an arbitrary function.

The core of Shor's factoring algorithm follows the same pattern, adapted for $f : \mathbb{Z}_k \to \mathbb{Z}_k$, where $f(x) = a^x \mod k$. The factored integer is k, and a is randomly selected to be tested for common factors with it, which can be derived by finding a period of f.

Remark. Cryptographers will recognize the first step of (1.1), transforming a boolean function into an involution, as a Feistel transformation, ubiquitous in block cyphers. Category theorists will recognize the second step of of (1.1) as a part of the obvious hom-embedding of sets and (partial) bijections into vector spaces and matrices, where X-dimensional vectors are viewed as functions $X \longrightarrow \mathbb{C}$, i.e. the vector spaces are in the form $\mathbb{C}^X$. Note that this embedding is not functorial for the functions in general. The relational composition is mapped to the matrix composition only for partial bijections. This is mentioned, e.g., in [**7**, Sec. 3].

1.2. Denoting classical data in quantum programs. A program generally describes a family of computations over a family of input data. The various input data to be computed with are denoted by variables. E.g., the polynomial $x^2 + x + 2$ can be construed as a program, describing the family of computations that can be performed for the various values of x. It is tacitly assumed that the possible values of x can be copied, so that one copy can be substituted for each occurrence of x in the polynomial $x^2 + x$; and that these data can also be deleted, if the polynomial is just 2, and x does not occur in it.

The first problem with quantum programming is that quantum data cannot be manipulated in this way: it is a fundamental property of quantum states that they generally cannot be copied [**57, 20**], or even deleted [**39, 3**]. So how do we write quantum programs? In particular, given a program $f(x)$ for a function f, what kind of a program transformation leads to the quantum program $U_f|x,y\rangle$, that we used to specify the unitary U_f above? The answer to this question is given in Sections 3 and 4. It turns out that the needed copying and deleting operations are closely related with the variable abstraction operations, on which λ-calculus and functional programming are based [**11, 6, 33, 46**].

---

[3]We ignore the renormalizing factors throughout.

On the other hand, copying, deleting and abstraction capabilities can be viewed as the characteristics of classical computation. In a quantum computer, a structure that supports copying, deleting and abstraction can be construed as its classical interface. This is what we call a *classical structure*. An early analysis of this structure was in [**12**]. In the meantime, there are several versions, and many applications [**14, 41, 17**]. In recent work, Coecke [**16**] uses the term *basis structures* for the same concept, because classical structures over finitely dimensional Hilbert spaces precisely correspond to choices of a basis [**19**], and can be viewed as a purely categorical, element-free version of this notion. While the simple basis intuitions are attractive, I stick here with the original terminology. One reason is that the correspondence of classical structures and the induced bases is not always as simple as it is in the category of finitely dimensional Hilbert spaces [**41**], and it is useful to keep the distinction. A more important reason is that classical structures express the fact that *classicality is relative* as an algebraic structure. The fact that classical data with respect to one classical structure may be entangled with respect to another one is the fundamental feature of quantum computation. This lies at the heart of the developments in [**15, 18**], where classical structures provide an algebraic framework for studying complementarity.

1.3. Outline of the paper. Section 2 provides a brief summary of the categorical prerequisites, notations and terminology. It gets us quickly *in medias res*, but the reader interested in is referred to [**13, 50**]. Section 3 recounts the relevant part of the story of [**40**] in the language of strings: the polynomial constructions, extended by Lambek from rings to (closed) cartesian categories [**32**], are further extended to (closed) monoidal categories. The upshot is that this tells us what will an abstraction calculus (such as the typed λ-calculus [**11**], which lies at the foundation of classical computation is based) have to look like in a quantum world:

- Thm. 3.4 tells that the abstraction operations in a monoidal category exactly correspond to commutative monoids in it.

Section 4 expands the story about abstraction from monoidal to dagger-monoidal categories:

- Thm. 4.3 and Corollary 4.5 tell that the abstraction operations in a dagger-monoidal category exactly correspond to (commutative) Frobenius algebras in it;
- Lemma 4.8 and Corollary 4.10 tell that the *normalized* abstraction operations in dagger-monoidal categories exactly correspond classical structures.

The upshot here is that classical structures thus arise from the need for abstraction over the classical data in the quantum world. This can be viewed as a rational reconstruction of classical structures (although it was actually the guiding idea behind their construction in [**12**]). The other way around, classical data can be recognized in the quantum world as those data that allow abstraction.

Section 5 provides a very brief overview of notion of complementarity with respect to a classical structure, and of the transformations to complementary bases. Complementarity is studied in [**14**] in much more detail. We recast in the present framework only as much as is needed to capture the transformations used in Simon's algorithm.

The final step of the described algorithm pattern, the measurement, is modeled in Section 6. This section refines and streamlines the ideas about measurements proposed in [**12**]. This categorical view captures not only the measurements as presented in the standard Hilbert space model, but it also provides interpretations in non-standard models. We spell out a relational interpretation, based on [**16, 41**]. In particular, Simon's algorithm turns out to have an effective *relational* implementation, assuming that a relational oracle effectively performs computations in an abelian group.

2. Preliminaries

2.1. Monoidal categories. We assume that the reader has some understanding of the basic categorical concepts and terminology [**37**], and work with symmetric monoidal categories $(\mathcal{C}, \otimes, I)$ [**27, 25**]. This is largely standard material, surveyed e.g. in [**53**].

2.1.1. *Strictness assumption.* For simplicity, and without loss of generality, we tacitly assume that each of our monoidal categories is *strictly associative and unitary*, i.e. that the objects form a monoid in the usual sense. This causes no loss of generality because every monoidal category is equivalent to a strictly associative and unitary one, along a monoidal equivalence. But note that the tensor symmetry cannot be "strictified" without essentially changing the category; the canonical isomorphisms $A \otimes B \xrightarrow{c} B \otimes A$ are thus generally *not* identities.

On the other hand, just like the tensors, we strictify functors: a *monoidal* functor F is always assumed to be strict, i.e. it preserves the monoidal structure on the nose: $F(A \otimes B) = FA \otimes FB$ and $FI = I$.

The arrows from I are sometimes called *vectors*, or *elements*. The abstract "vector spaces" are thus written $\mathcal{C}(X) = \mathcal{C}(I, X)$. When confusion is unlikely, we elide the tensor symbol and write XAf instead of $X \otimes A \otimes f$.

2.1.2. *String diagrams.* Calculations in monoidal categories are supported by a simple and intuitive graphical language: the string diagrams such as Figure 1. This language has its roots in Penrose's diagrammatic notation [**45**], and it has been formally developed in categorical *coherence theory*, and in particular in Joyal and Street's *geometry of tensor calculus* [**25**]. The objects are drawn as strings, and the morphisms as boxes attached on these strings. One can think that the information flows through the strings, and is processed in the boxes. A direction of this flow is chosen by convenience. We shall assume that the information flows up, so that the strings at the bottom of a box denote the domain of the corresponding morphism; the threads at the top the codomain. Drawing the strings A and B next to each other represents $A \otimes B$; similarly with the boxes. Drawing a thread from one box to another denotes the composition of the corresponding morphisms.

One of the salient features of this notation is that the associativity is implicit, and automatic, both of the tensor and of the composition. The tensor symmetry $c : B \otimes D \longrightarrow D \otimes B$ is denoted above by a circle. The circle is usually omitted, so that symmetry boils down to crossing the strings. The identity morphisms are the "invisible boxes", that can be placed on any thread. The tensor unit I is the "invisible thread", that can be added to any diagram. This means that a box representing a vector $a \in \mathcal{C}(I, AB)$ does not have any visible threads coming in from below. This is often emphasized by reducing the bottom of such a box to a point: e.g., the vector $I \xrightarrow{a} AB$ is denoted by a triangle. The box representing a

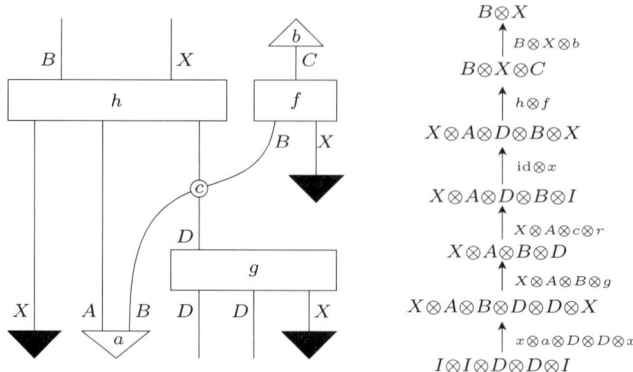

FIGURE 1. A String Diagram

covector $b \in \mathcal{C}(C, I)$ does not have any visible threads coming out, and boils down to a triangle pointing up. The black triangles denote the vector indeterminates $I \xrightarrow{x} X$, freely adjoined to monoidal categories to form polynomials. Such polynomial constructions will be discussed in Sec. 3.

2.1.3. *Monoids and comonoids.* A monoid in a monoidal category is a pair of arrows $X \otimes X \xrightarrow{\nabla} X \xleftarrow{\bot} I$ such that

$$\nabla \circ (\nabla \otimes X) = \nabla \circ (X \otimes \nabla)$$
$$\nabla \circ (\bot \otimes X) = \nabla \circ (X \otimes \bot) = \mathrm{id}_X$$

When the tensor is the cartesian product, this captures the usual notion of monoid.

A comonoid in a monoidal category is dual to a monoid: it is a pair of arrows $X \otimes X \xleftarrow{\Delta} X \xrightarrow{\top} I$ such that

$$(\Delta \otimes X) \circ \Delta = (X \otimes \Delta) \circ \Delta$$
$$(\top \otimes X) \circ \Delta = (X \otimes \top) \circ \Delta = \mathrm{id}_X$$

In string diagrams, we draw the monoid evaluations as trapezoids pointing up, whereas their units are little triangles pointing down. The comonoids are represented by the trapezoids and the little triangles in the opposite directions. E.g., the comonoid laws correspond to the following graph transformations

A monoid is *commutative* if $\nabla \circ c_{XX} = \nabla$. A comonoid is commutative if $c_{XX} \circ \Delta = \Delta$. In string diagrams, this means that the value of the output of ∇ does not change if the strings that come into it cross; and that the output of Δ does not change if the strings coming out of it cross.

2.1.4. *Cartesian categories.* A monoidal category $(\mathcal{C}, \otimes, I)$ is *cartesian* when it comes with natural transformations

$$X \otimes X \xleftarrow{\delta_X} X \xrightarrow{!_X} I$$

which make every object X into a comonoid. The naturality of this structure means that every morphism $X \xrightarrow{f} Y$ in $\mathcal{C}$ is a comonoid homomorphism. It is easy to see that this makes the tensor $X \otimes Y$ into a product $X \otimes Y$, such that any pair of arrows $A \xrightarrow{g} X$ and $A \xrightarrow{h} Y$ corresponds to a unique arrow $A \xrightarrow{\langle g,h \rangle} A \times B$, and the tensor unit I into the final object 1, with a unique arrow from each object. Cartesian structure is thus written in the form $(\mathcal{C}, \times, 1)$.

2.1.5. *Monads and comonads.* A *monad* on a category $\mathcal{C}$ can be defined as a functor $T : \mathcal{C} \longrightarrow \mathcal{C}$ together with a monoid structure $TT \xrightarrow{m} T \xleftarrow{h} \mathrm{Id}$ in the category of endofunctors on $\mathcal{C}$. With the corresponding monoid homomorphisms, monads form a category on their own [8]. Dually, *comonads* on $\mathcal{C}$ can be defined as comonoids in the category of endofunctors over $\mathcal{C}$, and accomodate similar developments.

The categories of algebras for a monad and coalgebras for a comonad, and in particular the Kleisli and the Eilenberg-Moore constructions that will be used below, are presented in detail in [**37**, **8**], and in many other books.

The following observation is the starting point for most of the constructions in this paper. The proof is left as an easy exercise.

PROPOSITION 2.1. *Every (co)monoid X in a monoidal category $\mathcal{C}$ induces a (co)monad $X \otimes (-) : \mathcal{C} \longrightarrow \mathcal{C}$. The corresponding Kleisli category $\mathcal{C}_{[X]}$ is monoidal if and only if the (co)monoid X is commutative.*

More precisely, the category of monoids in a monoidal category $\mathcal{C}$ is equivalent with the category of monads T on $\mathcal{C}$ such that $T(A \otimes B) = T(A) \otimes B$ and moreover $h_B = h_I \otimes B$ and $m_B = m_I \otimes B$ hold for all $A, B \in \mathcal{C}$. The dual statement holds for comonoids and comonads.

2.1.6. *Convolution and representation.* Any monoid $(X, \nabla, \bot)$ in a monoidal category $(\mathcal{C}, \otimes, I)$ induces the ordinary monoid $(\mathcal{C}(X), \bullet, \bot)$, whose operation

(2.1) $$a \bullet b \ = \ \nabla \circ (a \otimes b)$$

is often called *convolution*. A Cayley representation (or Yoneda embedding) of the monoid $(X, \nabla, \bot)$ is a map

(2.2) $$\begin{aligned} \widehat{(-)} : \mathcal{C}(X) &\longrightarrow \mathcal{C}(X, X) \\ \left(I \xrightarrow{a} X \right) &\longmapsto \left(X \xrightarrow{a \otimes X} X \otimes X \xrightarrow{\nabla} X \right) \end{aligned}$$

furthermore represents the vectors $a \in \mathcal{C}(X)$ as endomorphisms $\widehat{a} \in \mathcal{C}(X, X)$.

LEMMA 2.2. *(Cayley, Yoneda) The Cayley representation is a monoid isomorphism between the convolution monoid $(\mathcal{C}(X), \bullet, \bot)$ and the monoid $(\mathrm{Nat}(X, X), \circ, \mathrm{id}_X)$ of natural endomorphisms*

$$\mathrm{Nat}(X, X) \ = \ \{ f \in \mathcal{C}(X, X) \mid \forall ab \in \mathcal{C}(X).\ f \circ (a \bullet b) = (f \circ a) \bullet b \}$$

A comonoid structure on X induces a convolution monoid on $\mathcal{C}(X, I)$, with $c \bullet d = (c \otimes d) \circ \Delta$, and with a similar Cayley representation. In general, a convolution monoid can be defined over any hom-set $\mathcal{C}(X, Y)$, where X is a comonoid and Y a monoid, by setting $f \bullet g = \nabla_Y \circ (f \otimes g) \circ \Delta_X$.

2.1.7. *Scalars.* In general, the canonical isomorphism $I \otimes I \cong I$ makes the tensor unit I of $\mathcal{C}$ into a commutative monoid and comonoid, the tensor associativity is the associativity law of this (co)monoid. Our strictness assumption 2.1.1 makes the evaluation maps of these structures into identities. The tensor commutativity makes the (co)monoid commutative. The convolution monoid $(\mathcal{C}(I, I), \bullet, \mathrm{id}_I)$ is the abstract *scalar algebra* of the monoidal category $\mathcal{C}$. In general, the coherence conditions imply that there is only one monoid structure on I, hence $s \bullet t = s \circ t$ holds for all scalars $s, t \in \mathcal{C}(I, I)$. The strictness assumption implies even $s \otimes t = s \bullet t = s \circ t$.

Abusing notation, the scalar action $\bullet : \mathcal{C}(I, I) \times \mathcal{C}(A, B) \longrightarrow \mathcal{C}(A, B)$ is defined by $s \bullet f = s \otimes f$. If the tensor unit I is not strict, then $s \otimes f$ needs to be precomposed by $A \cong I \otimes A$ and postcomposed by $I \otimes B \cong B$.

2.2. Duals and daggers.

The abstract duality structures were first studied by Kelly and Laplaza in *compact closed categories* [28]. Abramsky with collaborators explained their computational content [4, 43], and recognized their role in abstract quantum mechanics [1]. Selinger clarified their subtle interactions with the daggers [48]. These rich structures allow a very succinct presentation.

Terminology: ioofs and embeddings. The daggers and many other categorical constructions lead to the evil Identity-On-the-Objects-Functors. I call them *ioofs*. If the reader finds this abbreviation objectionable, she is welcome to unfold each of its occurrences, and read out the full phrase[4].

In a similar development, the functors that are full and faithful are often called Full-and-Faithful-Functors. I call them *embeddings*. The reader may notice that every functor can be factored into an ioof followed by an embedding.

2.2.1. *Dualities.* A *duality* structure in a monoidal category $\mathcal{C}$ consists of two objects X and X^* and two arrows, the pairing $X \otimes X^* \xrightarrow{\varepsilon} I$ and the copairing $I \xrightarrow{\eta} X^* \otimes X$, such that

$$(\varepsilon \otimes X)(X \otimes \eta) = X \qquad (X^* \otimes \varepsilon)(\eta \otimes X^*) = X^*$$

A duality structure is written $(\eta, \varepsilon) : X \dashv X^*$. Note that $X^{**} = X$, because $(c\eta, \varepsilon c) : X^* \dashv X$ is also a duality structure. If every object $X \in \mathcal{C}$ has a chosen duality structure, then such choices induce a *duality functor* $* : \mathcal{C}^{op} \longrightarrow \mathcal{C}$, which maps $A \xrightarrow{f} B$ to

$$f^* : B^* \xrightarrow{\eta B^*} A^* A B^* \xrightarrow{A f B^*} A^* B B^* \xrightarrow{A^* \varepsilon} A^*$$

---

[4]For consistency, she may wish to do the same with POVMs, CW-complexes, and the SSH-protocol.

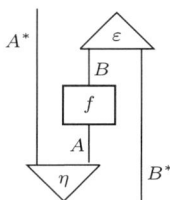

Using a duality $(\eta, \varepsilon): X \dashv X^*$, the abstract trace operators
$$\mathrm{Tr}_X^{AB} \;:\; \mathcal{C}(XA, XB) \longrightarrow \mathcal{C}(A, B)$$
as axiomatized in [**26**], can be defined as follows:
$$\frac{g: A \longrightarrow B}{\mathrm{Tr}_X^{AB} g: A \xrightarrow{\eta_X A} X^* X A \xrightarrow{X^* g} X^* X B \xrightarrow{\varepsilon_{X^*} B} B}$$

2.2.2. *Dagger-monoidal categories.* A *dagger* over a category $\mathcal{C}$ is an involutive ioof $\ddagger: \mathcal{C}^{op} \longrightarrow \mathcal{C}$. In other words, it satisfies $A^\ddagger = A$ on the objects and $f^{\ddagger\ddagger} = f$ on the arrows. This very basic structure turns out to suffice for some crucial concepts.

DEFINITION 2.3. A morphism $u \in \mathcal{C}(A, B)$ *unitary* if $u^\ddagger \circ u = \mathrm{id}_A$ and $u \circ u^\ddagger = \mathrm{id}_B$. An endomorphism $p \in \mathcal{C}(A, A)$ is a *projector* if $p = p^\ddagger = p \circ p$. A projector is *pure* if moreover $\mathrm{Tr}_A^{II}(p) = \mathrm{id}_I$.

Remarks. Note that the abstract trace operators, given above, require a monoidal structure in $\mathcal{C}$. The interactions between the dagger with the monoidal structure, and in particular with the duals, has been recognized and analyzed in [**1, 48, 49**]. A *dagger-monoidal* category $(\mathcal{C}, \otimes, I, \ddagger)$ is a dagger-category with a monoidal structure where all canonical isomorphisms, that form the monoidal structure, are unitary [**48**, Sec. 2.2]. When the monoidal structure is strict, this boils down to the requirement that the symmetry $c: A \otimes B \to B \otimes A$ is unitary.

In string diagrams, the morphism $f^\ddagger$ is represented by flipping the box f around its horizontal axis. The morphism boxes thus need to be made asymmetric to record this flipping: in [**48**], a corner of the box is filled; in [**12**], a corner is cut off.

2.2.3. *Abstract conjugates and reals.* Since the dagger and the duality functors $\ddagger, *: \mathcal{C}^{op} \longrightarrow \mathcal{C}$ commute commute in the obvious sense[5] of $f^{\ddagger *} = f^{*\ddagger}$ their composite defines the *conjugation* ioof $(-)_*: \mathcal{C}^{op} \longrightarrow \mathcal{C}$, which maps f to $f_* = f^{*\ddagger} = f^{\ddagger *}$.

---

[5]If functors are viewed as abstract arrows, it looks like these two functors are not composable, since the domain of one is different from the codomain of the other. But an ioof $F: \mathcal{C}^{op} \longrightarrow \mathcal{C}$ boils down to a specification of a map $F_{AB}: \mathcal{C}(A, B) \longrightarrow \mathcal{C}(B, A)$ for all $A, B \in \mathcal{C}$. The specifications of F and F^{op} are thus identical modulo renaming of the indices, and the equation $F = F^{op}$ is formally valid, e.g. in a Martin-Löf-style presentation of category theory. This is why "sorting out the types" of $*$ and $\ddagger$ and writing something like $f^{\ddagger(*)^{op}} = f^{*(\ddagger)^{op}}$ is not just in bad taste, but also wrong.

In the category of complex Hilbert spaces, the conjugation ioof corresponds is induced by the conjugation of the complex numbers. In the category of real Hilbert spaces, it degenerates into the identity functor.

DEFINITION 2.4. A morphism f is said to be *real* if $f = f_*$ (or equivalently $f^\ddagger = f^*$).

Remarks. Pursuing the Hilbert space intuitions, the arrows f and $f^\ddagger$ are sometimes thought of as each other's adjoints. On the other hand, in a completely different sense, the dual objects A and A^* are each other's adjoints, if the monoidal category is viewed as a bicategory with one object.

2.2.4. *Inner products and entanglement.* The dagger-monoidal structure has been proposed as a framework for categorical semantics of quantum computation [1, 48]. It turns out that this modest structure suffices for deriving many important notions:

- *inner product*

(2.3)
$$\langle -|-\rangle_A : \mathcal{C}(A) \times \mathcal{C}(A) \longrightarrow \mathcal{C}(I)$$
$$(I \xrightarrow{a,b} A) \longmapsto \left(I \xrightarrow{a} A \xrightarrow{b^\ddagger} I \right)$$

- *partial inner product*

(2.4)
$$\langle -|-\rangle_A^B : \mathcal{C}(A) \times \mathcal{C}(AB) \longrightarrow \mathcal{C}(B)$$
$$\left(I \xrightarrow{a} A, I \xrightarrow{b} AB \right) \longmapsto \left(I \xrightarrow{a} AB \xrightarrow{b^\ddagger \otimes B} B \right)$$

- *weakly entangled vectors* $\eta \in \mathcal{C}(A \otimes A)$, such that for all $a \in \mathcal{C}(A)$ holds

(2.5)
$$\langle a_* \mid \eta \rangle_A^A = a$$

Furthermore, an abstract version of strong entanglement can be defined as self-duality.

DEFINITION 2.5. A vector $\eta \in \mathcal{C}(X \otimes X)$ is said to be *(strongly) entangled* if $(\eta, \eta^\ddagger) : X \dashv X$ is a duality, i.e. satisfies $(\eta^\ddagger \otimes X)(X \otimes \eta) = X = (X \otimes \eta^\ddagger)(\eta \otimes X)$, and thus $X^* = X$.

PROPOSITION 2.6. *For every object X in a dagger-monoidal category $\mathcal{C}$ holds*

$$(a) \iff (b) \Longleftarrow (c)$$

where

(a) $\eta \in \mathcal{C}(X \otimes X)$ *is weakly entangled*
(b) $\eta^\ddagger \in \mathcal{C}(X \otimes X, I)$ *internalizes the inner product, as* $\langle a|b \rangle = \eta^\ddagger \circ (a_* \otimes b)$
(c) $\eta \in \mathcal{C}(X \otimes X)$ *is strongly entangled.*

The three conditions are equivalent if I generates $\mathcal{C}$, in the sense that whenever $fa = ga$ for all $a \in \mathcal{C}(X)$, then $f = g$.

A PROOF can be conveniently built by transforming the string diagrams corresponding to conditions *(a-c)*:

2.3. Notation and terminology. To describe relations on finite sets, we often find it convenient to use von Neumann's representation of ordinals, where $0 = \emptyset$ is the empty set, and $n = \{0, 1, \ldots, n-1\}$. Moreover, the pairs $\langle i, j \rangle \in n \times n$ are often abbreviated to $ij \in n \times n$.

When space is constrained and confusion unlikely, we often elide the tensors and write $AfXX$ instead of $A \otimes f \otimes X \otimes X$.

3. Polynomials and abstraction

In this section we formalize the program transformations needed to implement a classical function in a quantum computer. If a program is an arrow in a category, a program transformation is simply a functor out of it. But the problem with transforming a classical program into a quantum program is that classical data can be copied and deleted, whereas quantum data cannot. So the program transformation must map classical data to classical data, distinguished within a quantum universe. What does this mean? When the classical program $f'(x, y)$ was transformed into the corresponding quantum program $U_f|x, y\rangle$ in the Introduction, the classical inputs were denoted by the variables x, y, and mapped to the basis vector variables $|x, y\rangle$. The fact that the classical inputs can be copied and deleted was captured as a syntactical property of the variables.

If the data over which a program will compute are denoted by variables, then the program itself is a polynomial in some suitable algebraic theory. More precisely, a program is an *abstraction* over the as-yet-undetermined input data, and a computation is an *application* of the program. More generally, a program transformation can be viewed as a *substitution* into a polynomial. So we need functorial semantics of polynomial constructions, and of the abstraction and substitution operations. In the framework of cartesian (closed) categories, such a treatment goes back to Lambek and Scott's seminal work [32, 33]. It was extended to monoidal categories in [40]. Here we extend it to dagger-monoidal categories.

3.1. Polynomial constructions. Adjoining an indeterminate x to a ring R leads to the ring of polynomials $R[x]$. Its universal property is that every ring homomorphism $f : R \longrightarrow S$ extends to a unique ring homomorphism $f_a : R[x] \longrightarrow S$ for each choice of $a \in S$ to which x is mapped.

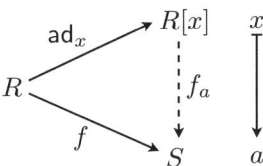

The same construction applies to other algebraic theories: e.g., one could form polynomial groups, or polynomial lattices. Categorically, for an arbitrary algebraic theory T, a polynomial T-algebra $A[x]$ can be viewed as the coproduct in the category of T-algebras of the T-algebra A and the free T-algebra over one generator, denoted x.

The polynomial construction also applies to algebraic structures over categories, such as cartesian, monoidal, or $*$-autonomous; polynomial categories can be built for any algebraic theory T over the category of categories. The polynomial category $\mathcal{S}[x]$ is then the free T-category obtained by freely adjoining a single generator x to the T-category $\mathcal{S}$; i.e. as the coproduct of $\mathcal{S}$ and the free T-category generated by x. However, categories are generated over graphs, rather than sets, so the question is what kind of a graph should x be. There seem to be two minimal choices:

(a) x is an object: a graph with one node and no edges; or

(b) x is an arrow: a graph with two nodes and an edge between them.

While case *(a)* leads to the constructions which do not involve the arrows, and thus largely boil down to the polynomial constructions of universal algebra, case (b) involves genuinely categorical aspects. These new aspects are isolated by assuming that *only* new arrows are adjoined to $\mathcal{S}$, and *no* new objects. More precisely, an indeterminate arrow $A \xrightarrow{x} B$ is freely adjoined between the extant objects A, B of $\mathcal{S}$. In other words, $\mathcal{S}[A \xrightarrow{x} B]$ can be viewed as the following pushout in the category of T-categories.

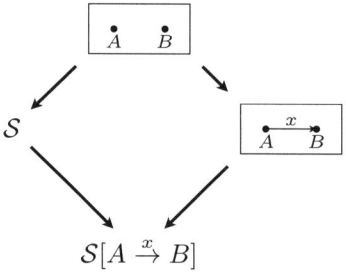

Lambek was the first to use polynomial categories in his interpretation of typed λ-calculus in cartesian closed categories [**32**]. The approach was elaborated in the book [**33**], from which categorical semantics branched in many directions. The terms containing a variable x of type X were represented as the arrows of the polynomial category $\mathcal{S}[x{:}X]$, built by adjoining to a cartesian closed category $\mathcal{S}$ an indeterminate arrow $1 \xrightarrow{x} X$, where X is an object of $\mathcal{S}$. The universal property of $\mathcal{S}[x{:}X]$ is the same as before: every structure preserving functor $F : \mathcal{S} \longrightarrow \mathcal{L}$ extends to a unique structure-preserving functor $F_a : \mathcal{S}[x] \longrightarrow \mathcal{L}$ by mapping $1 \xrightarrow{x} X$ to $1 \xrightarrow{a} FX$ in $\mathcal{L}$.

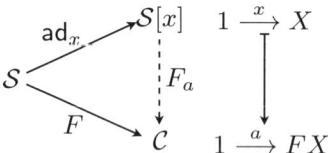

Just like a polynomial ring, the category $\mathcal{S}[x{:}X]$ can be constructed syntactically, i.e. by induction. However, the cartesian closed structure allows a more effective closed form presentation presentation of $\mathcal{S}[x{:}X]$.

THEOREM 3.1. [**32, 33**] *Let $\mathcal{S}$ be a cartesian category, $X \in \mathcal{S}$ an object and $\mathcal{S}[x{:}X]$ the free cartesian category generated by $\mathcal{S}$ and $1 \xrightarrow{x} X$. Then the inclusion functor* ad $: \mathcal{S} \longrightarrow \mathcal{S}[x{:}X]$ *has a left adjoint, the* abstraction *functor* ab $: \mathcal{S}[x{:}X] \longrightarrow \mathcal{S} : A \mapsto X \times A$

$$A \xrightarrow{\langle x, \mathsf{id}\rangle} X \times A \xrightarrow{f} B \qquad \mathcal{S}[x{:}X]\big(A, \mathsf{ad}(B)\big) \qquad A \xrightarrow{\varphi(x)} B$$
$$\Big\uparrow \qquad\qquad \Big(\Big) \qquad\qquad \Big\downarrow$$
$$X \times A \xrightarrow{f} B \qquad \mathcal{S}\big(\mathsf{ab}(A), B\big) \qquad X \times A \xrightarrow{\kappa x. \varphi(x)} B$$

and $\mathcal{S}[x{:}X]$ is equivalent with the Kleisli category for the comonad $X \times (-) : \mathcal{S} \longrightarrow \mathcal{S}$.

When $\mathcal{S}$ is cartesian closed, then $\mathcal{S}[x{:}X]$ is cartesian closed too. The Kleisli category for the comonad $X \times (-) : \mathcal{S} \longrightarrow \mathcal{S}$ is isomorphic with the Kleisli category for the monad $(-)^X : \mathcal{S} \longrightarrow \mathcal{S}$. The abstraction functor can now be viewed as a right adjoint of the inclusion $\mathsf{ad} : \mathcal{S} \longrightarrow \mathcal{S}[x{:}X]$

$$A \xrightarrow{\langle f, x\rangle} B^X \times X \xrightarrow{\varepsilon} B \qquad \mathcal{S}[x{:}X]\big(\mathsf{ad}(A), B\big) \qquad A \xrightarrow{\varphi(x)} B$$
$$\Big\uparrow \qquad\qquad \Big(\Big) \qquad\qquad \Big\downarrow$$
$$A \xrightarrow{f} B^X \qquad \mathcal{S}\big(A, \mathsf{ab}(B)\big) \qquad A \xrightarrow{\lambda x. \varphi(x)} B^X$$

This latter adjunction provides a categorical model of simply typed lambda-calculus.

On abstraction, copying and deleting of variables. Function abstraction is what makes programming possible. The first example of program abstraction were probably Gödel's numberings of primitive recursive functions [22]. Gödel's construction demonstrated that recursive programs, specifying entire families of computations (of the values of a function for all its inputs), can be stored as data. Von Neumann later explicated this as the fundamental principle of computer architecture. Kleene, on the other side, refined the idea of program abstraction into the fundamental lemma of recursion theory: the s-m-n theorem [29]. Church, finally[6] proposed the formal operations of function abstraction and data application as the driving force of all computation [11]. This proposal became the foundation of functional programming. Lawvere's observation that Church's λ-abstraction could be interpreted as an adjunction transposition [34] was a critical step towards categorical semantics of computation. Theorem 3.1 spells out this observation in terms of polynomial categories. Besides the familiar λ-abstraction, which uses the right adjoint of the inclusion $\mathsf{ad} : \mathcal{S} \longrightarrow \mathcal{S}[x{:}X]$ to transpose a polynomial into a function which outputs functions

$$\frac{\varphi(x) : A \to B}{\lambda x. \varphi(x) : A \to B^X}$$

the theorem points to an analogous abstraction operation which uses the *left* adjoint to the inclusion $\mathsf{ad} : \mathcal{S} \longrightarrow \mathcal{S}[x{:}X]$, and transposes polynomials into *indexed* families of functions

$$\frac{\varphi(x) : A \to B}{\kappa x. \varphi(x) : X \times A \to B}$$

This form of abstraction does not require higher-order types, and lifts from cartesian to monoidal categories [40]. In the present paper, we extend such abstraction operations to monoidal categories with enough structure to support the basic forms of quantum programming. — In this way, the usual quantum programming constructions can be viewed as a form of functional programming in Hilbert spaces.

---

[6] Although Church's paper appeared three years earlier than Kleene's, Church's proposal is the final step in the conceptual development of function abstraction as the foundation of computation.

But what kind of functional programming is it? The fundamental assumption of functional programming is that all data can be copied and deleted. Theorem 3.1 shows that this implies a canonical abstraction operation. The fundamental assumption of quantum programming is that some data —the quantum data— cannot be copied or deleted; but they can be entangled. Entanglement is then developed into a powerful computational resource. In-between the data that can be copied and deleted, and the data that can be entangled, there is a rich structure of diverse abstraction operations, that we shall now explore. The idea is that quantum programming can be "semantically reconstructed" a set of techniques for combining and interfacing quantum entanglement and classical abstractions.

3.2. Abstraction in monoidal categories. Given a monoidal category $\mathcal{C}$ and a chosen object A in it, we want to freely adjoin a variable arrow $I \xrightarrow{x} A$ and build the polynomial monoidal category $\mathcal{C}[x{:}X]$. Like before, $\mathcal{C}[x{:}X]$ can be built syntactically, as the free symmetric monoidal category over the graph spanned by $\mathcal{C}$ and $I \xrightarrow{x} A$, factored by the equations between the arrows of $\mathcal{C}$. Although this is not a very effective description, it does show that the polynomial category $\mathcal{C}[x{:}X]$ can in this case be quite complicated[7]. Moreover, in contrast with the cartesian (closed) case, the inclusion $\mathsf{ad} : \mathcal{C} \longrightarrow \mathcal{C}[x{:}X]$ does not have an adjoint in general, and thus does not support abstraction. The task is now to extend the polynomial construction to support abstraction. We follow, refine and strengthen the results from [40].

DEFINITION 3.2. Let $\mathcal{C}$ be a monoidal category, and E a set of well typed equations between some polynomial arrows in $\mathcal{C}[x{:}X]$. A *monoidal extension* is the monoidal category $\mathcal{C}[x{:}X; E] = \mathcal{C}[x{:}X]/E$ obtained by imposing the equations E on $\mathcal{C}[x{:}X]$, together with all equations that make it into a monoidal category. Every monoidal extension comes with the obvious ioof $\mathsf{ad} : \mathcal{C} \longrightarrow \mathcal{C}[x{:}X; E]$.

A *substitution functor* between monoidal extensions is a (strict) monoidal ioof $F : \mathcal{C}[x{:}X; E] \longrightarrow \mathcal{C}[y{:}Y; D]$.

We denote by $\mathsf{Ext}_\mathcal{C}$ the category of monoidal extensions of $\mathcal{C}$, with the substitution functors between them.

DEFINITION 3.3. A *(monoidal) abstraction* over a monoidal extension $\mathsf{ad} : \mathcal{C} \longrightarrow \mathcal{C}[x{:}X; E]$ is the adjunction $\mathsf{ab} \dashv \mathsf{ad}$ such that $\mathsf{ab}(A \otimes B) = \mathsf{ab}(A) \otimes B$, and the unit of the adjunction $h : \mathsf{Id} \longrightarrow \mathsf{ad} \circ \mathsf{ab}$ satisfies $h_A = x \otimes A$. We denote by $\mathsf{Abs}_\mathcal{C}$ the subcategory of $\mathsf{Ext}_\mathcal{C}$ spanned by the monoidal extensions that support abstraction.

Notation and terminology. Since the abstraction notation $\mathsf{ab} \dashv \mathsf{ad} : \mathcal{C} \longrightarrow \mathcal{C}[x{:}X; E]$ is generic, we often elide the structure and refer to an abstraction as $\mathcal{C}[x{:}X; E]$.

THEOREM 3.4. *The category $\mathsf{Abs}_\mathcal{C}$ of monoidal abstractions is equivalent with the category $\mathcal{C}_\times$ of commutative comonoids in $\mathcal{C}$. Each abstraction is isomorphic with the Kleisli adjunction for the comonad induced by the corresponding comonoid.*

PROOF (SKETCH). Given a commutative comonoid $(X, \Delta, \top)$ in $\mathcal{C}$, we construct the abstraction $\mathsf{ab} \dashv \mathsf{ad} : \mathcal{C} \longrightarrow \mathcal{C}[x{:}X; E]$ as follows. Let

$$E = E_{(\Delta, \top)}$$

---

[7]E.g., $\mathsf{Rel}[x]$ is not a locally small category.

be the set of equations

$$\underbrace{x \otimes x \otimes \cdots \otimes x}_{n \text{ times}} = \Delta_n \circ x \quad \text{for } n = 0, 1, 2 \ldots$$

where $\Delta_n : X \longrightarrow X^{\otimes n}$ is defined inductively:

$$\Delta_0 = \top \quad \Delta_1 = \mathrm{id}_X \quad \Delta_2 = \Delta$$
$$\Delta_{i+1} = (\Delta \times X^{\otimes i-1}) \circ \Delta_i$$

This determines the extension $\mathsf{ad} : \mathcal{C} \longrightarrow \mathcal{C}[x{:}X; E]$. Using the symmetry, it follows that every polynomial $\varphi(x) \in \mathcal{C}[x{:}X; E]$ must satisfy the equation

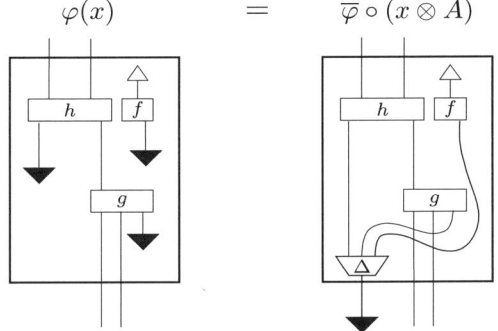

Setting $\kappa x.\varphi(x) = \overline{\varphi}$, define

$$\mathsf{ab} : \mathcal{C}[x{:}X; E] \longrightarrow \mathcal{C}$$
$$A \longmapsto X \otimes A$$
(3.1) $$\varphi(x) \longmapsto (X \otimes \kappa x.\varphi(x)) \circ (\Delta \otimes A)$$

The adjunction correspondence, with $\mathsf{ad}(B) = B$, is now

$$\mathcal{C}(\mathsf{ab}(A), B) \underset{\kappa x.}{\overset{(-)\circ(x \otimes A)}{\rightleftarrows}} \mathcal{C}[x{:}X; E](A, \mathsf{ad}(B))$$

$$\begin{aligned} (\kappa x.\, \varphi(x)) \circ (x \otimes A) &= \varphi(x) &&(\beta\text{-rule}) \\ \kappa x.\, (f \circ (x \otimes A)) &= f &&(\eta\text{-rule}) \end{aligned}$$

The other way around, given an abstraction $\mathsf{ab} \dashv \mathsf{ad} : \mathcal{C} \longrightarrow \mathcal{C}[x:X; E]$, the conditions from Def. 3.3 imply that $h(A) = x \otimes A$ and $\mathsf{ab}(A) = X \otimes A$. With the transposition κx as above, the comonoid structure must be

The arrow part of the claimed equivalence $\mathsf{Abs}_{\mathcal{C}} \simeq \mathcal{C}_\times$ follows in one direction from the fact that any comonoid homomorphism $f : Y \to X$ induces a unique ioof $F : \mathcal{C}[x{:}X] \longrightarrow \mathcal{C}[y{:}Y]$, mapping $\varphi(x)$ to $F\varphi(x) = \varphi(f \circ y)$. Since every structure-preserving functor F is easily seen to be induced by the comonoid homomorphism $f = \kappa y.\, Fx$ in this way, the bijective correspondence $\mathsf{Abs}\,(\mathcal{C}[x{:}X], \mathcal{C}[y{:}Y]) \cong \mathcal{C}_\times(X, Y)$ is established.

The isomorphism $\mathcal{C}[x : X] \cong \mathcal{C}_{[X]}$, where $\mathcal{C}_{[X]}$ is the Kleisli category for the comonoid X, is obtained by viewing the transpositions $\kappa x.(-)$ and $(-) \circ (x \otimes A)$ as functors. More precisely, this isomorphism is realized by the following ioofs:

$$K : \mathcal{C}[x{:}X] \longrightarrow \mathcal{C}_{[X]} \qquad\qquad H : \mathcal{C}_{[X]} \longrightarrow \mathcal{C}[x{:}X]$$
$$\varphi(x) \longmapsto \kappa x.\, \varphi(x) \qquad\qquad f \longmapsto f \circ (x \otimes A)$$

The fact that $H \circ K = \mathrm{id}$ is just the β-rule; the fact that $K \circ H = \mathrm{id}$ is the η-rule. Proving the functoriality of K and H, and the fact that they commute with the abstraction structure $\mathsf{ab} \dashv \mathsf{ad} : \mathcal{C} \longrightarrow \mathcal{C}[x : X; E]$ and the Kleisli adjunction $V \dashv G : \mathcal{C} \longrightarrow \mathcal{C}_{[X]}$ is an instructive exercise. $\square$

More on abstraction, copying and deleting. Theorem 3.1 had shown that the cartesian copying and deleting operations imply canonical abstraction operations over all types. In general, this implication is strictly one way, since some abstraction operations do not come about in this way: the examples include Geometry of Interaction [23] and reversible computing [2]. In contrast, Theorem 3.4 displays an *exact correspondence* between the general copying and deleting operations, provided by commutative monoids, and local abstraction operations. *All* abstraction operations thus arise from the copying and deleting operations, provided that drop the canonicity requirement from the abstraction operations on one hand, and the requirement that the copying and deleting operations are polymorphic (i.e. natural) on the other hand. Although technically rather straightforward, Theorem 3.4 has fairly deep and often missed conceptual repercussions, in particular with respect to the no cloning and no deleting conditions in quantum mechanics. The meaning of the polymorphism (naturality) of the copying operations in the quantum world was discussed in [3]. Here we proceed to study the meaning of the abstraction operations in the same framework.

Remarks. (a) The upshot of the preceding theorem is that the set of equations E in $\mathcal{C}[x{:}X; E]$ determines the comonoid structure $(\triangle, \top)$ over X; and *vice versa*: the comonoid structure $(\triangle, \top)$ determines the equations $E = E_{(\triangle, \top)}$, as in the above proof. Just like we often speak of a "comonoid X" and leave the actual structure $(\triangle, \top)$ implicit, we shall often elide E, and write $\mathcal{C}[x{:}X]$, or even $\mathcal{C}[x]$, whenever the rest of the structure is clear from the context. We shall also blur the distinction between the comonoid $(X, \triangle, \top)$ and the corresponding comonad, and denote both by X, writing $\mathcal{C}_{[X]}$ for the X-Kleisli category, the $\mathcal{C}^{[X]}$ for the X-Eilenberg-Moore category.

(b) The extension process can be iterated to construct $\mathcal{C}[x : X, y : Y] = \mathcal{C}[x : X][y{:}Y] \cong \mathcal{C}_{[X \otimes Y]}$, or $\mathcal{C}[x, y{:}X] \cong \mathcal{C}_{[X \otimes X]}$.

(c) The category $\mathcal{C}_\times$ of commutative comonoids is the cofree cartesian category over the monoidal category $\mathcal{C}$ [21]. The equivalence of categories established in 3.4 can be extended to an equivalence of 2-categories. The 2-cells of $\mathsf{Abs}_\mathcal{C}$ are the monoidal natural transformations. The 2-cells of $\mathcal{C}_\times$ can be obtained by dualizing the notion of natural transformations between the monoid homomorphisms. And the monoid homomorphisms are functors between categories with one object, so the usual notion of natural transformation just needs to be internalized. The reader may find it interesting to work this out.

(d) Recall (or see 2.1.3) that the tensor unit I carries a canonical structure of a commutative comonoid. Adjoining a variable $I \xrightarrow{y} I$ leads to $\mathcal{C}[y{:}I] \cong \mathcal{C}$, because $\Delta y = y \otimes y$ implies $y \cong \mathrm{id}_I$, i.e. that y is an isomorphism. When y is normalized, like in Sec. 4.2, $y = \mathrm{id}_I$ will hold.

COROLLARY 3.5. *In every extension $\mathcal{C}[x{:}X]$ that supports monoidal abstraction holds $\Delta x = x \otimes x$ and $\top x = \mathrm{id}_I$.*

PROOF. The first equation follows by postcomposing with x the equation $\Delta = \kappa x.\, x \otimes x$, which is the definition of Δ in $\mathcal{C}[x{:}X]$, and applying the β-rule. The second one is obtained by precomposing $\top = \kappa x.\, \mathrm{id}_I$ with x and applying the β-rule. □

COROLLARY 3.6. *If the extension $\mathcal{C}[x{:}X]$ supports abstraction, then X is generated by the tensor unit I. As a consequence, a weakly entangled vector $\eta \in \mathcal{C}[x{:}X](X \otimes X)$ is always strongly entangled.*

PROOF. By definition, I generates X in $\mathcal{C}$ if whenever $fa = ga$ for all $a \in \mathcal{C}(X)$, then $f = g$, for any $f, g \in \mathcal{C}(X, Y)$. But the η-rule implies that $fx = gx$ implies $f = g$. Hence the first claim. Furthermore, the same fact can be used to show that condition *(a)* implies condition *(c)* in Prop. 2.6. E.g., going back to the proof of 2.6, condition *(c)* can be obtained by composing the diagram for condition *(a)* and its dagger, after instantiating a to x. Condition *(c)* then follows by abstracting over x. □

3.2.1. *Substitutions.* But what does the variable x in the extension $\mathcal{C}[x]$ actually represent? What kind of vectors can be *substituted* for it?

DEFINITION 3.7. A *Substitution* for x in $\mathcal{C}[x{:}X]$ is a monoidal functor $\mathcal{C}[x{:}X] \longrightarrow \mathcal{C}$.

COROLLARY 3.8. *Substitutions $\mathcal{C}[x{:}X] \longrightarrow \mathcal{C}$ are in one-to-one correspondence with the comonoid homomorphisms $I \to X$, where X is the comonoid that induces the abstraction in $\mathcal{C}[x{:}X]$ as in Thm. 3.4.*

3.2.2. *Bases.* Only the vectors $\beta \in \mathcal{C}(X)$ that happen to be comonoid homomorphisms can thus be substituted for $x \in \mathcal{C}[x{:}X](X)$, leading to. In the category FHilb of finitely-dimensional Hilbert spaces, such vectors turn out to form a basis of the space X.

DEFINITION 3.9. A *basis vector* with respect to a comonoid $(X, \Delta, \top)$ in $\mathcal{C}$ is a comonoid homomorphism from I, i.e. an arrow $\beta : I \to X$ satisfying $\Delta \beta = \beta \otimes \beta$ and $\top \beta = \mathrm{id}_I$.

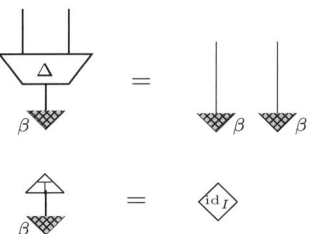

The *basis* of a comonoid is the set of its basis vectors.

In Hopf algebra theory, our basis vectors are sometimes called *set-like elements*. We shall see in the next section that, for a special family of comonoids that we call classical structures, the bases tend to form categories equivalent to the category of sets. The basis vectors of a type X in a monoidal category $\mathcal{C}$ are just the data that can be copied and deleted by a given comonoid structure on X.

Examples. Consider the monoidal category $(\mathsf{Rel}, \times, 1)$ of sets and relations. Every set X has a standard comonoid structure $X_1 = (X, \Delta, \top)$, induced by the cartesian structure of sets:

$$\Delta(x) = \{xx\} \qquad \top(x) = \{0\}$$

On the other hand, any monoid $(X, +, o)$ over the same underlying set induces a nonstandard comonoid $X_2 = (X, \widetilde{+}, \widetilde{o})$, where $\widetilde{r} : B \to A$ denotes the converse relation of $r : A \to B$, and thus

$$\widetilde{+}(u) = \{vw \mid u = v + w\} \qquad \widetilde{o}(u) = \{o\}$$

These different comonoids induce different monoidal extensions $\mathsf{Rel}[x:X; E_1]$ and $\mathsf{Rel}[x:X; E_2]$, with different abstraction operations. Both extensions have the same objects, and even the same arrows, but these arrows compose in different ways. Viewed in the Kleisli form, both categories consist of relations in the form $X \times A \longrightarrow B$. But the composites $X \times A \xrightarrow{r;s} C$ of $X \times A \xrightarrow{r} B$ and $X \times B \xrightarrow{s}$ will respectively be

$$\begin{aligned}(r;s)_1(u,a,c) &\iff \exists b.\ r(u,a,b) \wedge s(u,b,c) \\ (r;s)_2(u,a,c) &\iff \exists bvw.\ r(w,a,b) \wedge s(v,b,c) \\ &\qquad \wedge\ u = v + w\end{aligned}$$

As a consequence, each case allows substitution of different basis vectors. With respect to the standard comonoid $X_1 = (X, \Delta, \top)$, the basis vectors are just the singleton relations $\{u\} \in \mathsf{Rel}(X)$. The variable x in $\mathsf{Rel}[x : X; E_1]$ thus denotes an indeterminate element of the set X. On the other hand, with respect to the comonoid $X_2 = (X, \widetilde{+}, \widetilde{o})$, there is only one basis vector $\beta \in \mathsf{Rel}(X)$, which is the subset of X consisting of the invertible elements with respect to the monoid $(X, +, o)$. The variable x in $\mathsf{Rel}[x:X; E_2]$ thus denotes this one vector $\beta \in \mathsf{Rel}(X)$, since there is nothing else that can be substituted for x.

4. Daggers and classical structures

This section adds the dagger functor, and the dualities to the monoidal framework of abstraction (cf. 2.2.2). The abstraction now leads to classical structures, which were introduced in [12], albeit without discussing their conceptual origin in the abstraction operations. Here we show that these abstraction operations are formally equivalent to classical structures, and that moreover they display the copying and deleting capabilities as the usual operations on variables.

4.1. Dagger-monoidal abstraction.

DEFINITION 4.1. Let $\mathcal{C}$ be a dagger-monoidal category, and E a set of equations between some parallel arrows in the dagger-monoidal polynomial category $\mathcal{C}[x:X]$. A *dagger-monoidal extension* is the dagger-monoidal category $\mathcal{C}[x:X; E] =$

$\mathcal{C}[x:X]/E$, obtained by imposing the equations E on $\mathcal{C}[x:X]$, together with all equations that make it into a dagger-monoidal category. As all such constructions, it comes with the obvious ioof $\mathsf{ad} : \mathcal{C} \longrightarrow \mathcal{C}[x:X; E]$.

A *substitution functor* between the dagger-monoidal extensions is a monoidal ioof $F : \mathcal{C}[x:X; E] \longrightarrow \mathcal{C}[y:Y; D]$ which preserves the dagger, i.e. $F(\psi^\ddagger) = (F\psi)^\ddagger$.

We denote by $\ddagger\text{-}\mathsf{Ext}_\mathcal{C}$ the category of dagger-monoidal extensions of $\mathcal{C}$, with the substitution functors between them.

DEFINITION 4.2. A *dagger monoidal abstraction* over a dagger monoidal extension $\mathsf{ad} : \mathcal{C} \longrightarrow \mathcal{C}[x:X; E]$ is the adjunction $\mathsf{ab} \dashv \mathsf{ad}$, which satisfies the requirements of Definition 3.3, and moreover preserves the dagger, in the sense that $\mathsf{ab}.\varphi(x)^\ddagger = (\mathsf{ab}.\varphi(x))^\ddagger$.

We denote by $\ddagger\text{-}\mathsf{Abs}_\mathcal{C}$ the subcategory of $\ddagger\text{-}\mathsf{Ext}_\mathcal{C}$ where the abstraction is supported. Its objects are often called abstractions.

Thm. 3.4 established the correspondence between monoidal abstractions over X and the comonoid structures carried by X. The next theorem extends this correspondence to dagger monoidal categories: a monoidal abstraction corresponding to a comonoid structure preserves the dagger if and only if the Kleisli category, induced by the comonoid, is (equivalent with) the dagger monoidal extension itself.

THEOREM 4.3. *Let $\mathcal{C}$ be a dagger-monoidal category and $\mathsf{ad} : \mathcal{C} \longrightarrow \mathcal{C}[x:X; E]$ a dagger-monoidal extension. Suppose that it admits a monoidal abstraction $\mathsf{ab} \dashv \mathsf{ad}$ (as in Def. 3.3), with the induced comonoid $(X, \triangle, \top)$ (as in Thm. 3.4). Then the following statements are equivalent:*

(a) $\mathsf{ab} \dashv \mathsf{ad} : \mathcal{C} \longrightarrow \mathcal{C}[x:X; E]$ is a dagger-abstraction, i.e.

$$\mathsf{ab}.\varphi(x)^\ddagger \;=\; (\mathsf{ab}.\varphi(x))^\ddagger$$

(b) x is real, i.e. $x^* = x^\ddagger$

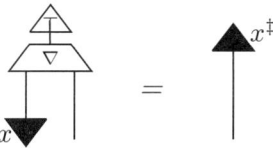

(c) $\mathsf{ab} \dashv \mathsf{ad} : \mathcal{C} \longrightarrow \mathcal{C}[x:X; E]$ is isomorphic with the Kleisli adjunction

$$V \dashv G \;:\; \mathcal{C} \longrightarrow \mathcal{C}_{[X]}$$

The following conditions provide further equivalent characterizations of (a-c), this time expressed in terms of the properties of the comonoid $(X, \triangle, \top)$ and its dual monoid $(X, \triangledown, \bot)$, where $\triangledown = \triangle^\ddagger$ and $\bot = \top^\ddagger$.

(i) $\eta = \triangle \circ \bot$ and $\varepsilon = \top \circ \triangledown$ make $X = X^*$ self-dual

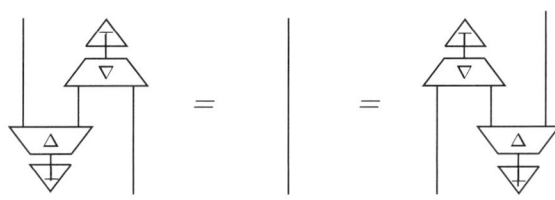

(ii) $(X \otimes \nabla) \circ (\eta \otimes X) = \Delta = (\nabla \otimes X) \circ (X \otimes \eta)$

(iii) $(X \otimes \nabla) \circ (\Delta \otimes X) = \Delta \circ \nabla = (\nabla \otimes X) \circ (X \otimes \Delta)$

Remark. Condition *(iii)* is the *Frobenius condition*, analyzed in [**10, 9, 30, 12**]. Condition *(ii)* is Lawvere's earlier version of the same [**35**]. In each of the last three conditions, the commutativity assumption makes one of the equations redundant. The equivalence of *(i-iii)*, however, holds without this commutativity.

PROOF. *(a⇒b)* Using the definition (3.1) of ab, condition *(a)* implies that $\nabla = (\mathsf{ab}.x)^\ddagger = \mathsf{ab}.x^\ddagger = (X \otimes \kappa x.x^\ddagger) \circ (\Delta \otimes X)$, or graphically

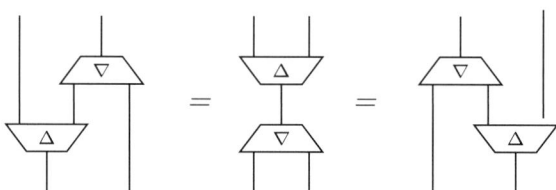

from which *(b)* follows by precomposing both sides with $(x \otimes X)$ and postcomposing with $\top$.

(b⇒i) Dualizing *(b)* gives $x = x_* = x^{\ddagger *}$, i.e.

Combining *(b)* and its dual gives

from which *(i)* follows, because the η-rule implies that $f \circ (x \otimes A) = g \circ (x \otimes A) \Longrightarrow f = g$

(i⇒ii) On one hand, if X is self-dual, then $X \otimes X$ is self-dual too, because

if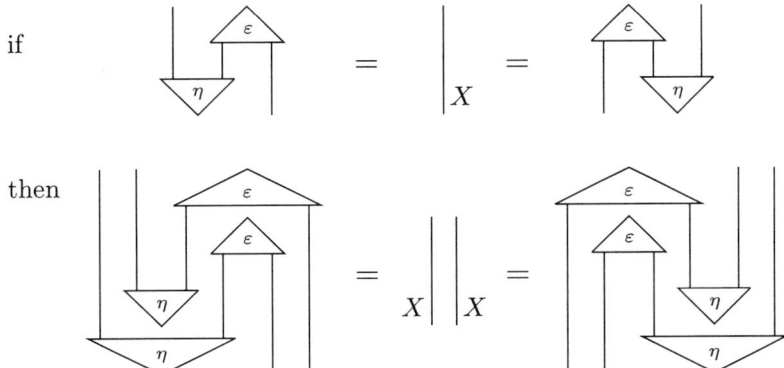

then

On the other hand, *(i)* also implies that $\triangledown^\ddagger = \triangledown^*$, and since $\Delta = \triangledown^\ddagger$ holds by definition, we have

(ii⇒iii) Using *(ii)* to expand Δ at the first step, and to collapse it at the last step, we get

(iii⇒i) follows in a way obvious from the diagrams, by precomposing the first equation of *(iii)* with $\bot \otimes X$ and postcomposing it with $X \otimes \top$; and by precomposing the second equation with $X \otimes \bot$ and postcomposing it with $\top \otimes X$.

(i⇒c) Using the self-duality of X, the dagger on $\mathcal{C}_{[X]}$ is defined by

Since this implies $\kappa x.\ \varphi(x)^\ddagger = (\kappa x.\ \varphi(x))^\ddagger$, it follows that the isomorphism $\mathcal{C}[x:X] \cong \mathcal{C}_{[X]}$, defined in the proof of Thm. 3.4, preserves the dagger.

(c⇒a) Since the dagger preservation under the isomorphism $\mathcal{C}[x:X] \cong \mathcal{C}_{[X]}$ means that the dagger in $\mathcal{C}_{[X]}$ must be as above, it follows

By (3.1), the left-hand side is **ab.** $\varphi(x)^\ddagger$, whereas the right-hand side is (**ab.** $\varphi(x))^\ddagger$. Hence *(a)*. □

DEFINITION 4.4. A *Frobenius algebra* in a monoidal category $\mathcal{C}$ is a structure $(X, \nabla, \Delta, \bot, \top)$ such that

- $(X, \nabla, \bot)$ is a monoid,
- $(X, \Delta, \top)$ is a comonoid, and
- the equivalent conditions (i-iii) of Thm. 4.3 are satisfied.

A *dagger-Frobenius algebra* in a dagger-monoidal category $\mathcal{C}$ is a Frobenius algebra where $\nabla = \Delta^\ddagger$ and $\bot = \top^\ddagger$.

Thm. 4.3 can now be summarized as follows.

COROLLARY 4.5. *The category of dagger-monoidal abstractions $\ddagger$-$\mathrm{Abs}_\mathcal{C}$ is equivalent with the category $\mathcal{C}_\Delta$ of commutative dagger-Frobenius algebras and comonoid homomorphisms in $\mathcal{C}$.*

Summary. The upshot of Thm. 4.3 is thus that a monoidal extension $\mathcal{C}[x:X]$, induced by a commutative comonoid X which also happens to be a dagger-Frobenius algebra, is necessarily a dagger-monoidal extension. The immediate corollary is the following.

COROLLARY 4.6. *The substitutions $\mathcal{C}[x:X] \longrightarrow \mathcal{C}$ of the basis vectors with respect to a Frobenius algebra X preserve not only the tensors and their unit, but also the daggers.*

Furthermore, since the basis vectors of the Frobenius algebra X are substituted for the variable x, which must be real, it is natural to expect, and easy to prove that

COROLLARY 4.7. *The basis vectors with respect to a dagger-Frobenius algebra are always real.*

Remark. This last statement may sound curious. There are many complex vectors in a complex Hilbert space, and each of them may participate in some basis. However, after a change of basis they may become real; and some vectors that were real will cease to be real.[8] The notion of reality depends on the choice of basis. However, just like people, the basis vectors themselves always satisfy their own notion of reality: they are in the form $\beta_1 = (1, 0, 0, \ldots, 0), \beta_2 = (0, 1, 0, \ldots, 0), \ldots, \beta_n = (0, 0, \ldots, 0, 1)$.

4.2. Classical structures. It turns out that Frobenius algebras with additional properties provide a purely algebraic characterization of the choice of a basis, e.g. in a Hilbert space. More generally, in an abstract quantum universe, we can thus distinguish classical data types, by means of algebraic operations. We begin by describing the additional property needed for this.

---

[8]Note that we are talking about vectors here, and not about quantum states. So the global phases are not factored out.

LEMMA 4.8. *Let $\mathcal{C}[x, y : X]$ be a dagger-monoidal extension induced by the Frobenius algebra $(X, \triangledown, \vartriangle, \bot, \top)$. Then the following conditions are equivalent:*

   (a) $\triangledown \circ \vartriangle = \mathrm{id}_X$
   (b) $\triangledown(x \otimes x) = x$
   (c) $\langle x|y \rangle^2 = \langle x|y \rangle$

and they imply

   (d) $\langle x|x \rangle = \mathrm{id}_I$

The equivalence of (a) and (b) is also valid for monoidal categories, with no dagger.

PROOF. $(a \Rightarrow b)$ $\triangledown(x \otimes x) = \triangledown \vartriangle x = x$, using Cor. 3.5.
$(b \Rightarrow c)$ $\langle x|y \rangle = x^\ddagger \circ y = x^\ddagger \circ \triangledown \circ (y \otimes y) = x^\ddagger \circ \vartriangle^\ddagger \circ (y \otimes y) = (x^\ddagger \otimes x^\ddagger)(y \otimes y) = (x^\ddagger \circ y) \otimes (x^\ddagger \circ y) = \langle x|y \rangle^2$, i.e.

$(c \Rightarrow a)$ $x^\ddagger \circ \triangledown \circ \vartriangle \circ y = (x^\ddagger \otimes x^\ddagger)(y \otimes y) = (x^\ddagger \circ y) \otimes (x^\ddagger \circ y) = \langle x|y \rangle^2 = \langle x|y \rangle = x^\ddagger \circ y$, and then use the η-rule.
$(b \Rightarrow d)$ Since by Thm. 4.3 $x^\ddagger = x^*$, and by Cor. 3.5 $\top x = \mathrm{id}_I$, we have $\langle x|x \rangle = x^\ddagger x = \top x = \mathrm{id}_I$.

□

Remark. Lemma 4.8(b) and Thm. 4.3 together say that a monoidal extension $\mathcal{C}[x : X]$ of a dagger monoidal category $\mathcal{C}$ is a classical extension if and only the variable x is real and idempotent, i.e. $x = x_* = x \bullet x$, where $a \bullet b = \triangledown(a \otimes b)$ is the convolution, mentioned in 2.1.6. Lemma 4.8(c) says that the idempotence of x is equivalent with the idempotence of the inner product $\langle x|y \rangle$ of any two variables of type X. (Idempotence with respect to which monoid? Recall from Sec. 2.1.6 that the convolution, the composition, and the tensor of scalars all induce the same monoid, since $s \bullet t = s \circ t = s \otimes t$ holds for all $s, t \in \mathcal{C}(I)$.)

Note that, by the η-rule, $\langle x|y \rangle = \langle x|z \rangle \Rightarrow y = z$. It follows that the monoid of scalars in a polynomial extension $\mathcal{C}[x, y, z : X]$ must have freshly adjoined elements, if $x \neq y \neq z$. Another interesting point is that the implication $\langle x|a \rangle = \langle x|b \rangle \Rightarrow a = b$, valid in $\mathcal{C}[x : X]$, is preserved under the substitutions *jointly*, provided that the basis vectors generate X: if $\langle \beta|a \rangle = \langle \beta|b \rangle$ holds for all basis vectors β, then $a = b$. Elaborating this, one could formulate the suitable soundness and completeness notions and for reasoning with polynomials and classical structures, but we shall not pursue this thread.

DEFINITION 4.9. A *classical structure* is a commutative Frobenius algebra satifsying 4.8*(a)*. A *classical extension* of $\mathcal{C}$ is a dagger-monoidal extension $\mathcal{C}[x:X]$ induced by a classical structure, i.e. satisfying 4.8(b-c).

Note that the above definition does not impose $\nabla = \Delta^\ddagger$ and $\bot = \top^\ddagger$, since it only requires that a classical structure is a commutative Frobenius algebra, and not necessarily a *dagger*-Frobenius algebra. In Sec. 4.2.2 below, we shall show that this always happens to be the case in classical structure nevertheless: any Frobenius algebra must be a dagger-Frobenius algebra, as soon as the conditions from Lemma 4.8 are satisfied.

But let us first spell out the upshot of Theorem 4.3 in terms of classical structures.

COROLLARY 4.10. *The category $\mathcal{B}$-Abs$_\mathcal{C} \subseteq \ddagger$-Abs$_\mathcal{C}$ of classical abstractions of $\mathcal{C}$ is equivalent with the category $\mathcal{C}_\mathcal{B}$ of classical structures and comonoid homomorphisms in $\mathcal{C}$.*

Note that the category $\mathcal{C}_\mathcal{B}$ is a cartesian subcategory of the category $\mathcal{C}_\times$ of commutative comonoids. While the forgetful functor $\mathcal{C}_\times \longrightarrow \mathcal{C}$ was couniversal for all monoidal functors from cartesian categories to $\mathcal{C}$, the forgetful functor $\mathcal{C}_\mathcal{B} \longrightarrow \mathcal{C}$ is couniversal for the conservative functors among them. The exactness properties of $\mathcal{C}_\mathcal{B}$, induced by the various properties of $\mathcal{C}$, were analyzed in [9]. If $\mathcal{C}$ is compact [28] and right exact with biproducts, then $\mathcal{C}_\mathcal{B}$ turns out to be a pretopos. In any case, if $\mathcal{C}$ represents a quantum universe, $\mathcal{C}_\mathcal{B}$ can be thought of as the category of classical data.

4.2.1. *Orthonormality of bases.* Definition 3.9 stipulated an abstract notion of a basis with respect to a comonoid. The notion of a classical structure now characterizes just those comonoids whose bases are *orthonormal*, in the sense of the following

DEFINITION 4.11. A vector $a \in \mathcal{C}(A)$ is *normalized* if $\langle a|b \rangle = \text{id}_I$. A pair of vectors $a, b \in \mathcal{C}(A)$ is *orthogonal* if $\langle a|b \rangle^2 = \langle a|b \rangle$. A set of vectors is *orthonormal* when each element is normalized, and each pair orthogonal.

Lemma 4.8 and Cor. 3.8 imply that

PROPOSITION 4.12. *The basis set of every classical structure is orthonormal.*

4.2.2. *Succinct classical structures.* The following lemma shows that being classical structure is a property of a comonoid, or of a monoid, rather than additional structure. In particular, whenever a monoid $(X, \nabla, \bot)$ and a comonoid $(X, \Delta, \top)$ satisfy conditions *(i-iii)* of Theorem 4.3, *and* the conditions of Lemma 4.8, then they must be dual, and thus any special Frobenius algebra must be a dagger-Frobenius algebra. A moment of thought shows that this boils down to the following lemma.

LEMMA 4.13. *The monoid and the comonoid part of a classical structure determine each other: e.g., $(X, \nabla, \Delta_1, \bot, \top_1)$ and $(X, \nabla, \Delta_2, \bot, \top_2)$ are classical structures, then $\Delta_1 = \Delta_2$ and $\top_1 = \top_2$.*

Since $(X, \nabla, \Delta, \bot, \top)$ is completely determined by $(X, \nabla, \bot)$ (and by $(X, \Delta, \top)$), it is justified to speak succinctly of the classical structure $(X, \nabla, \bot)$ (and of the classical structure $(X, \Delta, \top)$).

PROOF. It is enough to prove $\Delta_1 \circ \nabla = \Delta_2 \circ \nabla$, because this and $\nabla \circ \Delta_1 = \mathrm{id}_X$ give
$$\Delta_1 = \Delta_1 \circ \nabla \circ \Delta_1 = \Delta_2 \circ \nabla \circ \Delta_1 = \Delta_2$$
Here is a diagrammatic proof $\Delta_1 \circ \nabla = \Delta_2 \circ \nabla$:

□

4.2.3. Classifying classical structures.

PROPOSITION 4.14. **[19]** *In the category* (FHilb, $\otimes, \mathbb{C}, \ddagger$) *of finitely-dimensional complex Hilbert spaces and linear maps, the classical structures correspond to the orthonormal bases in the usual sense.* FHilb$_\mathcal{B}$ *is equivalent with the category* FSet *of finite sets and functions.*

PROPOSITION 4.15. **[41]** *In the category* $\left(\mathsf{Rel}, \times, 1, \widetilde{(-)}\right)$ *of sets and relations, the classical structures are just the biproducts (disjoint unions) of abelian groups.* Rel$_\mathcal{B}$ *is equivalent with the category* Set *of sets and functions.*

Each classical structure X in Rel decomposes as a disjoint union $X = \sum_{j \in J} X_j$ where each restriction $(X_j, \nabla_j, \perp_j)$ of $(X, \nabla, \perp)$ is an abelian group. A classical structure on X thus consists of (1) a partition $X = \sum_{j \in J} X_j$ and (2) an abelian group structure on each X_j. These partitions and group structures, and even the size of X are, however, indistinguishable by the morphisms of Rel$_\mathcal{B}$, because any two classical structures with the same number J of components are isomorphic.

Bases in Rel. The basis induced by the classical structure $X = \sum_{j \in J} X_j$ is in the form $\mathcal{B}(X) = \{X_j\}_{j \in J}$. While the bases with the same number of elements are indistinguishable in Rel$_\mathcal{B}$, they are the crucial resource for quantum computation in Rel. The bases induced by the *rectangular* structures $(\Xi_n, \Delta, \top)$, will be particularly useful, where

$$\begin{aligned}
\Xi_n &= \sum_n \mathbb{Z}_n = \{ij \mid 0 \leq i, j \leq n-1\} \\
\Delta(ij) &= \{\langle ik, i\ell \rangle \mid j = k + \ell\} \\
\top &= \{i0 \mid 0 \leq i \leq n-1\} \\
\mathcal{B}(\Xi_n) &= \{\beta_i = \{ij\} \mid 0 \leq i, j \leq n-1\}
\end{aligned}$$

4.3. Bases for Simon's algorithm.
Any bitstring function $f : \mathbb{Z}_2^m \to \mathbb{Z}_2^n$, considered in Simon's algorithm, can be viewed as a morphism $f \in \mathsf{FSet}_\wp(m, n)$ in the category of finite powersets and all functions between them. It is easy to see that this is a cartesian closed category, with $+$ as the cartesian product[9]. The program transformation from the function f to the corresponding Hilbert space unitary U_f is formalized as follows

$$\frac{f(x) = f \circ x \in \mathsf{FSet}_\wp[x{:}m](n)}{\frac{f'(x,y) = \langle x, y \oplus f(x)\rangle \in \mathsf{FSet}_\wp[x,y{:}m+n](m+n)}{U_f|x,y\rangle = \mathbb{B}^{\otimes f'(x,y)} \in \mathsf{FHilb}\left[|x,y\rangle{:}\mathbb{B}^{\otimes(m+n)}\right]\left(\mathbb{B}^{\otimes(m+n)}\right)}}$$

where $\mathbb{B} = \mathbb{C}^2$. The unitary U_f is thus the image of f' along the functor

$$\mathbb{B}^{\otimes(-)} \;:\; \mathsf{FSet}_\wp[x,y{:}m+n] \longrightarrow \mathsf{FHilb}\left[|x,y\rangle{:}\mathbb{B}^{\otimes(m+n)}\right]$$

which maps finite sets to the tensor powers of $\mathbb{B}$. Since $\mathbb{B}^{\otimes m} = \mathbb{C}^{(2^m)}$, any function $f : 2^m \to 2^n$ in $\mathsf{Set}_\wp$ is mapped to a linear operator $\mathbb{B}^{\otimes f} : \mathbb{B}^{\otimes m} \longrightarrow \mathbb{B}^{\otimes n}$ in FHilb, represented by the matrix $F = (F_{ij})_{2^n \times 2^m}$ where $F_{ij} = 1$ whenever $f(j) = i$, otherwise $F_{ij} = 0$. This determines a functor $\mathsf{FSet}_\wp \longrightarrow \mathsf{FHilb}$. It is extended to a substitution $\mathsf{FSet}_\wp[x,y{:}m+n] \longrightarrow \mathsf{FHilb}\left[|x,y\rangle{:}\mathbb{B}^{\otimes(m+n)}\right]$ by stipulating that the variables x, y are mapped to the variables $|x,y\rangle$.

The function $f \in \mathsf{FSet}_\wp(m, n)$ has a simpler, though nonstandard interpretation in the dagger-*premonoidal*[10] category $(\mathsf{Rel}_\wp, \otimes, 1, \ddagger)$, where $\mathsf{Rel}_\wp(m, n) = \mathsf{Rel}(2^m, 2^n)$ and $m \otimes n = m \times n$. The dagger is still just the relational converse. Like before, we define

$$\Xi^{\otimes(-)} \;:\; \mathsf{FSet}_\wp[x,y{:}m+n] \longrightarrow \mathsf{Rel}_\wp\left[|x,y\rangle{:}\Xi^{\otimes(m+n)}\right]$$

this time over the rectangular structure

$$\Xi = \Xi_2 = \{00, 01, 10, 11\}$$
$$\vartriangle(i0) = \{\langle i0, i0\rangle, \langle i1, i1\rangle\} \quad \vartriangle(i1) = \{\langle i0, i1\rangle, \langle i1, i0\rangle\}$$
$$\top = \{00, 10\}$$
$$\mathcal{B}(\Xi) = \{\beta_0 = \{00, 01\}, \beta_1 = \{10, 11\}\}$$

Note that this comonoid structure lifts from $(\mathsf{Rel}, \times, 1)$ to $(\mathsf{Rel}_\wp, \otimes, 1)$ because $\Xi \otimes \Xi = 2^2 \otimes 2^2 = 2^{2\times 2} = 2^{2+2} = 2^2 \times 2^2 = \Xi \times \Xi$. It furthermore lifts to any $\Xi^{\otimes m}$, since the commutative (co)monoid structures always extend to the tensor powers.

Since the underlying set of $\Xi^{\otimes m}$ is $2^{(2^m)}$, any function $f : 2^m \to 2^n$ in $\mathsf{Set}_\wp$, is mapped to a relation $\Xi^{\otimes f} : \Xi^{\otimes m} \longrightarrow \Xi^{\otimes n}$ in $\mathsf{Rel}_\wp$, represented by the matrix $F = (F_{ij})_{2^n \times 2^m}$ where $F_{ij} = 1$ whenever $f(j) = i$, otherwise $F_{ij} = 0$. The functor is extended into a substitution $\mathsf{Set}_\wp[x,y{:}m+n] \longrightarrow \mathsf{Rel}_\wp\left[|x,y\rangle{:}\Xi^{\otimes(m+n)}\right]$ like before. Mapping the polynomial $f'(x,y)$, constructed above, along this functor, we get a polynomial unitary relation $\Upsilon_f|x,y\rangle = \Xi^{\otimes f'(x,y)}$ on $\Xi^{\otimes(m+n)}$ in

---

[9]$\mathsf{FSet}_\wp$ is opposite to the Kleisli category for the $\wp\wp$-monad. Along the discrete Stone duality, $\mathsf{FSet}_\wp$ is thus dual to the category of free finite atomic Boolean algebras. Since Boolean algebras are primal, every function between them can be expressed as a polynomial.

[10]The tensor $m \otimes n = m \times n$ is functorial in each argument, but it is not a bifunctor. See [47] for a discussion about such structures. This has no repercussions for us, since the definition of the functor $\Xi^{\otimes(-)}$, spelled out explicitly below, makes no use of the arrow part of $\otimes$.

$\mathsf{Rel}_\wp\left[|x,y\rangle{:}\Xi^{\otimes(m+n)}\right]$. This polynomial can be viewed as a family of unitary relations indexed over the basis of $\Xi^{\otimes(m+n)}$; and each member of the family is a permutation on $\Xi^{\otimes(m+n)} = 2^{\left(2^{m+n}\right)}$.

5. Complementarity

5.1. Complementary classical structures.

DEFINITION 5.1. *A vector $a \in \mathcal{C}(X)$ is* unbiased *(or* complementary*) with respect to a classical structure $(X, \triangle, \top)$ if $\triangle a \in \mathcal{C}(X \otimes X)$ is strongly entangled (in the sense of Sec. 2.2.4). Two classical structures are* complementary *if every every basis vector with respect to one is complementary with respect to the other one, and* vice versa.

Remark. In the framework of Hilbert spaces, this definition is equivalent to the standard notion of complementary bases, used for describing the quantum uncertainty relations [**31, 56**]. Coecke, Duncan and Edwards [**14, 16**] have characterized complementary vectors in terms of their representations (cf. Sec. 2.1.6 (2.2)). The first part of the following proposition says that our definition is equivalent to theirs.

PROPOSITION 5.2. *With respect to a classical structure X, the representative $\widehat{b} \in \mathcal{C}(X, X)$ of $b \in \mathcal{C}(X)$ is*

(a) unitary if and only if b is unbiased;

(b) a pure projector if b is a basis vector.

The converse of (b) holds whenever the basis vectors generate X.

Recall from Sec. 2.2.2 that the usual definitions of projectors and unitaries lift to dagger-categories: a unitary is an endomorphism u such that $u^\ddagger = u^{-1}$, whereas a projector p satisfies $p = p^\ddagger = p \circ p$. For a pure projector over X we moreover require $\mathrm{Tr}(p) = \varepsilon \circ (X \otimes p) \circ \eta = \mathrm{id}_I$. The assumption that a set of vectors $\Gamma \subseteq \mathcal{C}(X)$ generates an object X means that for any $f \neq g \in \mathcal{C}(X, Y)$ there must be a basis vector $a \in \Gamma$ such that $fa \neq ga$.

PROOF OF 5.2. *(a)* Since $\triangledown$ is commutative, by the definition of $\widehat{b}$ in (2.2), $\widehat{b}^\ddagger = (\triangledown(b \otimes X))^\ddagger = (X \otimes b^\ddagger)\triangle$. The composites $\widehat{b} \circ \widehat{b}^\ddagger$ and $\widehat{b}^\ddagger \circ \widehat{b}$ can thus be viewed as the left-hand side and the right-hand side of the following diagram.

Both side diagrams can be transformed into the middle one by applying the Frobenius condition 4.3*(iii)*. Thus

$$\widehat{b} \circ \widehat{b}^\ddagger = \mathrm{id}_X \iff (X \otimes b^\ddagger \triangledown)(\triangle b \otimes X) = \mathrm{id}_X \iff \widehat{b}^\ddagger \circ \widehat{b} = \mathrm{id}_X$$

But by Defn. 2.5, the middle equation just says that $\triangle b$ is strongly entangled, i.e. that b is unbiased. Hence the claim.

(b) To begin from the easiest, first note that $\mathrm{Tr}(\widehat{b}) = \mathrm{id}_I \iff \top b = \mathrm{id}_I$, because $\mathrm{Tr}(\widehat{b}) = \top b$:

Secondly, we want to show that $\widehat{b} = \widehat{b}^{\ddagger} \iff b^* = b^{\ddagger}$, i.e.

The right-hand equation says that b is real, which is a property of every basis vector, according Cor. 4.7. The implication from left to right is obtained by postcomposing both sides of the left-hand equation with ⊤. The implication from right to left is obtained by tensoring by X on the right both sides of the right-hand equation, and then precomposing them with Δ. The left-hand equation is then obtained using 4.3(ii).

To complete the proof, we show that $\Delta b = b \otimes b$ implies $\widehat{b} \circ \widehat{b} = \widehat{b}$, by the following diagram:

□

5.2. Transforms. A given basis of a Hilbert space can be mapped into a complementary one using a Fourier transform. This is done in all HSP-algorithms: the basis vectors are entangled into one complementary vector, and the unitary U_f is then evaluated over that vector, thus computing all values of f in one sweep.

In order to complete the implementation of Simon's algorithm in $\mathsf{Rel}_\wp$, we need a pair of complementary bases for $\Xi^{\otimes(m+n)}$. As mentioned above, the classical structures of Ξ lift from Rel to $\mathsf{Rel}_\wp$. And in Rel in general, for a given classical structure $X = \sum_{j \leq m} X_j^1$ in Rel, a complementary vector is a set $\gamma \subseteq X$ such that $\gamma_j = \gamma \cap X_j^1$ is a singleton for every $j \leq m$. Another classical structure $X = \sum_{k \leq n} X_k^2$ over the same set is thus complementary if and only if $X_j^1 \cap X_k^2$ is a singleton for all $j \leq m, k \leq n$. Since X^1 and X^2 are partitions, it follows that all $\#X_j^1 = n$ and all $\#X_k^2 = m$. So X must decompose to m groups of order n, and to n groups of order n. In order to have an invertible transform from one basis to another, we need $m = n$. Unless we are interested in the various forms of entanglement engendered by the various group structures, we can thus restrict attention to rectangular structures from sec. 4.2.3. A simple transform mapping the basis vectors of Ξ_ℓ into a complementary basis is

$$H_\ell : \Xi_\ell \longrightarrow \Xi_\ell$$
$$ij \longmapsto ji$$

Using $H = H_2$ to transform $H^{\otimes m} : \Xi^{\otimes m} \longrightarrow \Xi^{\otimes m}$ we can now produce the superposition of all the basis vectors, representing the inputs of the function $f : \mathbb{Z}_2^m \to \mathbb{Z}_2^n$ from Simon's algorithm. The other way around, the H-image of any basis vector is the superposition of the complementary basis of $\Xi^{\otimes m}$. We can thus define the unitary polynomial $(H^{\otimes m} \otimes \mathrm{id}) \circ \Upsilon_f |x, y\rangle \circ (H^{\otimes m} \otimes \mathrm{id})$ on $\Xi^{\otimes (m+n)}$ in $\mathsf{Rel}_\wp \left[|x, y\rangle : \Xi^{\otimes (m+n)} \right]$ and evaluate it on the vector $|0, 0\rangle = \bot \in \mathsf{Rel}_\wp \left(\Xi^{\otimes (m+n)} \right)$, to get the outcome $S |x, y\rangle \in \mathsf{Rel}_\wp \left[|x, y\rangle : \Xi^{\otimes (m+n)} \right] \left(\Xi^{\otimes (m+n)} \right)$. To complete the execution of Simon's algorithm in $\mathsf{Rel}_\wp$, we just need to measure this outcome.

6. Measurements

So far, we have seen that the classical data in a quantum universe, represented by a dagger-monoidal category $\mathcal{C}$, can be characterized as just those data that can be annotated by the variables in $\mathcal{C}[x, y, \ldots]$, i.e. those data that support the abstraction operation κx. Quantum programs are thus viewed as polynomial arrows $\varphi(x, y, \ldots) \in \mathcal{C}[x, y, \ldots]$. In this respect, quantum programs are similar to classical programs: they specify that some operations should be applied to some input data, always classical, denoted by the variables. Semantics of computation is captured through abstractions and substitutions. Program execution, in particular, corresponds to substituting some input data for the variables, and evaluating the resulting expressions.

In classical computation, such evaluations yield the outputs. In quantum computation, however, there is more: the outputs need to be *measured*. The view of quantum programs as polynomials in dagger-monoidal categories needs to be refined to capture measurements. In the simplest case, a measurement will turn out to be just a projector in $\mathcal{C}[x{:}X]$.

DEFINITION 6.1. A morphism $X \otimes A \xrightarrow{\alpha} A$ in $\mathcal{C}$ on is an X-action A if $\alpha \circ (X \otimes \alpha) = \alpha \circ \triangledown$. An X-action is *normal* if moreover $\alpha \circ (\bot \times A) = \mathrm{id}_A$.

An X-equivariant homomorphism from $X \otimes A \xrightarrow{\alpha} A$ to $X \otimes B \xrightarrow{\beta} B$ is an arrow $f \in \mathcal{C}(A, B)$ such that $f \circ \alpha = \beta \circ (X \otimes f)$. The category of X-actions and X-equivariant homomorphisms is denoted $\mathcal{C}^{\{X\}}$.

The full subcategory of *normal* X-actions is $\mathcal{C}^{[X]} \hookrightarrow \mathcal{C}^{\{X\}}$.

Remark. Normal X-actions are the Eilenberg-Moore algebras for the monad $X \otimes (-) : \mathcal{C} \longrightarrow \mathcal{C}$. Equivalently, they are also actions of the monoid X, and this terminology tends to lead to less confusion.

LEMMA 6.2. *Let $(X, \triangle, \top)$ be a classical structure, $\alpha(x) : A \longrightarrow A$ an endomorphism in $\mathcal{C}[x{:}X]$ and $\alpha = \kappa x.\, \alpha(x) : X \otimes A \longrightarrow A$ its abstraction.*

(a) *The following conditions are equivalent:*
    (i) $\alpha(x) = \alpha(x) \circ \alpha(x)$, *i.e.* $\alpha(x)$ *is idempotent*
    (ii) $\alpha \circ (X \otimes \alpha) = \alpha \circ \triangledown$, *i.e.* α *is an X-action*

(iii) $\alpha \circ (X \otimes \alpha) \circ (\Delta \otimes A) = \alpha$, *i.e.* α *is idempotent as an endomorphism on* A *in* $\mathcal{C}_{[X]}$.

(b) On the other hand, the following conditions are also equivalent:
(i) $\alpha(x) = \alpha(x)^\ddagger$, *i.e.* $\alpha(x)$ *is self-adjoint*
(ii) $\alpha = (\varepsilon \otimes A) \circ \alpha^\ddagger$

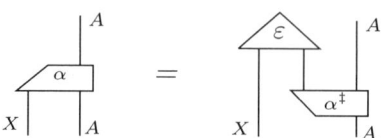

(iii) $(X \otimes \alpha) \circ (\Delta \otimes A) = (\nabla \otimes A) \circ (X \otimes \alpha^\ddagger)$

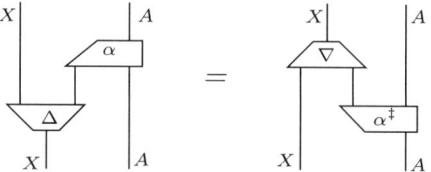

The PROOFS of the above equivalences are easy exercises with classical structure. The equivalence $(b)(ii \Leftrightarrow iii)$ can be viewed, and proven, in analogy with Thm. 4.3$(ii \Leftrightarrow iii)$.

DEFINITION 6.3. Let X be a classical structure in $\mathcal{C}$. An *X-measurement* over $A \in \mathcal{C}$ is a projector $\alpha(x) : A \longrightarrow A$ in $\mathcal{C}[x:X]$, i.e. a self-adjoint idempotent $\alpha(x) = \alpha(x)^\ddagger = \alpha(x) \circ \alpha(x)$.

A homomorphism $f : \alpha(x) \longrightarrow \beta(x)$, where $\alpha(x)$ is an X-measurement over A and $\beta(x)$ is an X-measurement over B, is an arrow $f \in \mathcal{C}(A, B)$ such that $f \circ \alpha(x) = \beta(x) \circ f$. The category of measurements in the classical structure $(X, \Delta, \top)$ is denoted by $\mathcal{C}\{x:X\}$.

Remark. Substituting a basis vector $\varphi \in \mathcal{B}(X)$ into a measurement $\alpha(x) \in \mathcal{C}[x:X](A,A)$ yields a projector $\alpha(\varphi) \in \mathcal{C}(A,A)$. The intuition is that this projector corresponds to an the outcome of the measurement α.

It is easy to see that $\mathcal{C}\{x:X\}$ is a dagger-monoidal category. The following two propositions show that this notion of a measurement is equivalent with the one from [**12**].

THEOREM 6.4. *Let X be a classical structure, and $\alpha(x) : A \longrightarrow A$ an endomorphism in $\mathcal{C}[x:X]$. Then (a)* $\Longleftrightarrow$ *(b)* $\Longrightarrow$ *(c).*

(a) $\alpha(x) : A \longrightarrow A$ *is a measurement*
(b) $\alpha = \kappa x.\, \alpha(x) : XA \longrightarrow A$ *is an X-action satisfying*

$$\alpha \circ (x \otimes A) = (x^\ddagger \otimes A) \circ \alpha^\ddagger$$

(c) α *is an X-action satisfying the following equivalent conditions*

(i) $(X \otimes \alpha) \circ (\triangle \otimes A) \quad = \quad \alpha^\ddagger \circ \alpha \quad = \quad (\triangledown \otimes A) \circ (X \otimes \alpha^\ddagger)$

(ii) $\alpha^\ddagger \circ \alpha \quad = \quad (X \otimes \alpha) \circ (c \otimes A) \circ (X \otimes \alpha^\ddagger)$

The converse $(c) \implies (a) \wedge (b)$ holds if the X-action α is normal. When this is the case, then also

$$\alpha \circ \alpha^\ddagger \quad = \quad \mathrm{id}_A$$

Remarks. The two equations in Thm. 6.4*(i)* imply each other by applying the dagger. They also imply that

- $X \otimes A \xrightarrow{\alpha} A$ is a retract of $X \otimes X \xrightarrow{\triangledown} X$ in the category of X-actions, along the restriction $\alpha^\ddagger : \alpha \rightarrowtail \triangledown$, and that
- $A \xrightarrow{\alpha^\ddagger} X \otimes A$ is a retract of $X \xrightarrow{\triangle} X \otimes X$ in the category of X-coactions, along the retraction $\alpha : \triangle \twoheadrightarrow \alpha^\ddagger$.

The Frobenius condition is the special case of both *(i)* and *(ii)*, since $\triangle$ and $\triangledown$ are just special actions.

PROOF. $(a \iff b)$ follows directly from Lemma 6.2. Part *(a)* of the lemma says that $\alpha(x)$ is idempotent if and only if α is an X-action. Part *(b)* says that $\alpha(x)$ is self-adjoint if and only if $\alpha = (\varepsilon \otimes A) \circ \alpha^\ddagger$, which is equivalent to $\alpha \circ (x \otimes) = (x^\ddagger \otimes A) \circ \alpha^\ddagger$ by the η-rule, using Thm. 4.3*(b)*.

$(a \implies ii)$ is proved as follows:

using Lemma 6.2, and the commutativity of Δ.

$(ii \Longrightarrow i)$ is a variation on the same theme:

Finally, if the X-action α is normal, then postcomposing *(i)* with $\top \otimes A$ gives condition 6.2*(b)*, and hence *(a)*.

$\alpha \circ \alpha^{\ddagger} = \mathrm{id}_A$ is left as an exercise. $\qquad\square$

PROPOSITION 6.5. *The category $\mathcal{C}\{x:X\}$ of measurements over X is equivalent with the category $\mathcal{C}^{\{X\}}$ of X-actions.*

6.1. Measuring the outcome. In general, the measurement outcome corresponding to a basis vector is the pure projector that represents it. In order to perform the measurement in the first component of $S|x, y\rangle$ from sec. 5, we use a partial representation of this vector.

LEMMA 6.6. $\sigma_y(x) = (\nabla_m \otimes \mathrm{id}_n) \circ S|x, y\rangle$ *is a measurement on* $\Xi^{\otimes(m+n)}$ *in* $\mathsf{Rel}_\wp \, [|y\rangle:\Xi^{\otimes n}]\{|x\rangle:\Xi^{\otimes m}\}$.

Substituting the basis vectors for x in $\sigma_y(x)$ gives the projectors on $\Xi^{\otimes(m+n)}$, from which the information about the period c is extracted like before.

7. Conclusions and future work

Simon's algorithm required three operations:

abstraction: : to represent classical functions and classical data in a quantum universe;

transform to a complementary basis: : to entangle classical data and make use of quantum parallelism;

measurement: : to extract the classical outcomes of quantum computation.

The abstraction operations shape the classical interfaces of quantum computers. Our analysis of the general abstraction operations uncovered a rich structure, that may be of interest beyond quantum computation. Are there other computational resources, besides entanglement, that provide exponential speedup when suitably combined with the general abstraction operations?

The other two operations that we formalized are typically quantum. Complementary bases provide access to entanglement, as the main resource of quantum computation, and thus enable quantum parallelism. The varied interactions among the different classical structures and with measurements give rise to the wealth of quantum algorithms that remain to be explored.

Our abstract model uncovered some abstract entanglement structures, and made them available for quantum computation in non-standard mathematical models. The algorithmic consequences of this semantical result need to be carefully explored. Some categorical tools for such explorations in nonstandard models are described in [**42**].

Acknowledgements. This work arose from conversations with Samson Abramsky and with Bob Coecke during their respective visits to Kestrel Institute in Palo Alto in 2005 and 2006. Bob asked me to present a part of it at his TANCL workshop in 2007, and Samson to present another part at his Clifford Lectures in 2008. Their encouragement is greatly appreciated. Some of the presented material was subsequently developed jointly with Bob, and presented in [**12, 17**]; but the guiding view of classical data as the data that can be denoted by the variables and that support abstraction, was not absorbed. This paper arose from my feeling that this view might nevertheless be useful for some developments ahead. Although I could not avoid reworking some of the structures that have in the meantime appeared and ramified into different forms, I am hoping that this step back towards the conceptual origins will not distract, but that it may help in the exciting new explorations [**18, 15, 5**]. The astute remarks, and even some instructive misunderstandings of the anonymous referee have helped me to improve the presentation.

References

1. S. Abramsky and B. Coecke, *A categorical semantics of quantum protocols*, Proceedings of the 19th Annual IEEE Symposium on Logic in Computer Science: LICS 2004, IEEE Computer Society, 2004, pp. 415–425.
2. Samson Abramsky, *A structural approach to reversible computation*, Theor. Comput. Sci. **347** (2005), 441–464.
3. Samson Abramsky, *No-cloning in categorical quantum mechanics*, Semantical Techniques in Quantum Computation (Simon Gay and Ian Mackie, eds.), Cambridge University Press, 2009.
4. Samson Abramsky, Simon Gay, and Rajagopal Nagarajan, *Interaction categories and the foundations of typed concurrent programming*, Proceedings of the 1994 Marktoberdorf Summer Sxhool on Deductive Program Design (M. Broy, ed.), Springer-Verlag, 1996, pp. 35–113.
5. Samson Abramsky and Chris Heunen, h^*-*algebras and nonunital Frobenius algebras: first steps in infinite-dimensional categorical quantum mechanics*, (2011), this volume.
6. H. P. Barendregt, *The lambda calculus : Its syntax and semantics*, North-Holland, 1981.

7. Michael Barr, *Algebraically compact functors*, Journal of Pure and Applied Algebra **82** (1992), no. 3, 211–231.
8. Michael Barr and Charles Wells, *Toposes, triples, and theories*, Grundlehren der mathematischen Wissenschaften, no. 278, Springer-Verlag, 1985.
9. Aurelio Carboni, *Matrices, relations, and group representations*, J. of Algebra **136** (1991), 497–529.
10. Aurelio Carboni and Robert F.C. Walters, *Cartesian bicategories, I*, J. of Pure and Applied Algebra **49** (1987), 11–32.
11. Alonzo Church, *A formulation of the simple theory of types*, The Journal of Symbolic Logic **5** (1940), no. 2, 56–68.
12. B. Coecke and D. Pavlovic, *Quantum measurements without sums*, Mathematics of Quantum Computing and Technology (G. Chen, L. Kauffman, and S. Lamonaco, eds.), Taylor and Francis, 2007, arxiv.org/quant-ph/0608035.
13. Bob Coecke, *Quantum picturalism*, Contemporary Physics **51** (2010), 59–83.
14. Bob Coecke and Ross Duncan, *Interacting quantum observables*, ICALP (2) (Luca Aceto, Ivan Damgård, Leslie Ann Goldberg, Magnús M. Halldórsson, Anna Ingólfsdóttir, and Igor Walukiewicz, eds.), Lecture Notes in Computer Science, vol. 5126, Springer, 2008, pp. 298–310.
15. _____, *Interacting quantum observables: categorical algebra and diagrammatics*, New Journal of Physics **13** (2011), no. 4, 043016.
16. Bob Coecke and William Edwards, *Toy quantum categories*, Proceedings of the 2008 QPL-DCM Workshop (Bob Coecke and Prakash Panangaden, eds.), Springer-Verlag, 2008, arXiv:0808.1037, pp. 25–35.
17. Bob Coecke, Éric Oliver Paquette, and Dusko Pavlovic, *Classical and quantum structuralism*, Semantical Techniques in Quantum Computation (Simon Gay and Ian Mackie, eds.), Cambridge University Press, 2009, pp. 29–69.
18. Bob Coecke and Aleks Kissinger, *The compositional structure of multipartite quantum entanglement*, Automata, Languages and Programming, 2010, pp. 297–308.
19. Bob Coecke, Dusko Pavlovic, and Jamie Vicary, *A new description of orthogonal bases*, Math. Structures in Comp. Sci. (2011), 13 pp., to appear, arxiv.org:0810.0812.
20. D. Dieks, *Communication by EPR devices*, Physics Letters A **92** (1982), no. 6, 271–272.
21. Thomas Fox, *Coalgebras and cartesian categories*, Comm. Algebra **4** (1976), no. 7, 665–667.
22. Kurt Gödel, *Über formal unentscheidbare Sätze der Principia Mathematica und verwandter Systeme*, I. Monatshefte für Mathematik und Physik **38** (1931), 173–198.
23. Esfandiar Haghverdi and Philip J. Scott, *Towards a typed geometry of interaction*, Mathematical Structures in Computer Science **20** (2010), no. 3, 473–521.
24. Sean Hallgren, *Polynomial-time quantum algorithms for Pellõs equation and the principal ideal problem*, Proceedings of the 34th ACM Symposium on Theory of Computing, ACM Press, 2002, pp. 653–658.
25. André Joyal and Ross Street, *The geometry of tensor calculus I*, Adv. in Math. **88** (1991), 55–113.
26. André Joyal, Ross Street, and Dominic Verity, *Traced monoidal categories*, Math. Proc. Cambridge Philos. Soc. **119** (1996), no. 3, 447–468.
27. Gregory M. Kelly, *Basic concepts of enriched category theory*, Cambridge University Press, 1982 (English), http://www.tac.mta.ca/tac/reprints/articles/10/tr10.pdf.
28. Gregory M. Kelly and Miguel L. Laplaza, *Coherence for compact closed categories*, J. of Pure and Applied Algebra **19** (1980), 193–213.
29. Stephen Cole Kleene, *Recursive predicates and quantifiers*, Transactions of the American Mathematical Society **53** (1943), no. 1, 41–73.
30. Joachim Kock, *Frobenius algebras and 2d topological quantum field theories*, London Mathematical Society Student Texts, vol. 59, Cambridge University Press, 2004.
31. K. Kraus, *Complementary observables and uncertainty relations*, Physical Review D **35** (1987), no. 10, 3070–3075.
32. Joachim Lambek, *From types to sets*, Adv. in Math. **36** (1980), 113–164.
33. Joachim Lambek and Philip J. Scott, *Introduction to higher order categorical logic*, Cambridge University Press, New York, NY, USA, 1986.
34. F. William Lawvere, *Adjointness in foundations*, Dialectica **23** (1969), 281–296.

35. _____, *Ordinal sums and equational doctrines*, Seminar on Triples, Categories and Categorical Homology Theory, Lecture Notes in Mathematics, vol. 80, Springer-Verlag, 1969, pp. 141–155.
36. Samuel J. Lomonaco and Louis H. Kauffman, *Quantum hidden subgroup algorithms: An algorithmic toolkit*, Mathematics of Quantum Computing and Technology (G. Chen, Louis Kauffman, and Samuel Lamonaco, eds.), Taylor and Francis, 2007.
37. Saunders Mac Lane, *Categories for the working mathematician*, Graduate Texts in Mathematics, no. 5, Springer-Verlag, 1971.
38. Michael A. Nielsen and Isaac L. Chuang, *Quantum computation and quantum information*, Cambridge University Press, October 2000.
39. A.K. Pati and S.L. Braunstein, *Impossibility of deleting an unknown quantum state*, Nature **404** (2000), 164–165.
40. Dusko Pavlovic, *Categorical logic of names and abstraction in action calculus*, Math. Structures in Comp. Sci. **7** (1997), 619–637.
41. _____, *Quantum and classical structures in nondeterministic computation*, Proceedings of Quantum Interaction 2009 (Peter Bruza, Don Sofge, and Keith van Rijsbergen, eds.), Lecture Notes in Artificial Intelligence, vol. 5494, Springer Verlag, 2009, arxiv.org:0812.2266, pp. 143–158.
42. _____, *Relating toy models of quantum computation: comprehension, complementarity and dagger autonomous categories*, E. Notes in Theor. Comp. Sci. **270** (2011), no. 2, 121–139, arxiv.org:1006.1011.
43. Dusko Pavlovic and Samson Abramsky, *Specifying interaction categories*, Category Theory and Computer Science '97 (E. Moggi and G. Rosolini, eds.), Lecture Notes in Computer Science, vol. 1290, Springer Verlag, 1997, pp. 147–158.
44. Dusko Pavlović and Martín Escardó, *Calculus in coinductive form*, Proceedings. Thirteenth Annual IEEE Symposium on Logic in Computer Science (V. Pratt, ed.), IEEE Computer Society, 1998, pp. 408–417.
45. Roger Penrose, *Structure of space-time*, Batelle Rencontres, 1967 (C.M. DeWitt and J.A. Wheeler, eds.), Benjamin, 1968.
46. Benjamin C. Pierce, *Types and programming languages*, MIT Press, 2002.
47. John Power and Edmund Robinson, *Premonoidal categories and notions of computation*, Mathematical. Structures in Comp. Sci. **7** (1997), no. 5, 453–468.
48. Peter Selinger, *Dagger compact closed categories and completely positive maps*, Electron. Notes Theor. Comput. Sci. **170** (2007), 139–163.
49. _____, *Idempotents in dagger categories: (extended abstract)*, Electr. Notes Theor. Comput. Sci. **210** (2008), 107–122.
50. _____, *A survey of graphical languages for monoidal categories*, New Structures for Physics (Bob Coecke, ed.), Springer Lecture Notes in Physics, vol. 813, Springer Verlag, 2011, pp. 289–355.
51. Peter W. Shor, *Polynomial-time algorithms for prime factorization and discrete logarithms on a quantum computer*, SIAM J. Comput. **26** (1997), no. 5, 1484–1509.
52. Daniel R. Simon, *On the power of quantum computation*, SIAM J. Comput. **26** (1997), no. 5, 1474–1483.
53. Ross Street, *Categorical Structures*, Handbook of Algebra (M. Hazewinkel, ed.), vol. 1, North Holland, 1996, pp. 528–576.
54. Vlatko Vedral, *Introduction to quantum information science*, Oxford University Press, 2007.
55. D.V. Widder, *An introduction to transform theory*, Pure and Applied Mathematics, vol. 42, Academic Press, New York and London, 1971.
56. W. K. Wootters, *Quantum measurements and finite geometry*, 2004, arXiv.org:quant-ph/0406032.
57. W.K. Wootters and W.H. Zurek, *A single quantum cannot be cloned*, Nature **299** (1982), 802–803.

Royal Holloway, Department of Mathematics, TW20 0EX, UK, and Universiteit Twente, Faculteit EEMCS, Postbus 217, 7500AE Enschede, The Netherlands

E-mail address: dusko.pavlovic@rhul.ac.uk

PSAPM/71